Jean-Marie Lehn

Supramolecular Chemistry

VCH

Related Titles from VCH

C. St. Cooper (ed.)
Crown Compounds.
Toward Future Applications
ISBN 3-527-28073-1

B. Dietrich, P. Viout, J.-M. Lehn
Macrocyclic Chemistry. Aspects of Organic
and Inorganic Supramolecular Chemistry
ISBN 3-527-28330-7

J. Fuhrhop, G. Penzlin
Organic Synthesis. Concepts, Methods,
Starting Materials
Second, Revised and Enlarged Edition
With a Foreword by E. J. Corey
Hardcover ISBN 3-527-29086-9
Softcover ISBN 3-527-29074-5

K.C. Nicolaou, E. J. Sorensen
Classics in Total Synthesis
Hardcover ISBN 3-527-29231-4
Softcover ISBN 3-527-29284-5

© VCH Verlagsgesellschaft mbH. D-69451 Weinheim (Bundesrepublik Deutschland), 1995

Distribution:
VCH, Postfach 10 11 61, D-69451 Weinheim (Bundesrepublik Deutschland)
Schweiz: VCH, Postfach, CH-4020 Basel (Schweiz)
United Kingdom und Irland: VCH (UK) Ltd., 8 Wellington Court, Cambridge CB1 1HZ
(England)
USA und Canada: VCH, 220 East 23rd Street, New York, NY 10010-4606 (USA)
Japan: VCH, Eikow Building, 10-9 Hongo 1-chome, Bunkyo-ku, Tokyo 113 (Japan)

ISBN 3-527-29311-6 (Softcover)   ISBN 3-527-29312-4 (Hardcover)

Jean-Marie Lehn

Supramolecular Chemistry

Concepts and Perspectives

A Personal Account

Built upon the

| George Fisher Baker Lectures in Chemistry at Cornell University | Lezioni Lincee Accademia Nazionale dei Lincei Roma |

VCH
Weinheim • New York
Basel • Cambridge • Tokyo

Prof. Dr. Jean-Marie Lehn
Institut Le Bel   and   Collège de France
Université Louis Pasteur   11, Place Marcelin Berthelot
4, rue Blaise Pascal   F-75005 Paris
F-67000 Strasbourg   France
France

Editorial Director: Dr. Ute Anton
Production Manager: Dipl.-Ing. (FH) Hans Jörg Maier

Cover Illustration: Creativity in art and science. The creative power in chemistry is expressed by the design of molecular species, which self-assemble into organized supramolecular assemblies. The cover illustration shows a synthetic double helix, the double stranded helicate containing a tris bipyridine ligand, held together by Cu(I) ions (see Section 9.3.1). The creative power in art is expressed by the sculpture, *la Main de Dieu*, by Auguste Rodin. Photograph by Bruno Jarret © ADAGP, Paris 1955.

Library of Congress Card No. applied for.

A catalogue record for this book is available from the British Library.

Deutsche Bibliothek Cataloguing-in-Publication Data:

Lehn, Jean-Marie:
Supramolecular chemistry : concepts and perspectives; a personal account / Jean-Marie Lehn. The George Fisher Baker Non-Resident Lectureship in Chemistry at Cornell University; Lezioni Lincee, Accademia Nazionale dei Lincei, Roma. – Weinheim; New York; Basel; Cambridge; Tokyo: VCH, 1995
ISBN 3-527-29311-6 brosch.
ISBN 3-527-29312-4 Gb.

Composition: Filmsatz Unger & Sommer GmbH, D-69469 Weinheim. Printing: betz-druck gmbh, D-64291 Darmstadt. Bookbinding: Wilhelm Osswald & Co., D-67433 Neustadt

Preface

The present text is intended to describe the concepts, the lines of development and the perspectives of the field of supramolecular chemistry that has been growing over the last twenty five years. It uses as main thread the work done in my own laboratories, complementing it with a selection of results obtained by various other groups. It adopts a synthetic view, painting a broad picture supported by specific illustrations, rather than giving a comprehensive description. The latter would anyway have been impossible to achieve here, if justice was to be done to the very diverse and extensive research conducted by a large and continuously increasing crowd of laboratories around the world, attracted by the multidisciplinary breadth and unifying power of supramolecular chemistry, as witnessed by the number of publications, reviews, monographs, lectures, seminars, schools and congresses that have been dedicated to this area over the years.

Comparable extension has been given to the research on molecular recognition, catalysis and transport processes (Chapters 2–6) on molecular and supramolecular devices (Chapter 8) and self-processes (Chapter 9). This was done to emphasize the more recently developed topics, despite the much larger volume of work on the former areas that have been described in many instances in the literature. The outline of Chapters 1–7 follows that of earlier reviews. Chapters 8 and 9 bring together approaches from various directions. Chapter 10 places the basic concepts into a broader perspective and is intended to make the score decidedly open ended.

The purpose of this text will be achieved if it succeeds in the intention of the author to inspire, stimulate and challenge the creative imagination of scientists not only in chemistry, but also in physics and biology, in basic as well as in applied fields, at the meeting point provided by the design and investigation of organised, informed and functional supramolecular architectures. It may seem appropriate that at the centennial of Emil Fischer's celebrated 1894 image of the Lock and the Key, supramolecular chemistry offers a fertile ground for taking up the challenge of "instructed" chemistry.

This is also the occasion to fulfil, although belatedly, the request for a write up of the Baker Lectures given at Cornell University in 1978 (!) and of the Lezioni Lincee held at the Accademia Nazionale dei Lincei and at the University of Rome in 1992. It is long overdue, but it certainly profits from the much firmer basis, the much stronger body and the much broader scope that the field of supramolecular chemistry has acquired. I thank my colleagues at these Institutions for their warm hospitality during my stay with them and for their patience!

I wish to express my gratitude to the numerous research groups with which we have enjoyed collaborating over the years and my deep appreciation to all those around the world who have contributed to make supramolecular chemistry what it has become and who will keep it alive and growing in the future.

I thank the Université Louis Pasteur, the Collège de France, the Centre National de la Recherche Scientifique and the Science Programmes of the European Community for providing the local intellectual environment and the financial support of our work.

I am much indebted to Dr. Bernard Dietrich for his competent and scrupulous eye in reading the draft and in checking the references and illustrations, to Doris Biltz for her expert and efficient typing as well as to Thierry Bataille, Serge Wechsler and Robert Weidmann for generating the drawings.

Finally, highest credit goes of course to my 200 or so coworkers from many countries who, over the years, have performed with skill and dedication the work realised and provided the stimulating international and cross-cultural atmosphere that is a major attraction of life as a scientist.

Strasbourg, April 1995   *Jean-Marie Lehn*

Contents

X   *Contents*

1 From Molecular to Supramolecular Chemistry

In the beginning was the Big Bang, and physics reigned. Then chemistry came along at milder temperatures; particles formed atoms; these united to give more and more complex molecules, which in turn associated into aggregates and membranes, defining primitive cells out of which life emerged.

Chemistry is the science of matter and of its transformations, and life is its highest expression. It provides structures endowed with properties and develops processes for the synthesis of structures. It plays a primordial role in our understanding of material phenomena, in our capability to act upon them, to modify them, to control them and to invent new expressions of them.

Chemistry is also a science of transfers, a communication centre and a relay between the simple and the complex, between the laws of physics and the rules of life, between the basic and the applied. If it is thus defined in its interdisciplinary relationships, it is also defined in itself, by its object and its method.

In its method, chemistry is a science of interactions, of transformations and of models. In its object, the molecule and the material, chemistry expresses its creativity. Chemical synthesis has the power to produce new molecules and new materials with new properties. New indeed, because they did not exist before being created by the recomposition of atomic arrangements into novel and infinitely varied combinations and structures [1.1].

For more than 150 years, since the synthesis of urea by *Friedrich Wöhler* in 1828 [1.2], molecular chemistry has developed a vast array of highly sophisticated and powerful methods for the construction of ever more complex molecular structures by the making or breaking of covalent bonds between atoms in a controlled and precise fashion.

Organic synthesis grew rapidly and masterfully, leading to a whole series of brilliant achievements, the great syntheses of the last 50 years, where elegance of strategy combined with feats of efficiency and selectivity. There is a long way from

$H_2N-\overset{O}{\overset{\|}{C}}-NH_2$

H_2NOC $CONH_2$
   $CONH_2$
H_2NOC
   N   R   N
      Co III
   H
      N   N
H_2NOC
   O
   $CONH_2$
   NH
   N
   N
   O   HO
   O   O
   $-O-P-O$ $R = Adenosyl$
   O CH_2OH

Wöhler's urea to the synthesis of Vitamin B_{12} by *Robert B. Woodward* [1.3] and *Albert Eschenmoser* [1.4] assisted by a hundred or so collaborators!

Molecular chemistry, thus, has established its power over the covalent bond. The time has come to do the same for non-covalent intermolecular forces. Beyond molecular chemistry based on the covalent bond there lies the field of *supramolecular chemistry*, whose goal it is to gain control over the intermolecular bond.

It is concerned with the next step in increasing complexity beyond the molecule towards the supermolecule and organized polymolecular systems, held together by non-covalent interactions.

It is a sort of molecular sociology! Non-covalent interactions define the inter-component bond, the action and reaction, in brief, the behaviour of the molecular individuals and populations: their social structure as an ensemble of individuals having its own organisation; their stability and their fragility; their tendency to associate or to isolate themselves; their selectivity, their "elective affinities" and class structure, their ability to recognize each other; their dynamics, fluidity or rigidity of arrangements and of castes, tensions, motions and reorientations; their mutual action and their transformations by each other.

Molecular interactions form the basis of the highly specific recognition, reaction, transport, regulation, etc. processes that occur in biology, such as substrate binding to a receptor protein, enzymatic reactions, assembling of multiprotein complexes, immunological antigen–antibody association, intermolecular reading, translation and transcription of the genetic code, regulation of gene expression by DNA binding proteins, entry of a virus into a cell, signal induction by neurotransmitters, cellular recognition, and so on. The design of artificial, abiotic systems capable of displaying processes of highest efficiency and selectivity requires the correct manipulation of the energetic and stereochemical features of the non-covalent, intermolecular forces (electrostatic interactions, hydrogen bonding, van der Waals forces, etc.)

within a defined molecular architecture. In doing so, the chemist finds inspiration in the ingenuity of biological events and encouragement in the demonstration that such high efficiencies, selectivities, and rates can indeed be attained. However, chemistry is not limited to systems similar to those found in biology, but is free to create unknown species and to invent novel processes.

Supramolecular chemistry is a highly interdisciplinary field of science covering the chemical, physical, and biological features of the chemical species of greater complexity than molecules themselves, that are held together and organized by means of intermolecular (non-covalent) binding interactions. This relatively young area has been defined, conceptualized, and structured into a coherent system. Its roots extend into organic chemistry and the synthetic procedures for molecular construction, into coordination chemistry and metal ion-ligand complexes, into physical chemistry and the experimental and theoretical studies of interactions, into biochemistry and the biological processes that all start with substrate binding and recognition, into materials science and the mechanical properties of solids. A major feature is the range of perspectives offered by the cross-fertilization of supramolecular chemical research due to its location at the intersection of chemistry, biology, and physics. Drawing on the physics of organized condensed matter and expanding over the biology of large molecular assemblies, supramolecular chemistry expands into a *supramolecular science*. Such wide horizons are a challenge and a stimulus to the creative imagination of the chemist.

Thus, supramolecular chemistry has been rapidly expanding at the frontiers of chemical science with physical and biological phenomena. Considering the great activity and the vast literature that have developed, there is no possibility here to do justice to the numerous results obtained, still less to provide an exhaustive account of this field of science [1.5, 1.6]. The present text will emphasize the conceptual framework, classes of species, types of processes and extend previous overviews (in particular [1.7–1.9]); illustrations will be taken mainly but not solely from our own work.

In any field of science, novelty is linked to the past. Where would the roots of supramolecular chemistry reach? It was *Paul Ehrlich* who recognized that molecules do not act if they do not bind (*"Corpora non agunt nisi fixata"*) thus introducing the concept of *receptor* [1.10]. But binding must be selective, a notion that was enunciated by *Emil Fischer* in 1894 [1.11] and very expressively presented in his celebrated "lock and key" image of steric fit, implying geometrical complementarity, that lays the basis of *molecular recognition*. Finally, selective fixation requires interaction, affinity between the partners, that may be related to the idea of *coordination* introduced by *Alfred Werner* [1.12], supramolecular chemistry being in this respect a generalization of coordination chemistry [1.13].

With these three concepts, fixation, recognition and coordination, the foundations of supramolecular chemistry are laid. Why did it take so long to come to life? Molecular associations have been recognized and studied for a long time [1.14] and the term *"Übermoleküle"* was even coined as early as the mid-1930's to describe entities of higher organisation (such as the dimer of acetic acid) resulting from the association of coordinatively saturated species [1.15]. Also supramolecular organisation has been known to play a major role in biology [1.16]. But the emergence and the rapid development of a new field require the conjunction of at least three factors, first the recognition of a new paradigm, of the significance of scattered and seemingly unrelated observations, data, results, and their organisation into a coherent whole with its conceptualisation; second, the availability of the tools for studying the objects of this field; there the development of the powerful physical methods of today for the analysis of structure and properties (IR, UV and especially NMR and mass spectroscopies, X-ray diffraction, etc.) has played a major role since the comparative lability of supramolecular species, due to the lower energy of the non-covalent interactions, makes them more difficult to characterize; third, the ripeness of the scientific scene, the resonance the novel field finds not only within its own scientific discipline but also in closely as well as distantly related ones. In brief, the novel field will impose itself, once revealed, by its "Eureka!" obviousness. Such is the case of supramolecular chemistry, as witnessed by the rate at which it has developed and permeated other fields of science over the last twenty five years.

Concepts and Language of Supramolecular Chemistry. *"Le langage est une législation, la langue en est le code"* wrote *Roland Barthes* [1.17]. For our present purpose, chemistry is the legislation and the words naming its objects form the code that gives access to a specific area and through which it communicates.

The intrinsic language of chemistry is formed by the vocabulary of chemical formulae and structural representations connected by the syntax of their interconversions. It describes a tangible reality; it is a "reification" of the word and the text; its signs are engraved into matter [1.18, 1.19].

As a novel field of science emerges, grows and matures, it evolves basic concepts and generates novel terminology to name the concepts that define it and to describe the objects that constitute it.

Such conceptualizing and naming play a very important role, not only for shaping the field but also by the ground they offer to the creative imagination. Indeed, one may let one's imagination be carried by the magic of the word and pulled by the evocative and stimulating power of the concept.

Definitions have a clear, precise core but often fuzzy borders, where interpenetration between areas takes place. These fuzzy regions in fact play a positive role since it is often there that mutual fertilization between areas may occur. This is certainly also true for the case at hand, the case of supramolecular chemistry and its language [1.20].

The field, as we now know it, started with the selective binding of alkali metal cations by natural [1.21–1.23] as well as by synthetic macrocyclic and macropolycyclic ligands, the crown ethers [1.24, 1.25] and the cryptands [1.26, 1.27]. The outlook broadened leading to the emergence and identification of molecular recognition [1.27] as a novel domain of chemical research that, by extension to intermolecular interactions and processes in general and by broadly expanding over other areas, grew into supramolecular chemistry. The chemistry of molecular recognition is also at the core of "host–guest chemistry" [1.28].

The concept and the term of supramolecular chemistry were introduced in 1978 [1.29] (see also [2.17]) as a development and generalization of earlier work in which the seed had been planted [1.27, 1.30]. It was defined in words, "Just as there is a field of *molecular chemistry* based on the covalent bond, there is a field of *supramolecular chemistry*, the chemistry of molecular assemblies and of the intermolecular bond", as well as in diagrammatic fashion (Fig. 1) [1.29].

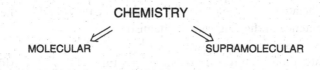

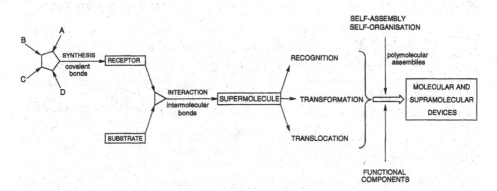

Fig. 1. From molecular to supramolecular chemistry: molecules, supermolecules, molecular and supramolecular devices.

It has since then been reformulated on various occasions, e.g., "Supramolecular chemistry may be defined as 'chemistry beyond the molecule', bearing on the organized entities of higher complexity that result from the association of two or more chemical species held together by intermolecular forces" [1.7].

The breadth and especially the unifying power of the concept became progressively more and more apparent, so that recent years have seen an explosive growth as measured by the increasing number of laboratories that join the field and whose work has been reported in a vast range of publications, books, journals, meetings and symposia (for a selection see Appendix).

Supramolecular chemistry has thus developed into a coherent and extremely lively body of concepts and objects, progressively generating and incorporating novel areas of investigation. A whole vocabulary, still incomplete however, has been developed and has become more and more widely accepted and used. Some of the names have been borrowed and adapted from other existing fields, in particular coordination chemistry and biology.

A chemical species is defined by its components, by the nature of the bonds that link them together and by the resulting spatial (geometrical, topological) features. The objects of supramolecular chemistry are "supramolecular entities, *supermolecules* possessing features as well defined as those of molecules themselves. One may say that supermolecules are to molecules and the intermolecular bond what molecules are to atoms and the covalent bond" [1.7, 1.8]. Thus, the supermolecule represents the next level of complexity of matter after the elementary particle, the nucleus, the atom and the molecule. Drawing a parallel with natural language, one might say that the atom, the molecule, the supermolecule would be the letter, the word and the sentence in the language of chemistry !

Supramolecular species are characterized both by the spatial arrangement of their components, their architecture or superstructure, and by the nature of the intermolecular bonds that hold these components together. They possess well-defined structural, conformational, thermodynamic, kinetic and dynamical properties. Various types of interactions may be distinguished, that present different degrees of strength, directionality, dependence on distance and angles: metal ion coordination, electrostatic forces, hydrogen bonding, van der Waals interactions, donor–acceptor interactions, etc. Their strengths range from weak or moderate as in hydrogen bonds, to strong or very strong for metal ion coordination. The former provide associations of stabilities comparable to enzyme–substrate species, whereas the latter makes available, by means of a single metal ion, binding strengths that lie in the domain of antigen–antibody complexes (or higher), where many individual interactions are involved. Intermolecular forces are however in general weaker than covalent bonds, so that supramolecular species are thermodynamically less stable, kinetically more labile and dynamically more flexible than molecules. Thus, supramolecular chemistry is concerned with soft bonds and represents a "soft chemistry".

In view of the many unspecific uses of the term *ligand* for either partner in a complex, the components of a supermolecule have been named *receptor* (ρ) and *substrate(s)* (σ) [1.27, 1.31], the substrate(s) being usually the smaller component(s) whose binding is being sought. This terminology conveys the relation to biological receptor–substrate interactions, with their highly defined structural and functional properties. Furthermore, it is easily converted from one language to another. The "inclusion compound" and "host-guest" designations also cover species (termed clathrates) [1.32, 1.33, A.18] that exist only in the solid state and are not strictly speaking discrete supermolecules, although they do represent supramolecular solid state species.

Selective binding of a specific substrate σ to its receptor ρ yields the super-molecule $\rho\sigma$ and involves a molecular recognition process. If, in addition to binding sites, the receptor also bears reactive functions, it may effect a chemical transformation on the bound substrate, thus behaving as a supramolecular reagent or catalyst. A lipophilic, membrane-soluble receptor may act as a carrier effecting the translocation of the bound substrate. Thus, *molecular recognition, transformation*, and *translocation* represent the basic functions of supramolecular species. More complex functions may result from the interplay of several binding subunits in a polytopic coreceptor. In association with organized polymolecular assemblies and phases (layers, membranes, vesicles, liquid crystals, etc.), functional supermolecules may lead to the development of *molecular* and *supramolecular devices*. This panorama is graphically represented in Figure 1. Recent lines of investigation concern *self-processes* (self-assembly, self-organization, replication) and the design of *programmed supramolecular systems*. All these features suppose the presence and operation of *molecular information*, a notion that has become the basic and crucial tenet of supramolecular chemistry. These and other concepts and terms will be described and illustrated in the following chapters.

One point to address concerns the use of the words *supra*molecular and *super*molecule. The concept of supramolecular chemistry has become a unifying attractor, in which areas that have developed independently have spontaneously found their place. The word "supramolecular" has been used in particular for large multiprotein architectures and organized molecular assemblies [1.16]. On the other hand, in theoretical chemistry, the computational procedure that treats molecular associations such as the water dimer as a single entity is termed the "supermolecule" approach [1.34, 1.35]. Taking into account the existence and the independent uses of these two words, one may then propose that "*supra*molecular chemistry" be the broader term, concerning the chemistry of all types of supramolecular entities from the well-defined supermolecules to extended, more or less organized, polymolecular associations. The term "*super*molecular chemistry" would be restricted to the specific chemistry of the supermolecules themselves.

Supramolecular chemistry may then be divided into two broad, partially overlapping areas concerning: (1) *supermolecules*, well-defined, discrete *oligo*molecular species that result from the intermolecular association of a few components (a receptor and its substrate(s)) following a built-in "Aufbau" scheme based on the principles of molecular recognition; (2) *supramolecular assemblies*, *poly*molecular entities that result from the spontaneous association of a large undefined number of components into a specific phase having more or less well-defined microscopic organization and macroscopic characteristics depending on its nature (such as films, layers, membranes, vesicles, micelles, mesomorphic phases, solid state structures, etc.). It thus covers the rational, coherent approach to molecular associations, from the smallest, the dimer, to the largest, the organized phase, and to their designed manipulation. Continuing along the comparison with natural languages made above, the polymolecular supramolecular entity would be the book!

The emergence of the concepts of supramolecular chemistry has led to the introduction of new terms or to the reappropriation of old ones, describing either classes of compounds or types of properties. As the field affirmed itself, the passage from molecular to supramolecular chemistry brought about a change in objects and goals from structures and properties towards systems and functions.

The naming of the objects of supramolecular chemistry, the supermolecules, poses some specific problems. Whereas the receptors and substrates themselves are covalent structures whose designations fall under the systematic naming of the Geneva Conference, there is a need for a coherent set of rules to describe the connectivity between the components in a supramolecular species. Conventions such as those used for coordination compounds of transition metal ions may represent a starting point. This is the case, for instance, for the notations κ (indicating the coordinating atom), *hapto* η (for coordinated groups) or μ (for bridging ligands) [1.36].

The description of the spatial localization of bound substrate(s) with respect to the receptor also requires some formalism. Whereas external addition complexes may be written as [A, B] or [A//B], the mathematical symbols of inclusion $\subset$ [1.27] and of intersection $\cap$ [1.37], respectively, have been introduced for the inclusion of σ into ρ, [$\sigma \subset \rho$], and for the partial interpenetration of σ and ρ, [$\sigma \cap \rho$].

The highly selective processes of molecular recognition are of course of stereochemical nature. Thus, a *supramolecular stereochemistry* may be defined that extends from supermolecules to polymolecular assemblies. Different spatial dispositions of the components of a supermolecule with respect to each other lead to *supramolecular stereoisomers*. Their eventual interconversion will depend on the properties of the interactions that hold them together, i.e., on the variation of the intermolecular interaction energy with distances and angles. There is thus an *intermolecular conformational analysis* like there is an intramolecular one. The time is ripe for putting together a coherent presentation that would found and describe this area on both experimental and theoretical bases. The specific aspects of *supramolecular chirality* concern chiral discrimination in receptor–substrate binding (see Sect. 2.5) as well as chirality features associated with the formation of supermolecules (e.g., supramolecular enantiomers) and supramolecular assemblies (see Sect. 9.7).

The active development of computational methods aimed at molecular design [1.38–1.41] provides a means for building up a *theoretical supramolecular chemistry* of not only explicative but also predictive power. Significant insight has already been gained and much more may be expected from the interplay of theory and experiment in the study of supramolecular features (interactions, structures, dynamics, association constants, medium effects, etc.) [1.35, 1.42–1.50, A.37].

The emergence and expansion of supramolecular chemistry make perceptible how concepts, terminology and actual research evolve in parallel, grow simultaneously and feed on each other, new results suggesting concepts and requiring terminology, new concepts and names inducing novel lines of investigation and carrying the creative imagination towards unexplored horizons.

Roland Barthes also noted that the tongue was oppressive: "*... toute langue est un classement et (que) tout classement est oppressif*", and "*un idiome se définit moins par ce qu'il permet de dire que par ce qu'il oblige à dire*" [1.17] ("... every tongue is a classification and every classification is oppressive"; "an idiom defines itself less by what it permits to say than by what it obliges to say"). Thus, the shaping of the language of a new field must avoid becoming a burden, bridling imagination, but rather build up from a fruitful interaction between living and lively chemical research and its codification in a nomenclature system.

At present the language of supramolecular chemistry is in the making, its vocabulary and syntax are progressively being developed, a body of concepts, of terms, of trivial names is being generated. It will become necessary to develop a systematic terminology in order to be able to designate supramolecular species in a comprehensive fashion and to allow their storage in and retrieval from data banks, as is required for the practice of chemical information science, in a way similar to molecular species. In particular, as the field grows and expands rules will be needed for designating the types, number and arrangement of partners, of binding sites and of interactions; the types of supramolecular architectures, etc., This also holds for confined species such as interlocked structures or polymolecular assemblies. One may recommend that the questions raised by supramolecular species be addressed so as to arrive at a coherent nomenclature of as general and univocal applicability as possible.

The unifying power and interdisciplinary nature of the concepts of supramolecular chemistry have attracted practitioners from various domains of chemical research and have led to the progressive incorporation of different areas that had developed independently. In order to retain its identity and not to lose its strength, the field of supramolecular chemistry must take care not to dilute itself into too large a body, however the attitude should be to keep an open house so that the family may be joined by all those who wish to become part of the common adventure!

2 Molecular Recognition

2.1 Recognition, Information, Complementarity

Molecular recognition is defined by the *energy* and the *information* involved in the *binding* and *selection* of substrate(s) by a given receptor molecule; it may also involve a specific *function* [1.27]. Mere binding is not recognition, although it is often taken as such. One may say that recognition is binding with a purpose, like receptors are ligands with a purpose. It implies a pattern recognition process through a structurally well-defined set of intermolecular interactions. Binding of σ to ρ forms a complex or supermolecule characterized by its (thermodynamic and kinetic) stability and selectivity, i.e., by the amount of energy and of information brought into operation.

Molecular recognition thus implies the (molecular) storage and (supramolecular) read out of molecular information. These terms have become characteristic of the language of supramolecular chemistry. Although the notions of recognition and information were used in connection with biological systems [2.1], they effectively pervaded the realm of chemistry only at the beginning of the 1970's in connection with, and as a generalization of, the studies on selective complexation of metal ions [1.27]. Since then molecular recognition has become a major area of chemical research and a very frequently used term.

Information may be stored in the architecture of the receptor, in its binding sites and in the ligand layer surrounding bound σ; it is read out at the rate of formation and dissociation of the supermolecule. In addition to size and shape, a receptor is characterized by the dimensionality, the connectivity and the cyclic order of its structural graph; these features have been used to define a ligand structural index L_T [1.27] (see Sect. 2.2; Fig. 2); conformation, chirality and dynamics also come into play. The binding sites are characterized by their electronic properties (charge, polarity, polarisability, van der Waals attraction and repulsion), their size, shape, number, and arrangement in the receptor framework as well as their eventual reactivity that may allow the coupling of complexation with other processes (such as protonation, deprotonation, oxidation or reduction). The ligand layer acts through its thickness, its lipo- or hydrophilicity and its overall polarity, being exo/endo–lipo/ polarophilic. In addition, stability and selectivity depend on the medium and result from a subtle balance between solvation (of both ρ and σ) and complexation (i.e., "solvation" of σ by ρ). Finally, for charged complexes medium-dependent cation–anion interactions influence markedly the binding stability and selectivity.

It has already been stressed that *information* is a key notion of supramolecular chemistry, in fact the most fundamental and general one, that constitutes the common thread running through the whole field. Indeed, in this respect supramolecular chemistry may be considered as a *chemical information science* or *molecular "informatics"* concerned with the molecular storage and the supramolecular reading and processing of the information via the structural and temporal features of molecules and supermolecules [1.7, 1.9]. These characteristics lead to the notion of programmed supramolecular systems (see Sect. 9.2). A more quantitative assessment of structural and interactional features in terms of information content represents a challenging endeavour that may be related to the evaluation of molecular complementarity or similarity [2.2a].

Recognition implies geometrical and interactional *complementarity* between the associating partners, i.e., optimal information content of a receptor with respect to a given substrate. This amounts to a generalized *double complementarity principle* extending over energetic features as well as over the geometrical ones represented by the "lock and key", steric fit concept of *Emil Fischer* [1.11].

It may be useful to devise a word for complementary partners. One may propose *pleromers* (from the greek πλήρωμα: complement μέροσ: part) i.e., parts that complement each other. Complementary interaction sites, binding subunits, molecular fragments or species could be described by the *bra-ket* notation < | and | > used in quantum mechanics to describe elements (vectors) belonging to conjugate spaces. Thus < A | B > would mean that A and B are either complementary entities, (pleromers) or complementary fragments or just complementary interaction sites [1.20].

High recognition by a receptor molecule ρ consists in a large difference between the binding free energies of a given substrate σ and of the other substrates. It results in a marked deviation from the statistical distribution. In order to achieve large differences in affinity several factors must be taken into account:

1)   *steric* (shape and size) *complementarity* between σ and ρ, i.e. presence of convex and concave domains in the correct location on σ and on ρ;

2)   *interactional complementarity*, i.e. presence of complementary binding sites (electrostatic such as positive/negative, charge/dipole, dipole/dipole, hydrogen bond donor/acceptor, etc.) in the correct disposition on ρ and σ so as to achieve complementary electronic and nuclear distribution (electrostatic, H-bonding and van der Waals) maps;

3)   *large contact areas* between ρ and σ so as to contain

4)   *multiple interaction sites*, since non-covalent interactions are rather weak compared to covalent bonds;

5)   *strong overall binding*; although high stability does in principle not necessarily imply high selectivity, this is usually the case; indeed, the differences in free energy of binding are likely to be larger when the binding is strong; high binding efficiency (i.e., high fraction of bound with respect to free σ) requires strong interaction; thus, in order to achieve efficient recognition, i.e., both high stability and high selectivity, strong binding of σ by ρ is required.

In addition, *medium effects* play an important role through the interaction of solvent molecules with ρ and σ as well as with each other; thus the two partners should present geometrically matched hydrophobic/hydrophobic or hydrophilic/hydrophilic (solvophobic, solvophilic, more generally) domains.

One may distinguish *positive* and *negative* recognition, depending on whether the discrimination between different substrates by a given receptor is dominated by attractive or repulsive (steric) interactions, respectively.

Affinity and recognition may be analysed along such qualitative or semi-quantitative considerations [1.27, 2.3, 2.4]. These are translated into quantitative factors through the interaction parameters and functions employed in molecular modelling methods [1.38–1.49].

The implementation of molecular recognition requires the search for a quantitative evaluation of complementarity or of its inverse, molecular similarity. Measurement of similarity rests on the definition and construction of molecular shape and potential maps [2.2a]. Fractal analysis may allow the characterization of the surfaces of molecules and solid materials [2.2b]. Such studies are of crucial importance for the design of artificial receptor molecules presenting high recognition. Enhanced recognition beyond that provided by a single equilibrium step may be achieved by proofreading mechanisms involving multistep recognition coupled to an irreversible reaction and non-equilibrium processes [2.5].

Biological molecular recognition represents the most complex expression of molecular recognition leading to highly selective binding, reaction, transport, regulation etc. It provides study cases, illustration and inspiration for the unravelling of basic principles and for the design of model systems as well as of abiotic receptors.

2.2 Molecular Receptors — Design Principles

Molecular receptors are defined as organic structures held by covalent bonds, that are able to bind selectively ionic or molecular substrates (or both) by means of various intermolecular interactions, leading to an assembly of two or more species, a supermolecule.

The design of molecular receptors amounts to expressing in an organic molecule the principles of molecular recognition.

Receptor chemistry, the chemistry of artificial receptor molecules, represents a *generalized coordination chemistry*, not limited to transition-metal ions but extending to all types of substrates: cationic, anionic, or neutral species of organic, inorganic, or biological nature [1.13].

The ideas of molecular recognition and of receptor chemistry have been penetrating chemistry more and more over the last twenty years, namely in view of their bioorganic implications, but more generally for their significance in intermolecular chemistry and in chemical selectivity.

In order to achieve high recognition, the factors mentioned above must be taken into account in the design of the receptor. In particular, complementarity depends on a well-defined three dimensional architecture with the correct arrangement of binding sites. Furthermore, ρ and σ will be in contact over a large area, if ρ is able to wrap around its guest so as to establish numerous non-covalent binding interactions and to sense its molecular size, shape, and architecture. This is the case for receptor molecules that contain intramolecular cavities, clefts or pockets into which the substrate may fit. In such *concave* receptors the cavity is lined with binding sites directed towards the bound species; they are endopolarophilic and *convergent,* and may be termed *endoreceptors. Exoreceptors* bearing externally oriented binding sites will be considered in Chapter 7.

Macropolycyclic structures are of special interest for designing artificial receptors: they are large (macro) and may therefore contain cavities of appropriate size and shape; they possess numerous branches, bridges, and connections (polycyclic) that allow the construction of a given architecture endowed with specific dynamic features; they provide means for the arrangement of structural groups, binding sites, and reactive functions. Binding of a substrate into the cavity will yield an inclusion complex, a *cryptate* [1.26, 2.6].

In addition to maximizing contact area, inclusion also leads to more or less complete solvent exclusion from the receptor site, thus minimizing the number of solvent molecules to be displaced by the substrate on binding.

The balance between rigidity and flexibility is of particular importance for the binding and the dynamic properties of ρ and of σ. Rigid, "lock and key"-type receptors are expected to present very efficient recognition, i.e., both high stability and high selectivity. On the other hand, flexible receptors that bind their substrate by an "induced fit" [2.7] process may display high selectivity but have lower stability, since part of the binding energy is used up in the change of conformation of the receptor.

Processes of exchange, regulation, cooperativity, and allostery require a built-in flexibility so that ρ may adapt and respond to changes. Flexibility is of great importance in biological receptor–substrate interactions, where adaptation is often required for regulation to occur. Such designed dynamics are more difficult to control than mere rigidity, and the developments in computer-assisted molecular design methods, allowing the exploration of both structural and dynamical features, may greatly help [1.38–1.50]. Receptor design thus covers both the static and dynamic features of macropolycyclic structures.

The stability and selectivity of σ binding result from the set of interaction sites in ρ and may be structurally translated into *accumulation* (or collection) and *organization* (or orientation) i.e., respectively bringing together binding sites and arranging them in a suitable pattern. Both processes cost energy, collection in fact appreciably more than orientation [2.8]. It is important to note that the corresponding intersite repulsions are being built into a polydentate ligand in the course of its synthesis.

Molecular receptors of extremely varied structural types have been conceived and investigated. Many of them are of macropolycyclic type. Their design has been guided by the goal of achieving structural control through planned organization, which increases with the cyclic order of the structure and depends on the nature of the fragments that it contains. It is the pursuit of such *preorganization* that led from acyclic to macrocyclic ligands (such as crown ethers [1.24, 1.25] or spherands [2.9, 2.10]), and to the macropolycyclic cryptands [1.26, 1.27].

Monotopic receptors (from the Greek τοπος: site, location) possess a single receptor unit and bind a single substrate. *Polytopic receptors* contain two or more discrete binding subunits [2.11]. These receptors are also usually *mesomolecules* [1.31], molecules of a size intermediate between the small molecules of organic chemistry and the large ones of macromolecular chemistry and of biology.

Receptor molecules belonging to numerous types of geometries can be imagined. A selection of such structures resulting from the combination of non-cyclic and cyclic elements is presented in Figure 2 ; they differ in dimensionality, cyclic order and connectivity and may be characterized by the molecular graph to which they belong and by the corresponding index L_T = connectivity + cyclic order + dimensionality [1.27] (see also [10.10]).

The size and shape of the binding cavity that they define and their rigidity or flexibility are determined by the nature of the structural subunits making up the branches of the graphical representations. Their overall shape and appearance has also led to the coining of a number of so-called trivial names that refer to a particular aspect of the structure. Proposals have been made to name and define different types of macrocyclic ligands and molecular receptors [2.12]. Without being exhaustive, one may mention here the acyclic *podands*, the macrocyclic *crown ethers*, *coronands* or *torands*, the *clathrochelates* and *coordinatoclathrates*, the macropoly-cyclic *cryptands* and *speleands*, the *spherands*, *cavitands* and *carcerands*, the *calixa-renes*, the *cyclophane* receptors, the *cryptophanes*, etc. (see below and Chapt, 3 and 4). All these creatures of a molecular bestiary have been brought to life as the chem-istry of molecular recognition developed. Such names, despite their apparent gratuitousness, nevertheless serve a real purpose of quick retrieval in ones memory of the class of structures they designate. They serve the same purpose as, for instance, cat and dog, words that immediately bring to mind the exact type of ani-mal, without having to detail all their features for conveying the information!

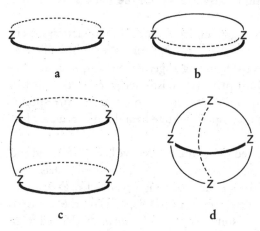

Fig. 2. Some macropolycyclic structures: (a) macrocyclic, (b) macrobicyclic, (c) cylindrical macrotricyclic, and (d) spherical macrotricyclic. L_T = 5, 8, 8 and 9, respectively [1.27].

We shall now discuss several recognition processes and describe the main prop-erties of the corresponding molecular receptors. Insofar as receptor molecules have cyclic geometries and contain cavities into which the substrate(s) may bind, the chemistry of molecular recognition also covers *macrocyclic chemistry* and *inclusion chemistry*. In view of the extensive literature concerning these domains, the reader is referred to specific monographs, reviews or original papers for more information (see Appendix).

2.3   Spherical Recognition — Cryptates of Metal Cations

The simplest recognition process is that of spherical substrates; these are either positively charged metal cations (alkali, alkaline-earth and lanthanide cations) or the negative halide anions (see Chapt. 3).

Although some scattered examples of binding of alkali cations (AC) were known (see [2.13, 2.14]) and earlier observations had suggested that polyethers interact with them [2.15], the coordination chemistry of alkali cations developed only in the last 30 years with the discovery of several types of more or less powerful and selective cyclic or acyclic ligands. Three main classes may be distinguished: 1) natural macrocycles displaying antibiotic properties such as valinomycin or the enniatins [1.21–1.23]; 2) synthetic macrocyclic polyethers, the crown ethers, and their numerous derivatives [1.24,1.25, 2.16, A.1, A.13, A.21], followed by the spherands [2.9, 2.10]; 3) synthetic macropolycyclic ligands, the cryptands [1.26, 1.27, 2.17, A.1, A.13], followed by other types such as the cryptospherands [2.9, 2.10].

In fact, it is from these initial studies of selective AC binding that emerged what eventualy turned into the field of molecular recognition. More specifically, it is the design and investigation of alkali cryptates that started our own work, which by progressive generalization was to become later on supramolecular chemistry. These origins have been briefly described [1.7].

Numerous studies have been performed that are reported in many papers and summarized in reviews and books [A.1, A.13, A.21, A.22]. For instance, valinomycin **1** gives a strong and selective complex **2** in which a K^+ ion is included

1   2

in the macrocyclic cavity [1.21, 2.13, A.2]. Similar inclusion takes place in the complexes of crown ethers, such as **4**, the complex of the Rb^+ ion with dibenzo-18-crown-6 **3**, the bis-annelated derivative of the parent 18-crown-6 (or $[18]-O_6$) macrocycle **5** [2.13, A.2]. Many macrocyclic polyethers have been synthesized by a variety of routes including an efficient high pressure approach [2.19, A.13]; the structures of ligands and metal complexes have been determined [2.13, 2.14, A.2] as well as the thermodynamics and kinetics of the complexation process [1.27, 2.16, 2.18, 2.20]. The stability and selectivity of the complexes depend on the size of the polyoxyethylene crown-ether ring, the best bound cation being that which fits best into the cavity [2.16, 2.18, A.13]. In lariat ethers [2.18d], the side chains influence the binding properties. The ability of macrocycles to form more stable and selective complexes than acyclic ligands has been termed the *macrocyclic effect* [2.16, 2.21–2.23].

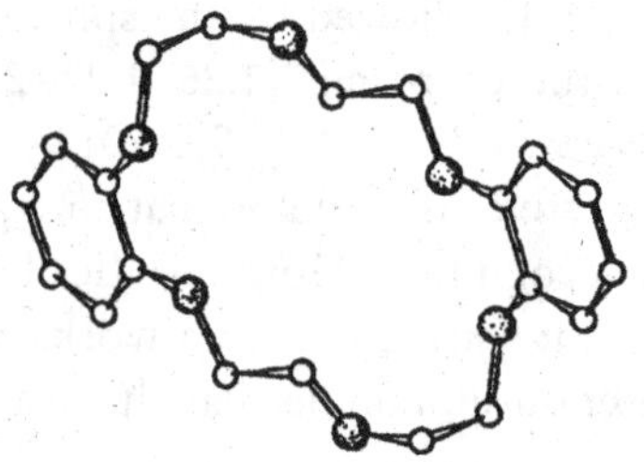

3

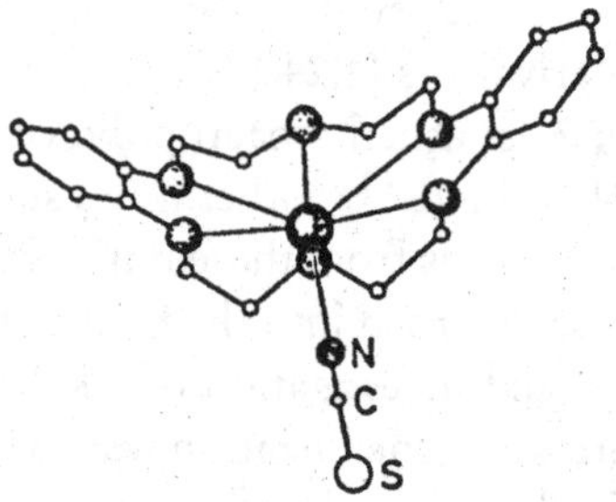

4

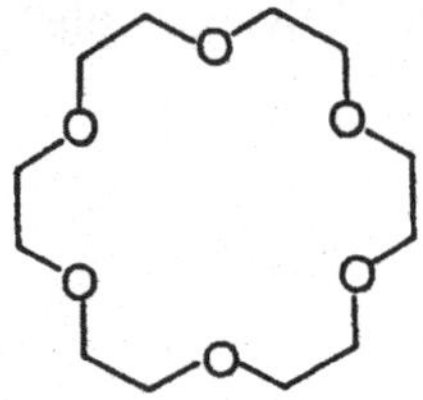

5

Whereas macrocycles define a two-dimensional, circular hole, macrobicycles define a three-dimensional, spheroidal cavity, particularly well suited for binding the spherical alkali cations (AC) and alkaline-earth cations (AEC).

Indeed, macrobicyclic ligands such as **7–9** form cryptates $[M^{n+} \subset \text{cryptand}]$, **10**, by inclusion of a metal cation inside the molecule [1.26, 1.27, 2.17, 2.24–2.26]. The optimal cryptates of AC and AEC have stabilities several orders of magnitude higher than those of either the natural or synthetic macrocyclic ligands. They show pronounced selectivity as a function of the size complementarity between the cation

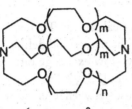

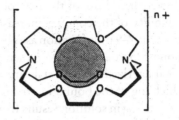

6   m = n = 0
7   m = 0, n = 1
8   m = 1, n = 0
9   m = n = 1

10

and the intramolecular cavity, a feature termed *spherical recognition*. As the bridges of the macrobicycle are lengthened from cryptand 7 to 9, the size of the cavity increases gradually and the most strongly bound ion becomes, respectively, Li^+, Na^+, and then K^+. Thus, these ligands present peak selectivity, being able to discriminate against cations that are either smaller or larger than their cavity. Plateau selectivity is observed for more flexible cryptands, which contain longer chains and therefore larger, more adjustable, cavities; like macrocyclic antibiotics they have high K^+/Na^+ selectivity but weak K^+, Rb^+, Cs^+ discrimination [1.27, 2.17, 2.27]. The cavity size criterion is more of operational than explicative nature, the binding features in solution presenting enthalpic and entropic features more complex than simple steric fit [2.28a].

Proton cryptates are obtained by internal protonation, the cavity concealing the protons very efficiently especially in the case of the protonated forms 11 and 12 of the smaller cryptand 6 [2.29, 2.30].

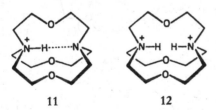

11   12

The macrobicyclic cryptands also bind AEC very strongly. Ligand 9 displays a unique and very high preference for Sr^{2+} and Ba^{2+} over Ca^{2+}. Suitable structural modifications allow control over the M^{2+}/M^+ selectivity from preference for AEC to preference for AC binding [2.31].

Lanthanide cryptates presenting interesting photophysical properties have been obtained [2.32, 2.33] (see also Sect. 8.2).

The cation *exchange kinetics* of the cryptates [1.27] are several orders of magnitude slower than those of the macrocyclic complexes [2.20b] and usually decrease as the stability of the complex increases.

Cryptands 7–9 thus function as receptors for spherical cations. Their special complexation properties result from their macropolycyclic nature and define a *cryptate effect* characterized by high stability and selectivity, slow exchange rates, and efficient shielding of the bound ion from the environment [2.17, 2.27].

Since binding in solution results from a compromise between interaction with the ligand and solvation, new insights into the origin of the cation recognition process and of the macrocyclic and cryptate effects can be gained from experimental gas phase studies [2.34, 2.35] as well as from computer modelling calculations in vacuo or in a solvent [1.35b, 1.42, 1.43, 1.45, 2.36, 2.37, A.37]. In particular, molecular dynamics calculations indicate that complementarity is reflected in restricted motion of the ion in the cavity [1.45, 2.36].

AC and AEC complexation is also effected efficiently by other macrocyclic ligands such as the spherands 13, cryptospherands 14 [2.9, 2.10], calixarenes [2.38, A.6, A.23], torands [2.39], etc., some of them, for instance the spherands displaying particularly high stabilities. A special case is represented by the endohedral complexes of fullerenes in which the cation (Sr^{2+}, Ba^{2+}, lanthanides) is locked inside the closed carbon framework [2.40].

13   14

Polyaza-, polythia-ligands. Recognition of transition metal ions. Replacing the oxygen sites with nitrogen or sulphur yields macrocycles and cryptands that show marked preference for transition metal ions and may also allow highly selective complexation of toxic heavy metals such as cadmium, lead and mercury [2.41–2.44, A.14].

In fact the marked complexation properties of peraza-macrocycles were known before those of the peroxa crown ethers. A large number of aza-macrocycles, that may also contain sulphur and oxygen sites, and polythiamacrocycles [2.45–2.48] have been synthesized, as well as polyaza [2.49, 2.50] and polythia [2.51] macro-polycyclic cryptands such as **15** and **16** [2.52, 2.53]. Closely related to the latter are the sepulchrate (see **17**) [2.54] and analogous macrobicyclic cage ligands, that encap-sulate the bound cation very tightly. Macrocyclic and macrobicyclic ligands containing internally directed functionalized units (such as catechols, see **18** [2.55]) form very strong complexes with various transition metal, lanthanide and actinide ions [2.55–2.57].

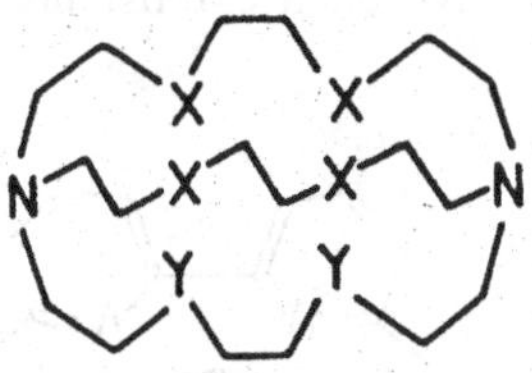

X, Y = O, NR, S

15

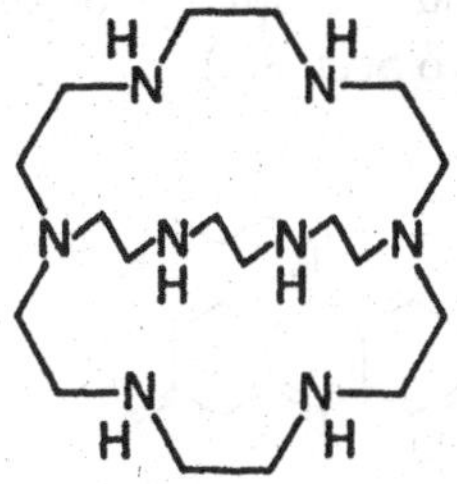

16

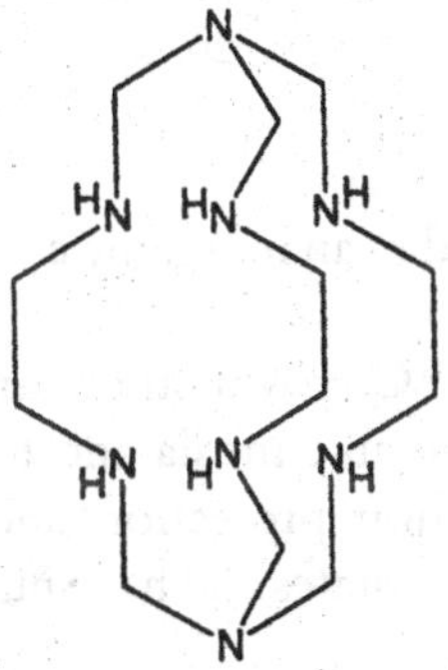

17

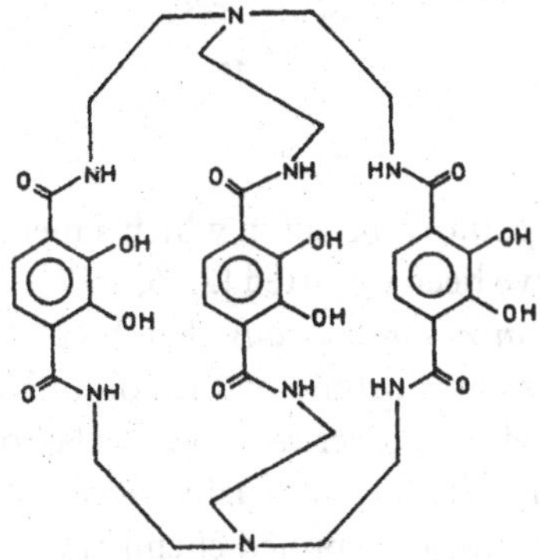

18

The synthetic routes may often involve template directed condensations, a widely used reaction being the (carbonyl + amine) to imine condensation that efficiently leads to a variety of Schiff-base macrocycles [2.58–2.60, A.7, A.14], macrobicyclic cryptands [2.61–2.63] and lacunar cyclidene ligands [2.60, 2.64].

All these ligands present a very rich complexation chemistry with transition metal ions that depends on their specific structure. Thus, they can perform *transition metal ion recognition* by selecting metal ions [2.65, 2.66] according to the coor-dination features of the ion and the nature, number and disposition of the binding sites, as well as to their rigidity. By imposing an unusual, more or less distorted coordination geometry they may markedly affect the spectral, redox and magnetic properties of the bound ion [2.67, 2.68, A.7, A.16], and also activate it in an "entactic state" fashion [2.69]. This is especially true for the transition metal cryptates in which the ion is trapped in a closed cavity.

Numerous macrocyclic and macropolycyclic ligands featuring subheterocyclic rings such as pyridine, furan or thiophene have been investigated [2.70] among which one may, for instance, cite the cyclic hexapyridine torands (see 19) [2.39] and the cryptands containing pyridine, 2,2'-bipyridine (bipy), 9,10-phenanthroline (phen) etc. units [2.56, 2.57, 2.71-2.73]. The [Na$^+$ $\subset$ tris-bipy] cryptate 20 [2.71] and especially lanthanide complexes of the same class have been extensively studied [2.74, 2.75] (see also Sect. 8.2).

19   20

Inorganic cryptates in which a metal cation is enclosed in an inorganic cage structure have been reported [2.76, 2.77].

Chemical applications. The formation of AC and AEC crown ether complexes and cryptates promotes the solubilization of salts in organic media and has three major effects, decreased cation/anion interaction, cation protection and anion activation. These are usually more pronounced for the cryptates and have numerous uses in pure and applied chemistry.

Whereas in crown ether complexes the bound cation is still accessible for ion pairing from the open "top" and "bottom" faces, this is not or less the case for the cryptates. Cryptate formation transforms a small metal cation into a very large, spheroidal, charge-diffuse organic cation about 10 Å in diameter, a sort of super-heavy AC or AEC [2.78]. The stability of the cryptates and the large distance imposed by the organic ligand shell separating the enclosed cation from the environment (both the anion and the solvent) have many physical and chemical consequences. If the anion associated with such a cation is an electron, the resulting species could be considered as a very large alkali-metal atom of extremely low ionization potential that in the isolated state would have analogies with a Rydberg atom [2.78].

Of particular significance in this respect has been the ability to prepare, characterize and study most intriguing species, the *alkalides* [2.79, 2.80] and the *electrides* [2.80, 2.81] containing an alkali metal anion and an electron, respectively, as counterion of the complexed cation. Thus, cryptates are able to stabilize species such as the sodide {[Na$^+$ ⊂ 9]Na$^-$} and the electride {[K$^+$ ⊂ 9]e$^-$}. They have also allowed the isolation of anionic clusters of the heavy post-transition metals, as in ([K$^+$ ⊂ cryptand]$_2$ Pb$_5^{2-}$) [2.82].

Between the two extremes, the atom and the electride, one may envisage (as suggested earlier [2.78]) a species in which the electron would be located on acceptor sites incorporated into the ligand shell surrounding the metal cation bound in the molecular cavity of a spheroidal cryptate, thus yielding a self-contained neutral entity. This represents a way of expanding the atom by stabilizing the electron on a ligand orbital rather than on a metal-centered one. It may be described either as a radical contact ion pair, containing a more or less delocalized electron, or, more intriguingly, as a sort of molecular "expanded atom". Such a species has been obtained by the electroreductive crystallisation of the macrobicyclic tris(bipy) sodium cryptate **20** [2.83] (Fig. 3). It represents the first member of the *cryptatium* family of potential other such molecular "elements" and has been termed sodiocryptatium. In this compound the counter electron is located on the bpy acceptors, localized in the solid state [2.83] and apparently delocalized in solution [2.84b]. Other related species have been generated in solution by electroreduction of the corresponding Ca^{2+} and La^{3+} cryptates [2.84a].

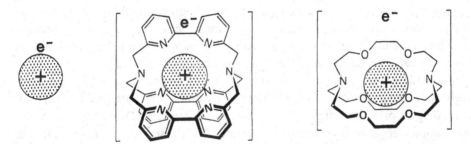

Fig. 3. Schematic representation of the series: (from the left to right) alkali-metal atom, cryptatium (expanded atom or radical contact-ion pair {[Na$^+$ ⊂ tris-bipy cryptand]e$^-$}), electride (Rydberg atom or electron salt) {[K$^+$ ⊂ 9]e$^-$}.

In a broader view one may conjecture that there is here a whole new field to be developed, having to do with large delocalized anionic entities obtained by injecting electrons into extended organic π systems. Cryptatium species represent one class of such compounds having a central positive potential. Many others of either spheroidal or non-spheroidal shape may be envisaged [2.84c].

Cryptation (as well as crown ether complexation) promotes the dissociation of ion pairs resulting in strong *anion activation* [2.78, 2.85]. It markedly affects numerous reactions, such as those involving the generation of strong bases, nucleophilic substitutions, carbanion reactions, alkylations, rearrangements, anionic polymerizations; it may even change their course. Extensive use has also been made of crown ethers and cryptands in *phase transfer catalysis* [2.78, 2.86–2.88]. Conversely, cryptate formation inhibits reactions in which cation participation (electrostatic catalysis) plays an important role. Thus, cryptands are powerful tools for studying the mechanism of ionic reactions that involve complexable metal cations. Their effect on a reaction is a criterion for ascertaining the balance between anion activation and cation participation under a given set of conditions [2.78].

Crown ethers and cryptands, either alone or fixed on a polymer support [2.89], have been used in many processes, including selective extraction of metal ions, solubilization, isotope separation [2.90], decorporation of radioactive or toxic metals [2.17, 2.49], and cation-selective analytical methods [2.89, 2.91, 2.92] (see also Sect. 8.2.2 and 8.4.5). A number of patents have been granted for such applications.

2.4 Tetrahedral Recognition by Macrotricyclic Cryptands

Selective binding of tetrahedral substrates requires the construction of a receptor molecule with a tetrahedral recognition site. This may be achieved by positioning four suitable binding sites at the corners of a tetrahedron and linking them with six bridges. Such a structure has been realized in the spherical macrotricyclic cryptand **21**, which contains four nitrogens located at the corners of a tetrahedron and six oxygens located at the corners of an octahedron, as shown by **22** [2.93].

Indeed, **21** binds the tetrahedral NH_4^+ cation exceptionally strongly and selectively (as compared with K^+), forming the ammonium cryptate [$NH_4^+ \subset$ **21**], **23**. This complex presents a high degree of structural (shape and size) and interaction site complementarity between the substrate, NH_4^+, and the receptor, **21**. The ammonium ion fits into the cavity of **21**, and is held by a tetrahedral array of $^+N-H\cdots N$ hydrogen bonds and by electrostatic interactions with the six oxygens [2.94, 2.95]. The strong binding in **23** results in an effective pK_a for bound NH_4^+ that is about six units higher than that of free NH_4^+; this brings to light how large may be the changes in substrate (or in receptor as well) properties brought about by binding. Similar changes may take place when substrates bind to enzyme active sites and to biological receptors.

The macrobicycle **9** also binds NH_4^+, forming cryptate **24**. The dynamic properties of **24** compared with **23** reflect the receptor–substrate binding complemen-

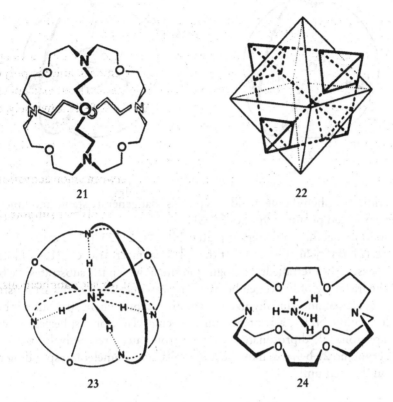

21   22

23   24

tarity: whereas NH_4^+ is firmly held inside the cavity in **23**, it undergoes internal rotation in **24** [2.95].

The remarkable protonation features of **21** led to the formulation of the diprotonated species as the water cryptate, $[H_2O \subset$ **21**, $2H^+]$ **25**, in which the water molecule accepts two $^+N\text{-}H\cdots O$ bonds from the protonated nitrogens and donates two $O\text{-}H\cdots N$ bonds to the unprotonated ones [2.17, 2.96]. The second protonation of **21** is facilitated by the substrate; it represents a "positive cooperativity" effect, mediated by H_2O, in which the first proton and the effector molecule water set the stage both structurally and energetically for the fixation of a second proton. When **21** is tetraprotonated it forms the chloride cryptate cryptate $[Cl^- \subset$ **21**,$4H^+]$ **26**, in which the included anion is bound by four $^+N\text{-}H\cdots X^-$ hydrogen bonds [2.97] (see also Chapt. 3).

Molecular mechanics and dynamics calculations have shown that the preorganization of **21** for cation or anion binding and recognition results from its high connectivity and its preformed cavity with converging binding sites and fairly fixed size [1.45, 2.36b].

Unsymmetrical derivatives of **21** display notably perturbed NH_4^+ binding, with a marked loss in both stability and selectivity [2.94]. Replacing the oxygen atoms in

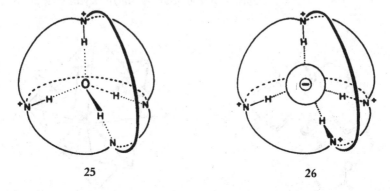

25   26

21 by pyridine or phenyl units yields cryptands that generate stable and kinetically inert metal ion and proton cryptates [2.98].

Considering together the three cryptates [NH_4^+ $\subset$ 21] 23, [H_2O $\subset$ (21, 2H^+)] 25, and [Cl^- $\subset$ (21,4H^+)] 26, it is seen that the spherical macrotricycle 21 is a molecular receptor possessing a tetrahedral recognition site in which the substrates are bound in a tetrahedral array of hydrogen bonds. It represents a state of the art illustration of the molecular engineering involved in abiotic receptor chemistry. Since it binds a tetrahedral cation NH_4^+, a bent neutral molecule H_2O, or a spherical anion Cl^- when unprotonated, diprotonated, or tetraprotonated, respectively, the macrotricyclic cryptand 21 behaves like a sort of molecular chameleon responding to pH changes in the medium!

2.5   Recognition of Ammonium Ions and Related Substrates

In view of the important role played by substituted ammonium ions in chemistry and in biology, the development of receptor molecules capable of recognizing such subtrates is of special interest. It is based on the use of polar $^+$N–H$\cdots$X hydrogen bond interactions [2.99]. Macrocyclic polyethers bind primary ammonium ions [1.24] by anchoring the NH_3^+ group into their circular cavity via three $^+$N–H$\cdots$O hydrogen bonds as shown in 27 [2.100–2.105] however, they complex alkali cation such as K^+ more strongly. Selective binding of R–NH_3^+ may be achieved by extending the results obtained for NH_4^+ complexation by 21 and making use of the oxa-aza macrocyles [2.105, 2.106]. Indeed, the symmetrical triazatrioxamacrocycle [18]-N_3O_3, which forms a complementary array of three $^+$N–H$\cdots$N bonds, 28, selects R–NH_3^+ over K^+ and is thus a receptor unit for this functional group [2.106].

A great variety of macrocyclic polyethers have been shown to bind R–NH_3^+ molecules with structural and chiral selectivity [2.100–2.103, 2.107]. Extensive

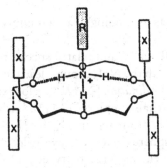

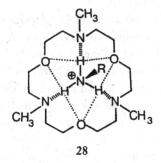

28

a)   X = H
b)   X = CO$_2$⁻
c)   X = CONYY'
d)   X = (L)–CH–CO$_2$Me
          |
        CH$_2$SH

27

29

studies of chiral recognition have been performed with optically active macrocyclic receptors containing subunits such as binaphthyl (see 29) [2.101, 2.102, 2.107] spiro-bifluorenyl [2.108] or sugar derivatives [2.102]. Although the introduction of the binaphthyl group markedly decreases the binding constants, it leads to pronounced chiral discrimination between enantiomeric substrates and also allows the positioning of functional groups around the binding site.

Particularly strong binding is shown by the optically active tetracarboxylate 27b, derived from tartaric acid, which conserves the desirable, [18]-O$_6$ ring and adds strong electrostatic interactions (see also [2.18e]), thus forming the most stable metal-ion and ammonium complexes of any polyether macrocycle [2.109a]. It displays marked selectivity in favour of primary against more highly substituted ammonium ions (central discrimination). Of special interest is its selective binding of biologically active ions such as noradrenalin and norephedrine with respect to their N-methylated derivatives, adrenalin and ephedrine.

Varying the side groups X in 27b affects both the stability and selectivity of the complexes (lateral discrimination), and allows the receptor–substrate interactions in biological systems to be modelled, for instance, the interaction between nicotin-amide and tryptophan [2.109b]. One may attach to 27b amino acid residues (leading to "parallel peptides" [2.109] as in 27c), nucleic acid bases or nucleosides, saccharides, etc. The structural features of 27 and its remarkable binding properties make it an attractive unit for the construction of macropolycyclic multisite receptors, molecular catalysts, and carriers for membrane transport. Such extensions require separate handling of the side groups, leading to face- and side-discriminated derivatives of **27** [2.110].

Gas phase studies reveal size selective effects on binding of ammonium cations to crown ethers and acyclic analogs [2.111].

Binding of metal–amine complexes $[M(NH_3)_n]^{m+}$ and related species to macrocyclic polyethers via N–H$\cdots$O interactions with the NH_3 groups, leads to a variety of supramolecular species by *second-sphere coordination* [2.112]. Cyclodextrins also act as second-sphere ligands for transition metal complexes [2.113]. Such "supercomplexes" may present properties resulting from the interaction between the bound species and the groups of the second-sphere ligand (energy or electron transfer, luminescence quenching, modification of chemical reactivity, etc.) [2.114].

R = CO_2^-

30

Receptor sites for secondary and tertiary ammonium groups are also of interest. $R_2NH_2^+$ ions bind to the [12]-N_2O_2 macrocycle via two hydrogen bonds [2.115]. The case of quaternary ammonium ions is considered in Section 4.3. The guanidinium cation binds to [27]-O_9 macrocycles through an array of six H-bonds [2.116, 2.117] yielding a particularly stable complex **30** with a hexacarboxylate receptor derivative, which also complexes the imidazolium ion [2.116]. Three-point binding of primary amines occurs in polyether macrocycles containing a boron site [2.118].

2.6 Binding and Recognition of Neutral Molecules

The binding and recognition of neutral molecules make use of electrostatic, donor-acceptor and especially of hydrogen bonding interactions [2.119–2.123]. Polar organic molecules such as malonodinitrile form weak complexes with crown ethers and related ligands [2.120].

Of special interest is the use of hydrogen bonding between polar sites for substrate binding. Thus, substrate recognition results from the formation of specific hydrogen bonding patterns between complementary subunits, in a way reminiscent of base pairing in nucleic acids [2.121, 2.124, 2.125]. Such groups have been positioned in acyclic [2.126–2.128] or macrocyclic [2.128] receptors, defining respectively clefts or cavities into which binding of substrate(s) of complementary structure has been shown to take place. Structures **31** and **32** illustrate respectively the complexation through hydrogen bonding of adenine in a cleft [2.126, 2.128] and of barbituric acid in a macrocyclic receptor [2.128]. Extensions towards the binding of peptides have been pursued [2.129, 2.130]. Hydrogen bonding plays a major role in the recognition of nucleic acid sequences by specially designed synthetic molecules [2.131a] or by proteins [2.131b,c] (see also Sect. 5.2) and in the recognition of oligosaccharides by proteins [2.132].

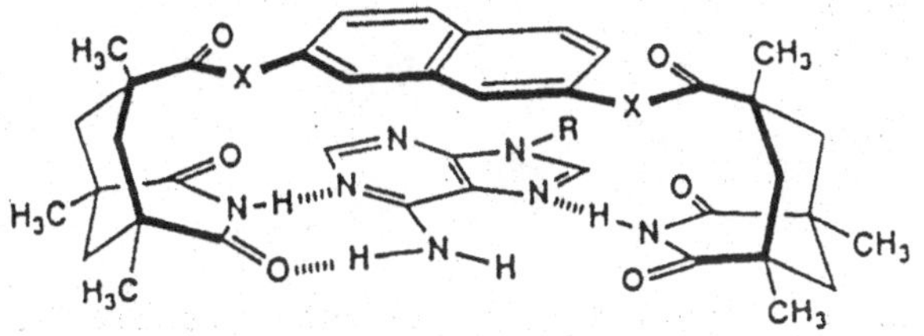

31

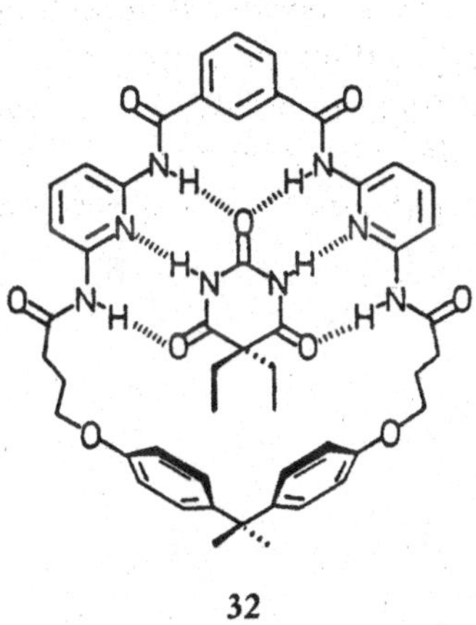

32

The variety of hydrogen bonding patterns that may be envisaged makes these interactions a highly versatile tool for the recognition and orientation of molecules for both biomimetic and abiotic purposes. Approaches to the analysis and quantitative evaluation of such molecular associations have been made [2.133, 2.134]. The precise design of the hydrogen bonding units also benefits from theoretical studies of interaction patterns [2.135]. The binding of neutral substrates into large molecular cavities and recognition at surfaces and in the solid state will be considered in Sections 4.3 and 7.2, respectively.

3 Anion Coordination Chemistry and the Recognition of Anionic Substrates

Although anionic species play a very important role in chemistry and in biology their binding features went unrecognized, whereas the complexation of metal ions and, more recently, of cationic molecules was extensively studied. The coordination chemistry of anions may be expected to yield a great variety of novel structures and properties of both chemical and biological significance. To this end, anion receptor molecules and binding subunits for anionic functional groups have to be devised. Research has been increasingly active along these lines in recent years and anion coordination chemistry is progressively building up as a new area in coordination chemistry [1.8, 1.29, 2.17, 2.25, 2.97, 3.1–3.5].

Anionic substrates have specific features. They are large, compared with cations; they possess a range of geometries, spherical (halides), linear (N_3^-, OCN^-, etc.), planar (NO_3^-, $R-CO_2^-$, etc.), tetrahedral (SO_4^{2-}, ClO_4^-, phosphates, etc.), octahedral ($M(CN)_6^{n-}$).

Positively charged or neutral electron-deficient groups may serve as interaction sites for anion binding. Ammonium and guanidinium units, which form $^+N-H\cdots X^-$ bonds, have mainly been used, but neutral polar hydrogen bonds (e.g., with –NHCO– or –COOH functions), electron-deficient centers (boron, tin, mercury, [3.6, 3.7] as well as perfluoro crown ethers and cryptands [3.8], etc.), or metal-ion centres in complexes also interact with anions.

Polyammonium macrocycles and macropolycycles have been studied most extensively as anion receptor molecules. They bind a variety of anionic species (inorganic anions, carboxylates, phosphates, etc.) with stabilities and selectivities resulting from both electrostatic and structural effects.

Spherical recognition of halide ions is displayed by protonated macropolycyclic polyamines. Thus, macrobicyclic diamines yield katapinates [3.9]. *Anion cryptates* are formed by the protonated macrobicyclic **16**–6H$^+$ [2.52] and macrotricyclic **21**–4H$^+$ [2.97] polyamines, with preferential binding of F$^-$ and Cl$^-$ respectively in an octahedral and in a tetrahedral array of hydrogen bonds.

21–4H$^+$ binds Cl$^-$ very strongly and very selectively compared with Br$^-$ and other types of anions, giving the [Cl$^-$ ⊂ (21–4H$^+$)] cryptate 26. Quaternary ammonium derivatives of oxygen free macrotricycles of type 21 also bind spherical [3.10a] and other [3.10b] anions.

Linear recognition is displayed by the hexaprotonated form of the ellipsoidal cryptand bis-tren 33, which binds various monoatomic and polyatomic anions and extends the recognition of anionic substrates beyond the spherical halides [3.11, 3.12]. The crystal structures of four such anion cryptates [3.11b] provide a unique series of anion coordination patterns (Fig. 4). The strong and selective binding of the linear, triatomic anion N$_3^-$ results from its size, shape and site complementarity to the receptor 33–6H$^+$. In the [N$_3^-$ ⊂ 33–6H$^+$] 34 formed, the substrate is held inside the cavity by two pyramidal arrays of $^+$N–H···N$^-$ hydrogen bonds, each of which binds one of the two terminal nitrogens of N$_3^-$.

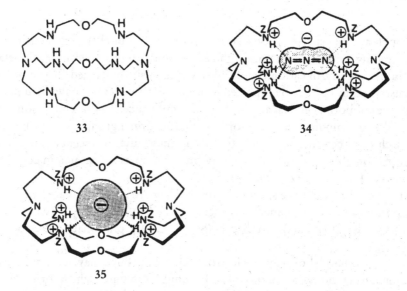

33

34

35

The non-complementarity between the ellipsoidal 33–6H$^+$ and the spherical halides results in much weaker binding and appreciable distortions of the ligand, as seen in the crystal structures of the cryptates 35 where the bound ion is F$^-$, Cl$^-$, or Br$^-$. In these complexes, F$^-$ is bound by a tetrahedral array of hydrogen bonds whereas Cl$^-$ and Br$^-$ display octahedral coordination (Fig. 4). Thus, 33–6H$^+$ is a molecular receptor for the recognition of linear triatomic species of a size compatible with the size of the molecular cavity [3.11].

A cryptate effect is observed for anion complexes as is the case for cation complexes (Section 2.3). In general, an increase in cyclic order from acyclic to macrocyclic to macrobicyclic significantly increases the stability and selectivity of the anion complexes formed by polyammonium ligands.

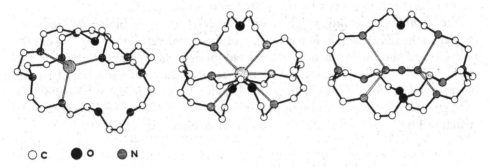

○ C   ● O   ◓ N

Fig. 4. Crystal structures of the anion cryptates formed by the hexaprotonated receptor molecule 33-6H⁺ with fluoride (left), chloride (centre), and azide (right) anions.

Furthermore, in **33** the receptor is built from two protonated tripodal subunits of the tren type, $N(CH_2CH_2NH_2)_3$, located at each pole of the molecule, which cooperate in substrate binding. This is a feature of coreceptor molecules, which will be discussed below (Chapter 4).

Carboxylates and *phosphates* bind to polyammonium macrocycles of various ring sizes (for instance **36–38**) with stabilities and selectivities determined by the structure and the charge of the two partners [3.1–3.5, 3.11–3.16]. The design of receptor units for these functional groups is of much interest since they serve as anchoring sites for numerous biological substrates. Thus, strong complexes are obtained with macrobicyclic polyammonium pockets in which carboxylate (formate, acetate, oxalate, etc.) and phosphate groups interact with several ammonium sites [3.11, 3.12]. The strong complexation of adenosine mono-, di-, and triphosphates (AMP, ADP, and ATP) is particularly significant in view of their role in bioenergetics. It offers the possibility of devising molecular catalysts and carriers for these substrates (see Chapter 5). Crystal structure determination and molecular dynamics calculations on **38**–6H⁺, 6Cl⁻ and on a related macrocycle give information about receptor conformation and chloride ion binding in the solid state and in solution [3.13b]. Substances like **36** and **37** are cyclic analogues of biological polyamines and could thus interact with biomolecules; indeed several macrocyclic polyamines induce efficient polymerization of actin [3.17].

36   37   38

Expanded porphyrins display anion binding in solution and in the solid, a new facet of the chemistry of porphyrin-related macrocycles [3.18].

The *guanidinium group*, which serves as interaction site in biological receptors (as well as the similar amidinium group), is of special interest for the binding of carboxylate and phosphate functions and related species, since it may form two chelating H-bonds with these anionic units. It has been introduced into acyclic [3.19a] and macrocyclic [3.19b] structures. In particular, chiral acyclic receptors containing a rigid guanidinium binding subunit perform chiral recognition of carboxylates (see structure **39**) and of nucleotides through multiple interactions [3.20].

39

The binding of complex anions of transition metals such as the hexacyanides $M(CN)_6^{n-}$ yields second-coordination-sphere complexes, *supercomplexes*, [3.13a] and affects markedly their electrochemical [3.21, 3.22] and photochemical [3.23] properties.

Cascade-type binding [3.24] of anionic species occurs when a ligand first binds metal ions, which then serve as interaction sites for an anion. Such processes occur for instance in lipophilic cation–anion pairs [1.31] and with Cu(II) complexes of bis-tren **33**, of macrocyclic polyamines [3.25, 3.26] and of calixarenes [3.27].

40   **41**

Heteronuclear NMR studies give information about the electronic effects induced by anion complexation as found for chloride–cryptates [3.28].

The complexation of various molecular anions by other types of macrocyclic ligands has been reported [3.1–3.4] in particular with cyclophane-type compounds. Two such receptors are represented by the protonated forms of the macropolycycles **40** [3.29] and **41** [3.30]. Quaternary polybipyridinium compounds also bind anionic substrates [3.31]. Progress is also being made towards the developments of neutral anion receptor molecules [3.32]. The thermodynamic and kinetic data for anion complexation by macrocyclic receptors have been reviewed [2.18c].

Theoretical studies may be of much help in the design of anion receptors and in the a priori estimation of binding features, as recently illustrated by the calculation of the relative affinity of **21**–4H$^+$ for chloride and bromide ions [3.33].

Anion coordination chemistry has thus made very significant progress in recent years. The development of other receptor molecules possessing well-defined geometrical and binding features will make it possible to further refine the requirements for anion recognition, so as to yield highly stable and selective anion complexes with characteristic coordination patterns.

4 Coreceptor Molecules and Multiple Recognition

Once recognition units for specific groups have been identified, one may consider combining several of them within the same macropolycyclic architecture. This leads to *polytopic coreceptor molecules* containing several discrete binding subunits which may cooperate for the simultaneous complexation of several substrates or of a multiply bound (polyhapto) polyfunctional species. Suitable modification would yield *cocatalysts* or *cocarriers* performing a reaction or a transport on the bound substrate(s). Furthermore, because of their ability for multiple recognition and because of the mutual effects of binding subunit occupation, such coreceptors provide entries into higher forms of molecular behaviour such as cooperativity, allostery and regulation, as well as communication or signal transfer if a species is released or taken up [2.11] (see also Chapters 8 and 9).

One may distinguish *cosystems* for which the binding of several substrates is commutative, and *cascade systems,* for which the substrate binding steps are non-commutative but follow a given sequence. In the case of the binding of two substrates σ_1 and σ_2 to a ditopic coreceptor, cascade binding follows the sequence $\rho + \sigma_1 \rightarrow [\rho\sigma_1] \xrightarrow{\sigma_2} [\rho\sigma_1\sigma_2]$ whereas in a cosystem binding is commutative, either σ_1 or σ_2 may bind first to ρ followed by the other substrate.

Polytopic receptors have been called *homotopic* or *heterotopic* depending on whether they contain identical or different subunits respectively [2.11]; one might prefer the words *autotopic* and *allotopic* respectively, in order to avoid confusion with terminology used for molecular chirality. Each receptor subunit , like a monotopic receptor, contains one or several binding sites (corresponding to groups of atoms such as functional groups) each one involving elementary interaction sites (electrostatic, hydrogen bonding, etc.). Restricting the structural elements to chelating (X) and macrocyclic (M) subunits, combination of m macrocycles yields m-topic M^m-macro ($m+1$) cycles; x chelates and m macrocycles yield an $(x+m)$-topic X^x–M^m-macro-$(m+1)$ cycle; if the subunits are different, this may be noted by using primes or indices, e.g., MM' or M_1M_2 for two different macrocycles. The *topicity* is equal to the total number of subunits and the cyclic order to $(m+1)$, where m is the number of macrocyclic subunits.

Substrate binding by either type of receptor (homo- or heterotopic) may occur at a given subunit with different localization and orientation in space, for instance inside or outside a cavity yielding supramolecular *endo/exo isomers*; this holds also for monotopic receptors. Heterotopic coreceptors present in addition *haptoselectivity*, i.e., preferential fixation of a given substrate at a given subunit, resulting from local complementarity of size, shape and binding interactions.

With respect to number and nature of the bound substrates the resulting supermolecule may be *mono-* or *polynuclear* and *homonuclear* or *heteronuclear*, respectively if the substrates are identical or different.

The mode of substrate binding may be *monohapto* or *polyhapto* (η^n), depending on whether fixation occurs via a single or via multiple (substrate group-binding subunit) association(s) (see also [2.11]).

The simplest class of coreceptors is that containing two binding subunits, ditopic coreceptors, which may belong to different structural types. The combination of chelating, tripodal, and macrocyclic fragments yields macrocyclic, axial or lateral macrobicyclic, or cylindrical macrotricyclic structures (Fig. 5). Depending on the nature of these units, the resulting coreceptors may bind metal ions, organic molecules, or both.

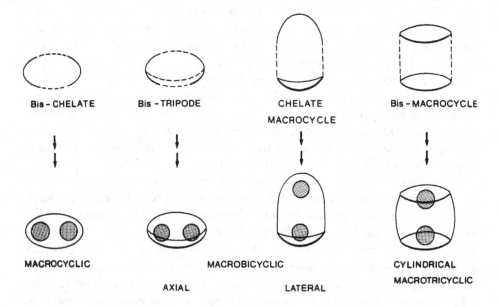

Fig. 5. Some dinuclear cryptates of macropolycyclic cryptands resulting from connection of chelating, tripodal, and macrocyclic subunits [3.24].

4.1 Dinuclear and Polynuclear Metal Ion Cryptates

Macropolycyclic ligands of coreceptor type incorporating two or more binding subunits for metal ions, form dinuclear or polynuclear cryptates in which the distance and arrangement of the cations held inside the molecular cavity may be controlled through ligand design. They allow the study of cation–cation interactions (magnetic coupling, electron transfer, redox, and photochemical properties) as well as the inclusion of bridging substrates to yield cascade complexes, which are of interest for bioinorganic modelling and multicentre–multielectron catalysis.

Depending on the nature and number of binding subunits and of connecting bridges used as building blocks, a variety of macropolycyclic structures may be envisaged. Ditopic ligands that contain two chelating, tripodal, or macrocyclic units bind two metal ions to form dinuclear cryptates of various types (Fig. 5). Combining three or four such groups leads to tritopic and tetratopic metal ion receptors. Dissymmetric ligands that contain subunits with "hard" and "soft" binding sites yield complexes in which the bound ions act as either Lewis acid or redox centres. Many representatives of these types of ligands and complexes have been obtained and studied. Only a few will be illustrated here [3.24, 4.1].

Numerous dinucleating macrocyclic and macrobicyclic ligands have been synthesized, in particular by the versatile amine + carbonyl → imine reaction; they form dinuclear metal complexes as well as cascade complexes with bridging groups [2.58–2.63, 3.24–3.27, 4.1–4.4], for instance in dicobalt complexes that are oxygen carriers [3.26].

Cascade type dinuclear copper(II) cryptates of macrocyclic ligands (e.g., **38**) or macrobicyclic ligands (e.g., **33**) containing bridging groups (imidazolato, hydroxo, or azido) may display antiferromagnetic or ferromagnetic coupling between the ions [4.4] and bear relation to dinuclear sites of copper proteins (see, for instance, **42**) [4.2]. Macrobicyclic hexaimine structures, produced in a one-step multiple condensation reaction, form dinuclear and trinuclear cryptates such as the bis-Cu(I) **43** [2.61], tris-Ag(I) **44** [4.5] and strongly coupled dicopper [2.63b] complexes.

42   43

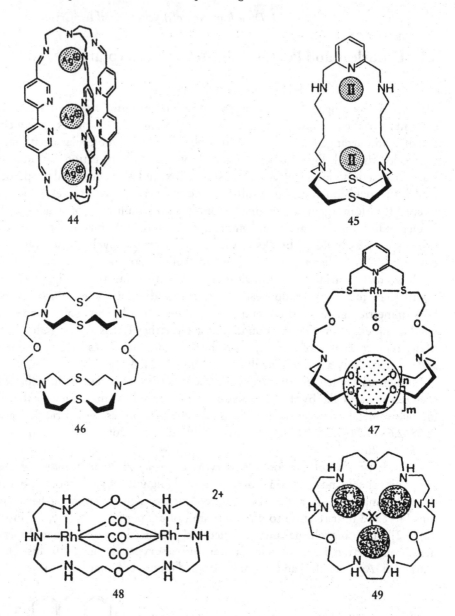

44

45

46

47

48

49

Lateral macrobicycles are dissymmetric by design; thus, monoelectronic reduction of the Cu(II) ion bound to the [12]-N$_2$S$_2$ macrocyclic subunit in the bis-Cu(II) cryptate **45**, gives a mixed valence Cu(I)–Cu(II) complex [4.6]. Macrotricycle **46** forms a dinuclear Cu(II) cryptate that acts as a dielectronic receptor and exchanges two electrons in a single electrochemical wave [4.7]. Complexes of type **47** combine a redox centre and a Lewis acid centre for the potential activation of a bound substrate [4.8].

Polytopic receptors have the ability to assemble metal ions and bridging species within their molecular cavity to form *"cluster cryptates"*. The bis-chelating macrocycle **38** gives complex **48**, in which a triply bridged [Rh(CO)$_3$Rh]$^{2+}$ unit is built inside the ligand cavity [4.9]. A trinuclear complex **49** containing a [tris-Cu(II), bis μ_3-hydroxo] group in the cavity is formed by a tritopic, tris-ethylene diamine macrocyclic ligand [4.10]. Modelling of biological iron-sulphur cluster sites may employ acyclic assembling ligands [4.11] or inclusion into appropriate macrocyclic cavities [4.12]. These few examples may at least have shown how rich the field of polynuclear metal cryptates is, both in structures and properties. Their chemical reactivity and use in catalysis have barely been studied up to now.

4.2 Linear Recognition of Molecular Length by Ditopic Coreceptors

Receptor molecules possessing two binding subunits located at the two poles of the structure will complex preferentially substrates bearing two appropriate functional groups at a distance compatible with the separation of the subunits. This distance complementarity amounts to a recognition of molecular length of the substrate by the receptor. Such *linear recognition* by ditopic coreceptors has been achieved for both dicationic and dianionic substrates, diammonium and dicarboxylate ions respectively; it corresponds to the binding modes illustrated in **50** and **51**.

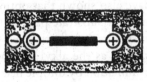
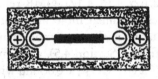

50   51

Incorporation of [18]-N$_2$O$_4$ macrocyclic subunits that bind $-NH_3^+$ groups into cylindrical macrotricyclic [4.13] and macrotetracyclic [4.14] structures yields ditopic coreceptors that form molecular cryptates such as **52** with terminal diammonium cations $^+H_3N-(CH_2)_n-NH_3^+$. In the resulting supermolecules the substrate is located in the central molecular cavity and anchored by its two NH$_3^+$ groups in the macrocyclic binding sites, as shown by the crystal structure **53** of the receptor in **52** (R = NA) with its complementary substrate (A = (CH$_2$)$_5$) [4.15]. Changing the length of the bridges R in **52** modifies the binding selectivity in favour of the substrate of complementary length. Other macrocyclic units and branches may be incorporated into macropolycyclic receptors leading to specific recognition prop-

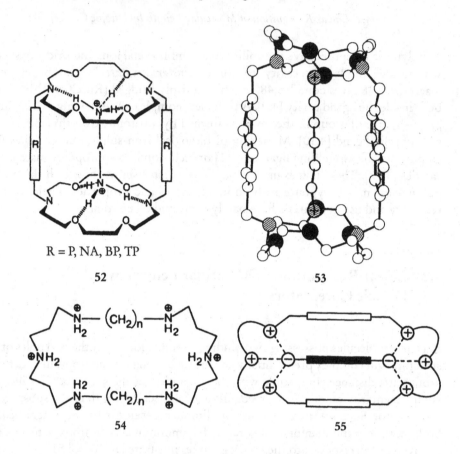

R = P, NA, BP, TP

52   **53**

54   **55**

erties [2.105, 4.16, 4.17]. NMR relaxation data have also shown that optimal partners present similar molecular motions in the receptor substrate pair. Thus, complementarity in the supramolecular species expresses itself in both steric and dynamic fit (see Section 4.5). *Dianionic substrates*, such as the alkyl dicarboxylates
$^-O_2C-(CH_2)_n-CO_2^-$, are bound with length discrimination by ditopic macrocycles
such as **54**. These coreceptors contain two triammonium groups as binding subunits
interacting with the terminal carboxylate functions via a pattern schematically
shown in **55** [4.18].

The crystal structure **57** of the strong and selective complex formed by the
terephthalate dianion with a hexaprotonated macrobicyclic polyamine shows that it
is a molecular cryptate **56** with the dianion tightly enclosed in the cavity and held
by formation of three hydrogen bonds between each carboxylate and the ammonium groups [4.19]. Both structures **53** and **57** illustrate nicely what supermolecules
really are; they show two covalently built molecules bound to each other by a set of
non-covalent interactions to form a well-defined novel entity of supramolecular
nature. Acyclic [4.20a,b] and macrobicyclic [4.20c] hydrogen bonding receptors

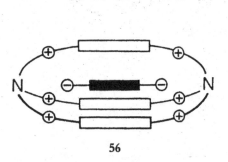

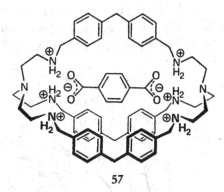

56   57

also bind dicarboxylic acids and dicarboxylates, a helicene-derived ligand effecting high diastereoselective recognition [4.20b].

Thus, for both the terminal diammonium and dicarboxylate substrates, selective binding by the appropriate receptors describes a linear recognition process based on length complementarity in a ditopic binding mode. Important biological species, such as polyamines, amino acid and peptide diamines, and dicarboxylates [4.18] may also be bound selectively. Recognition is achieved by multiple coordination to metal ions in dinuclear bis-macrocyclic coreceptors that complex selectively complementary bis-imidazole substrates of compatible length [4.21].

Numerous variations in the nature of the binding subunits or of the bridges linking them are conceivable and may be tailored to specific complexation properties (see, for instance, [4.16, 4.17, 4.22]).

4.3 Heterotopic Coreceptors — Cyclophane Receptors, Amphiphilic Receptors, Large Molecular Cages

The combination of binding subunits of different nature yields heterotopic receptors that may complex substrates by interacting simultaneously with cationic, anionic and neutral sites, making use of electrostatic and van der Waals forces as well as of donor–acceptor and solvophobic effects. Whereas homotopic coreceptors complex dicationic or dianionic substrates, heterotopic coreceptors may allow the binding of two different substrates, of ion pairs or of zwitterionic species [4.23, 4.24]. Enantioselective and diastereoselective molecular recognition is achieved by chiral coreceptors [4.25a]. A particularly interesting example of the latter is the binding of aromatic amino acids with high enantioselective recognition by an acyclic tritopic receptor that contains a guanidinium, a macrocyclic and a naphthalenic unit for simultaneous interaction with, respectively, the carboxylate, the ammonium

and the aromatic groups of the substrate (see structure **58**) [4.25b]. Hydrogen bonding and stacking or ionic forces have been used for the recognition of amino acids
and nucleotides [3.20], as well as of neutral heterocyclic molecules through interactions of base-pairing type [2.126–2.128].

58

A short zwitterionic recognition sequence of much biological importance is the
"signal" peptide RGD (ArgGlyAsp); its binding to specific biological receptors by
interaction through its guanidinium and carboxylate side chains plays a critical role
in various processes involving cell adhesion [4.26].

Fitting the macrocyclic polyamine **38** with a side chain bearing a 9-aminoacridine
group yields a coreceptor capable of both anion binding, via the polyammonium
subunit, and stacking interaction via the intercalating dye. It interacts with both the
triphosphate and the adenine groups of ATP and provides in addition a catalytic site
for its hydrolysis (see structure **82** and Section 5.2) [4.27].

The positioning of suitable binding subunits and the shaping of a hydrophobic
cavity for inclusion of organic substrates require receptors presenting large and
more or less rigidly connected architectures of macrocyclic or cage-like [2.56, 2.57]
nature.

The naturally occurring *cyclodextrins* were the first receptor molecules whose
binding properties towards organic molecules were recognized and extensively
studied, yielding a wealth of results on physical and chemical features of molecular
complexation [1.32, 4.28, 4.29].

Numerous types of synthetic macrocyclic receptors that contain various organic
groups and polar functions have been developed in recent years. They complex both
charged and uncharged organic substrates. Synergetic operation of electrostatic and
hydrophobic [4.30] effects may occur in *amphiphilic receptors* combining charged
polar sites with organic residues that shield the polar sites from solvation and
increase electrostatic forces. Although the results obtained often describe mere
binding rather than actual recognition, they have provided a large body of data that
make it possible to analyse the basic features of molecular complexation and the

properties of structural fragments to be used in receptor design. We give here only some illustrative examples and describe some of our own results in this area, referring the reader to specific reviews of this vast subject.

α-cyclodextrin

Large hydrophobic cavities have been obtained by means of macrocyclic or of three-dimensional macropolycyclic receptors of *cyclophane* type, which can be made water soluble by appropriate polar groups [2.56, 4.31–4.34, A.11]. Many such molecules have been obtained and some of them present remarkable binding properties towards various substrates. They make use of a variety of structural components and of interactions and play a very important role in both the basic and applied aspects of molecular recognition of organic molecules. Macrocyclic species based on condensed rings and forming belts, collars, etc. represent interesting structures for the development of rigid cyclophane type receptors [4.35]. Another class of compounds, the *calixarenes*, has been very actively investigated. A great number of derivatives of various types have been synthesized; they display a variety of substrate binding and reactivity features [4.36, A.6, A.23]. The *cavitands* [4.37] present enforced cavities as do the rigid and hollow cucurbituril type receptors [2.103c]. Large cyclic structures have been obtained based on porphyrin [4.38a,b] and steroid [4.32b, 4.38c] subunits as well as on extended phenylacetylene scaffoldings [4.38d].

The spherically shaped *cryptophanes* are of much interest in particular for their ability to bind derivatives of methane, achieving for instance chiral discrimination of CHFClBr; they allow the study of recognition between neutral receptors and substrates, namely the effect of molecular shape and volume complementarity on selectivity [4.39]. The efficient protection of included molecules by the *carcerands* [4.40] makes possible the generation of highly reactive species such as cyclobutadiene [4.41a] or orthoquinones [4.41b] inside the cavity. Numerous container molecules [A.38] capable of including a variety of guests have been described. A few representative examples of these various types of compounds are shown in structures **59** (cyclophane) **60** (cubic azacyclophane [4.34]), **61a**, **61b** ([4]- and [6]-calixarenes), **62** (cavitand), **63** (cryptophane), **64** (carcerand).

59

60

61a

61b

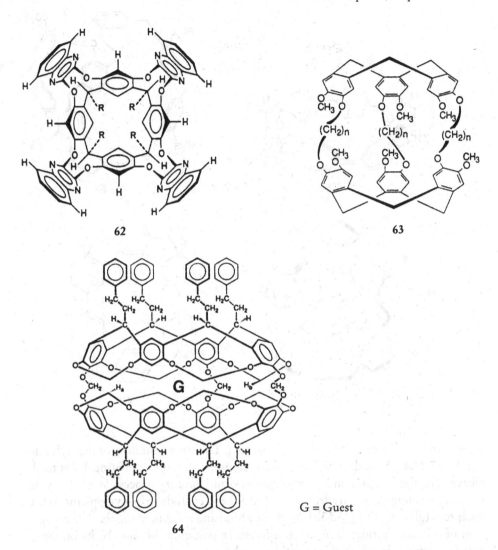

62

63

G = Guest

64

   Macropolycyclic structures containing polar binding subunits maintained by
apolar shaping components, termed *speleands*, yield molecular cryptates (*speleates*)
by substrate binding [4.42]. Thus, the macrocyclic **65** [4.43] and macrobicyclic **66**
[4.44] cyclophane type receptors, incorporating carboxylate groups and diphenyl-
methane units form stable complexes with various ammonium ions, and in particular strongly bind quaternary ammonium substrates. Similar effects operate in other
anionic receptors complexing quaternary ammonium cations and other substrates
[4.45–4.52]. Of special significance is the complexation of *acetylcholine*, which provides information about the type of interactions that may play a role in biological
acetylcholine receptors, such as the combination of negative charge with hydro-

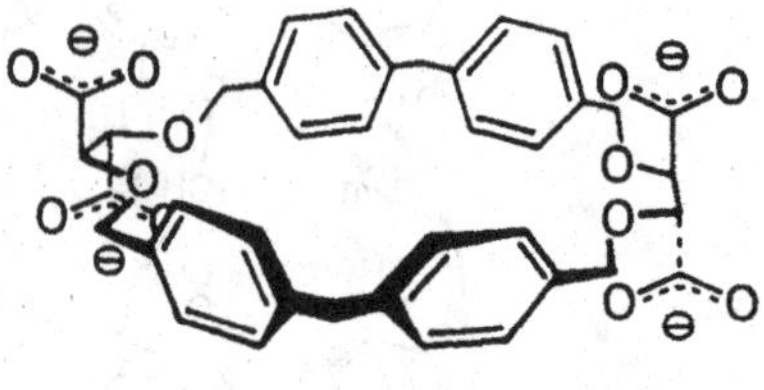

65

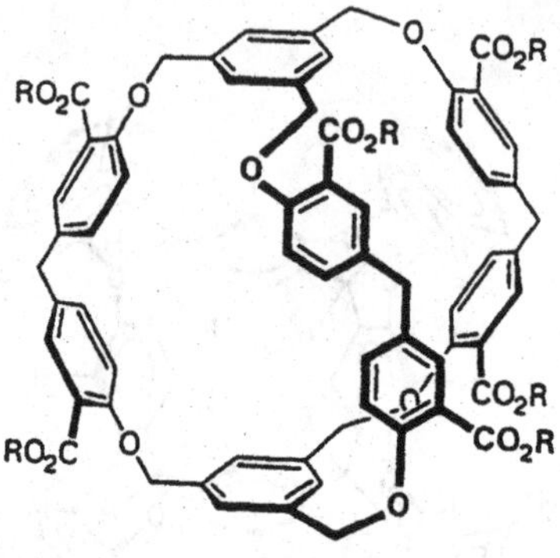

66

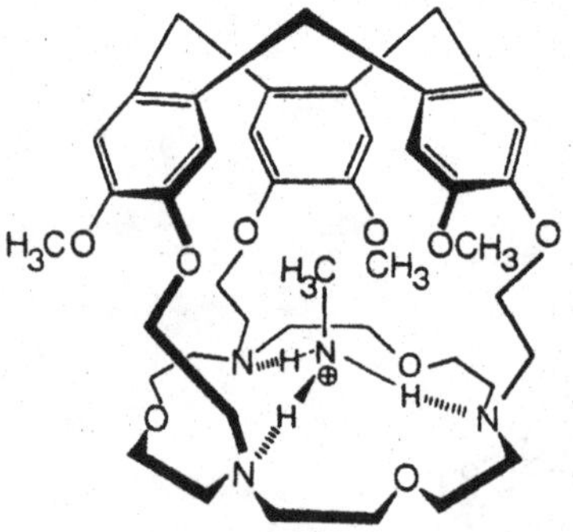

67

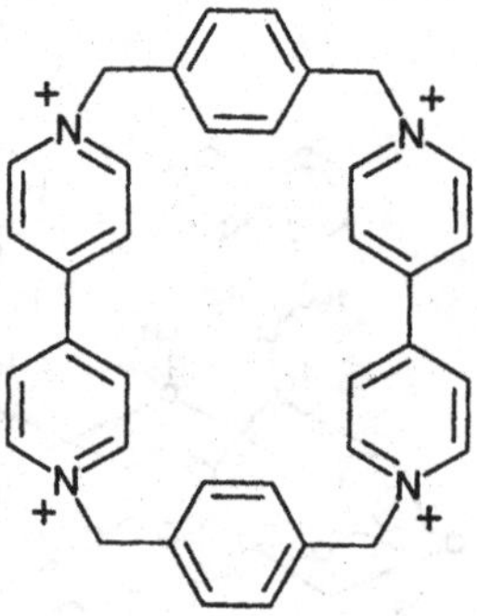

68

phobic walls and cation–π interactions [4.45c]. The strong binding of methylviologen by **65** [4.43a] yields a photosensitive complex that on irradiation leads to the cleavage of the receptor in a sort of *photosuicide binding* process [4.43b]. Cyclophanes containing a polyhydroxo core [4.53a] and cyclodextrin–cyclophane hybrid receptors (glycophanes) [4.53b] are able to bind and solubilize sugars. The recognition of oligosaccharides is of much interest, in particular because of its biological role for instance in cell–cell interaction [4.53c].

The CH_3–NH_3^+ cation forms a selective speleate **67** by binding to the [18]-N_2O_3 subunit of a macropolycycle maintained by a cyclotriveratrylene shaping component. The tight intramolecular cavity efficiently excludes larger substrates [4.42, 4.54].

Amphiphilic binding also occurs for molecular anionic substrates [4.31, 4.32, 4.34]. Charged heterocyclic ring systems, such as those derived from the pyridinium group, represent an efficient way to introduce simultaneously electrostatic interactions, hydrophobic effects, structure, and rigidity into a molecular receptor; in addition, they may be electroactive and photoactive. 4,4′-Bipyridinium groups have

been used with particular success as interaction sites and as components in molecular receptors, such as the rigid macrocyclic box-like structure 68 [4.55].

Planar units such as diazapyrenium dications bind flat organic anions remark-
ably well in aqueous solution, through electrostatic interactions as well as hydrophobic stacking [4.56a]. They also interact strongly with   and effect efficient photochemical cleavage of nucleic acids, in particular at guanine sites [4.56b]. π-Stacking
and hydrogen bonding sites may be used simultaneously for binding substrates such
as nucleobases or related heterocycles [2.128, 4.57–4.59]. Tweezer-like receptors
containing two subunits of intercalating type complex flat organic molecules
through double π-stacking (π-sandwiching) and may also contain hydrogen bonding sites [4.57, 4.58].

Receptors of *cyclointercaland* type, which incorporate intercalating units into a
macrocyclic system, are of interest both for the binding of small molecules and for
their eventual interaction with nucleic acids. A cyclo-bisintercaland has been found
to form an intercalative molecular cryptate in which a nitrobenzene molecule is in-
serted between the two planar subunits [4.60a]. Such receptors are well suited for
the recognition of substrates presenting flat shapes through π-stacking, for instance
when electron deficient intercalating dyes are incorporated. They may also make use
of additional hydrogen bonding [4.58b]. The macrobicyclic bis-intercaland 69 and
related receptors bind nucleosides, nucleotides and planar anionic substrates
[4.60b–e]. Such compounds may also interact with nucleic acids as shape selective
structural probes [4.60f].

X = O, A = — (CH$_2$)$_6$—

69

Although much work has already been performed and many interesting results
have been obtained, there is still a very open future for the design of receptor molecules especially tailored towards the recognition of more or less complex organic
molecules, such as the C$_{60}$ fullerene which forms a supramolecular association with

γ-cyclodextrin [4.61a], with calixarenes [4.61b] and in the solid state [4.61c]. Combination of structural units whose features have already been defined and construction of novel architectures offer wide perspectives.

4.4 Multiple Recognition in Metalloreceptors

Metalloreceptors are heterotopic coreceptors that are able to bind both metal ions and organic molecules by means of substrate-specific units. Cascade complexes form by first binding metal ions, which then serve as interaction site for another substrate (see above Section 4.2). Such is also the case, for instance, for the binding of neutral guests (e.g., urea) onto an electrophilic metal centre (e.g., UO_2^{2+}) bound in a macrocyclic cavity or in a cleft [4.62], for the recognition of nucleobases by simultaneous coordination to a complexed metal ion and hydrogen bonding to oxygen sites [4.63a], for the binding of the hydrazinium ion to an organopalladium crown ether [4.63b] or of phen ligands to three Cu(I) centres held inside a large macrobicyclic cavity [4.63c]. Substrate bound on the metal sites serves as template for the synthesis of macrocycles containing several porphyrin complexes as subunits [4.64]. Suitably arranged porphyrin subunits may present novel properties, such as the catalysis of multi-electron redox reactions by cofacial metallo-diporphyrins [4.65].

Porphyrin and α, α'-bipyridine (bipy) groups have been introduced as metal-ion binding units into macropolycyclic coreceptors (see also [4.38]) containing in addition macrocyclic sites for anchoring NH_3^+ groups [4.66a]. These receptors yield mixed-substrate supermolecules by simultaneously binding metal ions and diammonium cations as shown in 70. The bis-porphyrin receptor present in 70 binds to

70

polynucleotides, displaying higher affinity for the single-stranded than for the double-stranded species [4.66b].

Metalloreceptors, and the supermolecules that they form, thus open up a vast area for the study of interactions and reactions between simultaneously co-bound organic and inorganic species [4.67, 4.68]. In view of the number of metal-ion complexes known and of the various potential molecular substrates, numerous types of metalloreceptors may be imagined that would be of interest as abiotic chemical species or as bioinorganic model systems.

4.5 Supramolecular Dynamics

A receptor–substrate supermolecule is characterized by its geometric (structure, conformation), its thermodynamic (stability, enthalpy and entropy of formation) and its kinetic (rates of formation and of dissociation) features. Numerous physico–chemical studies have provided quantitative data on the thermodynamic and kinetic properties of supramolecular species; this holds for metal ion complexes of crown ethers, of cryptands and of a great variety of macrocyclic and macropolycyclic ligands, as well as for anion binding by various receptors (see references in Sections 2.3–2.5, 4.1–4.4, and Chapter 3). Similar data have also been obtained for numerous supermolecules formed by interaction of neutral molecules with receptor species [2.18, 4.69, 4.70]. All these studies give information about the free energies, the enthalpies and the entropies of formation of the supramolecular species as well as on their rates of association and dissociation, revealing relationships between the nature of the components, the energetic properties of the non-covalent interactions (interaction functions, thermochemistry [4.71]) and the recognition features. The latter are markedly affected by the medium, solvent effects being expressed in the enthalpies and entropies of formation, such as the negative entropies of formation of alkali cryptates in aqueous solution [2.28] and the increasing exothermicity of the binding to cyclophane receptors as the polarity of the solvent increases [4.72].

The medium may have a marked effect on the shape of receptor molecules itself. Shape modifications could strongly influence their substrate binding properties, for instance in the case of amphiphilic cyclophane receptors subjected to hydrophobic–hydrophilic factors in aqueous solution. Such medium effects in action are visualized by the solid state structures of two different forms of the water-soluble hexasodium salt of the macrobicyclic cyclophane 66, which could be crystallized in two very different shapes: an inflated cage structure 71 building up cylinders disposed in a hexagonal array; and a flattened structure 72 stacked in molecular layers separated by aqueous layers in a lamellar arrangement [4.73]. These two

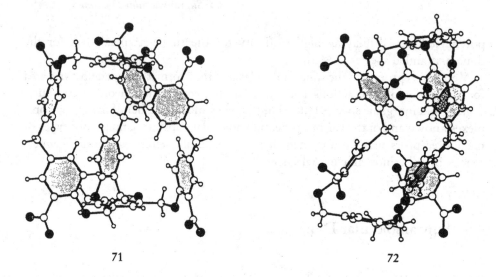

71   72

structures clearly illustrate how medium effects perturb molecular shape and how this shape is related to the supramolecular organisation of the medium. The occurrence of two highly different structures for the same molecule **66** may be compared with the behaviour of amphiphilic molecules that can organize into different periodic structures depending on the conditions [4.74, 4.75]. The ability of molecules such as **66** to selectively bind given substrates is expected to be influenced by medium effects, e.g. hydrophobic effects, that will deform the receptor shape. Such factors may be expected to come into play when functional molecules are incorporated into membrane phases as well as in the determination of the form and function of biomolecules.

In addition to its classical features (geometry, thermodynamics, kinetics) a receptor–substrate supermolecule is also characterized by its internal dynamics, its *dynamic cohesion*, i.e., by the coupling between the molecular motions of the two (or more) spatially co-confined entities of which it is composed. Nuclear relaxation data provide information about *overall* molecular motions and insight into *local* molecular motions, which may be divided into overall reorientations and internal segmental motions [4.76, 4.77]. Extension of these methods to the investigation of molecular association has shown that donor–acceptor complexes may be extremely short lived, with a lifetime comparable to the overall reorientation rates for the supramolecular fluorene–trinitrobenzene pair [4.78]. This raises the question about the existence criterion of a complex or of a supermolecule, i.e. what minimal lifetime it should possess to exist as a discrete species, such as a lifetime at least comparable to its reorientation time [4.78]. On the other hand, the relaxation data for the free and complexed species, indicate that complexes of α-cyclodextrin present weak dynamic coupling between substrate and receptor (anisodynamic ρ, σ pair), the bound substrate displaying faster molecular motions than the receptor in the complex (low coupling coefficient) [4.79]. Potential energy calculations have shown that there are small or no barriers to rotation of a substrate inside the cavity of α-, β- or γ-cyclodextrin [4.80].

Extension of such studies to the supermolecules formed by the binding of diammonium ions to macrotricyclic receptors, as in **52**, showed that structurally compatible species present strong dynamic coupling (isodynamic ρ, σ pair), both partners having similar molecular motions, as indicated in **73** (correlation times given in picoseconds); this results from the combination of *steric fit* and *dihapto-binding*, that anchors σ by its two end groups into the receptor. For substrates that are either too short or too long the cohesion is weakened with a concomitant decrease in the coupling coefficient [4.81].

73

The dynamic character of the ρ, σ pair in the diammonium cryptates **52** represents the extent of *dynamic fit*. Thus, complementarity in the supermolecular species expresses itself in both steric and dynamic fit, strong or weak motional coupling between ρ and σ depending on the degree of complementarity.

Dynamic features of supermolecules correspond on the intermolecular level to the internal conformational motions present in molecules themselves and define molecular recognition processes by their dynamics in addition to their structural aspects. They add a further important facet to the behaviour of these species and may influence their functional features in reactions and transport processes as well as in polymolecular assemblies.

5 Supramolecular Reactivity and Catalysis

The design of highly efficient and selective reagents and catalysts is one of the major goals of research in chemistry, the science of matter and of its transformations.

The particularly remarkable features displayed in this respect by the natural catalysts, the enzymes, has provided major stimulus and inspiration for the development of novel catalysts by either manipulating the natural versions or by trying to devise entirely artificial catalysts that would nevertheless display similar high efficiencies and selectivities. In his Nobel award lecture in 1902, *Emil Fischer* has shown remarkable prescience when he said: *"I can foresee a time in which physiological chemistry will not only make greater use of natural enzymes but will actually resort to creating synthetic ones"* [5.1].

Since enzymatic reactions involve binding of and reaction with a precisely defined substrate they have the characteristics of a supramolecular process. On the other hand, reactivity and catalysis represent major features of the functional properties of supramolecular systems [1.7, 1.29]. Molecular receptors bearing appropriate reactive groups in addition to binding sites may complex a substrate (with given stability, selectivity, and kinetic features), react with it (with given rate, selectivity, and turnover), and release the products, thus regenerating the reagent for a new cycle (Fig. 6).

Supramolecular reactivity and catalysis thus involve two main steps: *binding*, which selects the substrate, and *transformation* of the bound species into products within the supermolecule formed. Both steps take part in the molecular *recognition* of the *productive* substrate and require the correct molecular information in the reactive receptor. Compared to molecular reactivity, a binding step is involved that precedes the reaction itself. Catalysis additionally comprises a third step, the release of the substrate.

The selection of the substrate is not the only function of the binding step. In order to promote a given reaction, the binding should strain the substrate [5.2] so as to bring it toward the transition state of the reaction; thus, efficient catalysts should

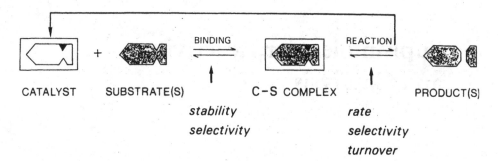

Fig. 6. Schematic representation of the supramolecular catalysis process.

bind the transition state more strongly than the free state of the substrate in order to lower the free energy of activation [5.3, 5.4]. However, the design of catalysts capable of optimal transition state stabilisation does not consist in searching for strongest binding of strict transition state analogues (TSA) of the substrate, but rather of (TSA–X), i.e., of TSA minus X, where X represents the atom(s) of the functional group(s) in the catalyst that react(s) with the bound substrate.

A major role is played by the existence of strong interactions between the substrate and the receptor site of the catalyst. They may be used to facilitate the reaction in several ways, such as: a thermodynamic effect, strong binding forcing the substrate into contact with the reactive groups; a steric effect, fixation of the substrate being able to distort it towards its transition state geometry; an electrostatic (electronic, protonic, ionic) effect, consisting of a possible activation of the functional groups of the catalyst (and possibly of the substrate also) by modification of their physico–chemical properties (pK_a, polarity, etc.) as a consequence of substrate fixation, which may perturb charge distributions in both the substrate and the catalyst with respect to their free, unbound state; such an activation of the catalyst by the substrate itself is a sort of *suicidal* behaviour, the substrate facilitating its own consumption (see below). The effects induced by the fixation of the substrate play a role in the reactions catalysed by either artificial species or by enzymes and may result in a sort of *mutual activation* of the two partners.

The design of efficient and selective supramolecular reagents and catalysts may give mechanistic insight into the elementary steps of catalysis, provide new types of chemical reagents, and reveal factors contributing to enzymatic catalysis [5.4–5.7]. This has led to numerous investigations, which have made use of reagents based on various receptor frameworks.

One must note that in a number of cases the processes described amount to facilitation of a given reaction in a catalytic cycle but not to full catalysis, due to slow regeneration steps, product inhibition, etc. The term catalysis will be applied to examples mentioned below with this restriction in mind.

The first class of substances to be used as substrate binding reagents were the *cyclodextrins*. Following the earlier pioneering studies [1.32], numerous investigations on substrate binding and transformation have been performed using natural or variously substituted cyclodextrins with the aim of developing artificial enzyme-like processes and often with biomimetic intentions [5.8–5.11]. Our own work has been concerned mainly with reagents based on polyoxa and polyaza macrocycles [1.7, 5.12, 5.13].

5.1 Catalysis by Reactive Macrocyclic Cation Receptor Molecules

The ability of [18]-O_6 macrocyclic polyethers to bind primary ammonium ions (Section 2.5) opens the possibility to induce chemical transformations on such substrates. *Activation* and *orientation* by binding was observed for the hydrolysis of *O*-acetylhydroxylamine, which forms such a stable complex **27b** (R = CH_3COO) with the macrocyclic tetracarboxylate receptor that it remains protonated and bound even at neutral pH, despite the low pK_a (ca. 2.15) of the free species. As a consequence, its hydrolysis is accelerated and exclusively gives acetate and hydroxylamine, whereas in the presence of K^+ ions, which displace the substrate, it yields also acetylhydroxamic acid, $CH_3CONH–OH$ (ca. 50%) [5.14]. Thus, strong binding may be sufficient for markedly accelerating a reaction and affecting its course, a result that also bears on enzyme-catalysed reactions.

Chemical transformations may be induced by reaction between a bound substrate and functional groups borne by the macrocyclic receptor unit, as illustrated in structure **74**.

Ester cleavage processes have been most frequently investigated in enzyme model studies. Macrocyclic polyethers fitted with side chains bearing thiol groups

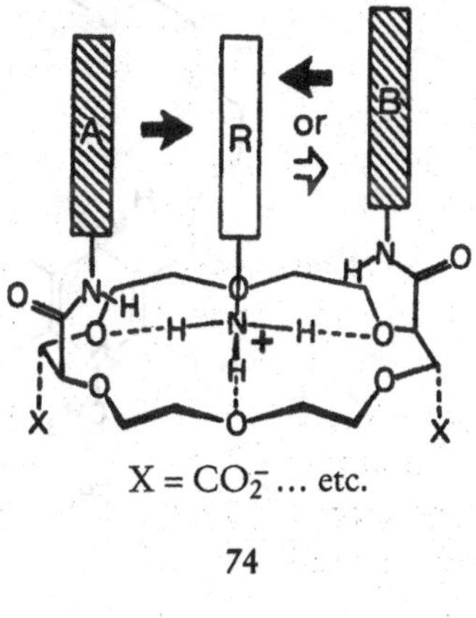

$X = CO_2^- \ldots$ etc.

74

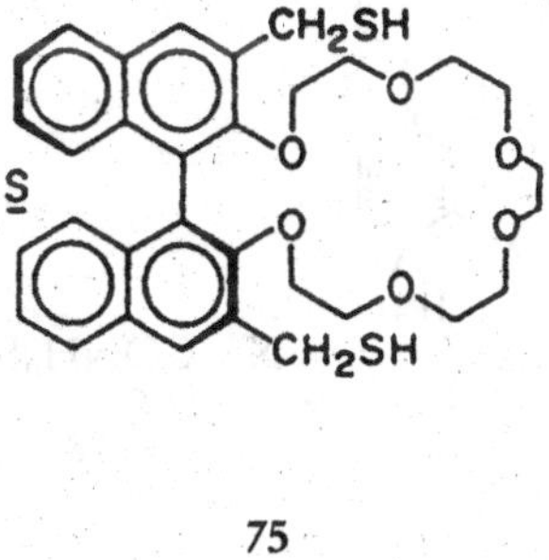

75

cleave activated esters with marked rate enhancements and chiral discrimination between optically active substrates [5.12, 5.13, 5.15–5.17]. The optically active binaphthyl reagent **75** performs thiolysis of activated esters of amino acids with pronounced acceleration and chiral recognition (see **76**) [5.15]. The tetra-L-cysteinyl macrocycle **27d** binds *p*-nitrophenyl (PNP) esters of amino acids and peptides, and reacts with the bound species, releasing *p*-nitrophenol as shown in **77** [5.16]. The reaction displays (1) substrate selectivity in favour of dipeptide ester substrates, with (2) marked rate enhancements, (3) inhibition by complexable metal cations that displace the bound substrate, (4) high chiral recognition between enantiomeric dipeptide esters, and (5) slow but definite catalytic turnover. An interesting effect was the finding that binding of the substrate probably activates the thiol groups by increasing their acidity, an illustration of the suicidal effect noted above, that may also be

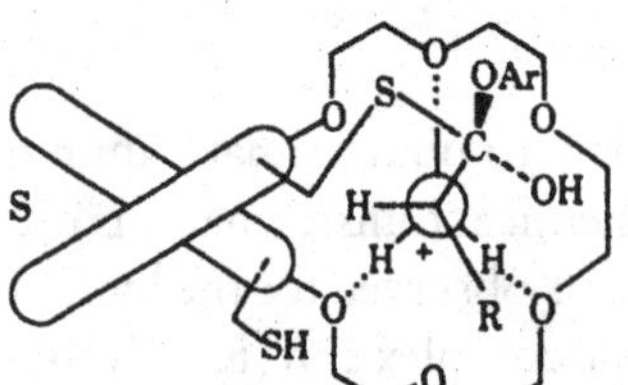

76

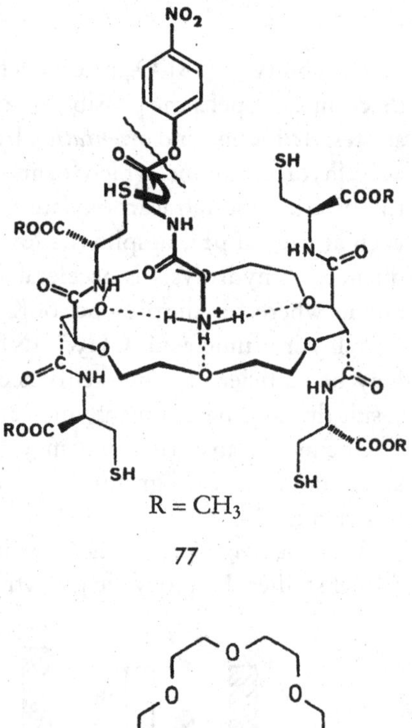

$R = CH_3$

77

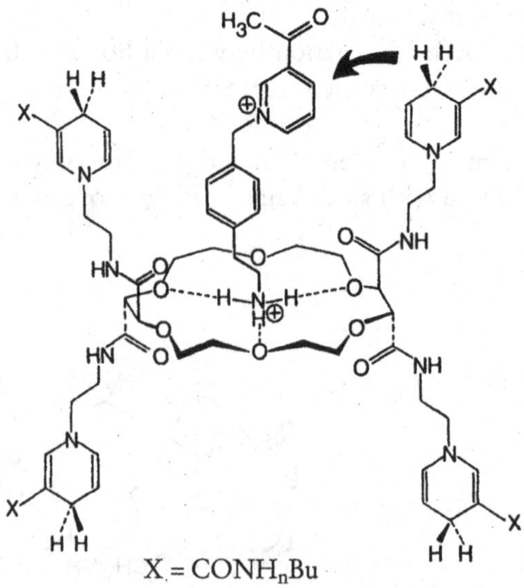

X = CONH$_n$Bu

78

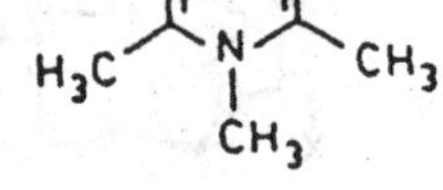

79

of importance in enzymatic reactions. A partial mimic of a transacylase has been synthesized and shown to accelerate substantially the transacylation of amino ester salts [5.18].

Hydrogen transfer has been induced with macrocyclic receptors bearing 1,4-dihydropyridyl (DHP) groups. Bound pyridinium substrates are reduced by hydrogen transfer from DHP side chains within the supramolecular species **78**; the first-order intracomplex reaction is inhibited and becomes bimolecular on displacement of the bound substrate by complexable cations [5.19]. Reactions with carbonyl or sulphonium substrates have been performed with other DHP containing macrocycles, such as **79** [5.20].

5.2 Catalysis by Reactive Anion Receptor Molecules

The development of anion coordination chemistry and anion receptor molecules has opened up the possibility to perform molecular catalysis on anionic substrates of chemical and biochemical interest, such as adenosine triphosphate. The catalysis of *phosphoryl transfer* is of particular interest, namely in view of the crucial role of such processes in biology and of the numerous enzymes that catalyse them.

ATP hydrolysis was found to be catalysed by a number of protonated macrocyclic polyamines. In particular, [24]-N_6O_2 **38**, strongly binds ATP and markedly accelerates its hydrolysis to ADP and inorganic phosphate over a wide pH range [3.15, 5.21–5.23]. The reaction presents first-order kinetics and is catalytic with turnover. It proceeds via initial formation of a complex between ATP and protonated **38**, followed by an intracomplex reaction that may involve a combination of acid, electrostatic, and nucleophilic catalysis. Structure **80** represents a possible binding mode in the ATP–X complex and indicates how cleavage of the terminal phosphoryl groups might take place. A transient intermediate, identified as phosphoramidate **81**, is formed by phosphorylation of the macrocycle by ATP and is subsequently hydrolysed. In this process, catalyst **38** presents ATPase type activity.

80   81

Multiple recognition and catalysis in ATP hydrolysis with increased ATP/ADP selectivity has been achieved with a multifunctional anion receptor containing a macrocyclic polyamine as anion binding site, an acridine group as stacking site and a catalytic site for hydrolysis (structure 82) [4.27]. Phosphoryl transfer is accelerated by other types of hydrogen-bonding receptors [5.24a].

82

An important goal is the development of efficient catalysts for *phosphodiester hydrolysis*, in view of the design of selective DNA and RNA cleavage reagents. Work towards this end is being pursued using in particular bis-guanidinium receptors [5.24b,c]. Intercalating units bearing imidazole groups may effect the cleavage of RNA in a process mimicing ribonuclease A [5.25a,b]. Purine–acridine combinations act as artificial endonucleases cleaving DNA at apurinic sites [5.25c].

Molecular recognition of the structural and base sequence features coupled with the selective cleavage of nucleic acids is an area of major interest both for basic reasons as well as for its biological implications. Processes involving single or double strand recognition by pairing, intercalation, groove binding, triple helix formation, using thermal or photochemical reagents of peptidic, oligonucleotidic or inorganic nature, have been actively investigated and highly selective reactions have been developed (for a few references see [5.26–5.31]).

Enolase type activity is displayed in the efficient supramolecular catalysis of H/D exchange in malonate and pyruvate bound to macrocyclic polyamines [5.32]. Other processes that have been studied comprise for instance the catalysis of nucleophilic aromatic substitution by macrotricyclic quaternary ammonium receptors of type 21 [5.33], the asymmetric catalysis of Michael additions [5.34], the selective functionalization of doubly bound dicarboxylic acids [5.35] or the activation of reactions on substituted crown ethers by complexed metal ions [5.36].

5.3   Catalysis with Cyclophane Type Receptors

A number of studies have made use of functionalized cyclophanes for developing supramolecular catalysts and enzyme models [4.31–4.34, 5.37, 5.38]. Their catalytic behaviour is based on the implementation of electrostatic, hydrophobic and metal coordination features for effecting various reactions in aqueous media.

Hydrophobic species bearing hydrocarbon chains present vitamin B_{12} or vitamin B_6 type activity [5.37]. Such systems lend themselves to inclusion in membrane or micellar media. They thus provide a link with catalysis in more or less organized media such as membranes, vesicles, micelles, polymers [5.39–5.41] (see Section 7.4). Water soluble cyclophanes showing, for example, transaminase [5.42], acetyl transfer [5.43], pyruvate oxidase [5.44] or nucleophilic substitution [5.45] activity have been described.

Cyclophane catalysts offer a rich playground for developing novel reactions and enzyme models in view of the variety of their structural types, the large cavities they contain and the possibility to attach several functional groups.

5.4   Supramolecular Metallocatalysis

Supramolecular metallocatalysts consist in principle of the combination of a recognition subunit (such as a macrocycle, a cyclodextrin, a cyclophane, etc.) that selects the substrate(s) and of a metal ion, bound to another subunit, that is the reactive site. Complexed metal ions presenting free coordination positions may present a variety of substrate activation and functionalization properties. Heterotopic coreceptors such as 70 bind simultaneously a substrate and a metal ion bringing them into proximity, thus potentially allowing reaction between them.

Approaches towards the development of artificial metalloenzymes [5.46–5.58] have been made, based on cyclodextrins [5.46, 5.47, 5.49] or macrocycles [5.50, 5.51] and involving various metal ions, such as Zn(II), Cu(II), Co(III), for facilitating hydrolysis [5.47–5.52], epoxidation [5.53], hydrogen transfer [5.20], etc.

Metalloporphyrins have been used for epoxidation and hydroxylation [5.53] and a phosphine–rhodium complex for isomerization and hydrogenation [5.54]. Cytochrome P-450 model systems are represented by a porphyrin-bridged cyclophane [5.55a], macrobicyclic transition metal cyclidenes [5.55b] or β-cyclodextrin-linked porphyrin complexes [5.55c] that may bind substrates and perform oxygenation reactions on them. A cyclodextrin connected to a coenzyme B_{12} unit forms a potential enzyme–coenzyme mimic [5.56]. Recognition directed, specific DNA cleavage reagents also make use of the reactivity features of various complexed metallic sites [5.26–5.29, 5.57]. The remarkable facilitation of amide hydrolysis by dinuclear copper complexes [5.58a] and related processes may open routes towards metallo-cleavage of proteins [5.58b]

Selective metalloprocesses, in particular asymmetric reactions utilizing external chiral ligands [5.59], such as hydrogenation, epoxidation, hydroxylation, etc. are of great value for organic synthesis and are being actively investigated.

Supermolecular metallocatalysts, by combining a substrate recognition unit with a catalytic metallic site, offer powerful entries to catalysts presenting shape, regio- and stereoselectivity.

5.5   Cocatalysis: Catalysis of Synthetic Reactions

A further step in supramolecular catalysis lies in the design of systems capable of inducing *bond formation* rather than bond cleavage processes, thus effecting synthetic as compared to degradative reactions. To this end, the presence of several binding and reactive groups is essential. Such is the case for coreceptor molecules in which subunits may cooperate for substrate binding and transformation [2.11]. They should be able to perform *cocatalysis* by bringing together substrate(s) and cofactor(s) and mediating reactions between them within the supramolecular architecture (Fig. 7).

Processes of this type have been realized in supramolecular phosphorylation reactions. Indeed, the same [24]-N_6O_2 macrocycle 38 as that already used in the studies of ATP hydrolysis was also found [5.60] to mediate the synthesis of pyrophosphate from acetylphosphate (AcP). Substrate consumption was accelerated and catalytic with turnover following the steps: (1) substrate AcP binding by the protonated molecular catalyst 38; (2) phosphorylation of 38 within the supramolecular complex, giving the phosphorylated intermediate PN 81; (3) binding of the substrate HPO_4^{2-} (P); (4) phosphoryl transfer from PN to P with formation of pyrophosphate PP (Fig. 8); (5) release of the product and of the free catalyst for a new cycle [5.60]. PP is also formed in the hydrolysis of ATP in the presence of divalent metal ions [5.61].

The fact that 38 is a ditopic coreceptor containing two diethylenetriamine subunits is of special significance for both PN and PP formation. These subunits may cooperate in binding AcP and activating it for phosphoryl transfer via the ammonium sites, in providing an unprotonated nitrogen site for PN formation and in mediating phosphoryl transfer from PN 81 to P. Thus 38 would combine electrostatic and nucleophilic catalysis in a defined structural arrangement suitable for PP synthesis via two successive phosphoryl transfers, displaying kinase type activity (Fig. 8)

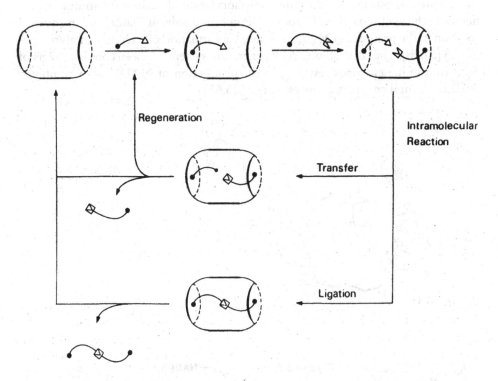

Fig. 7. Schematic illustration of cocatalysis processes: group transfer and ligation reactions occuring within the supramolecular complex formed by the binding of substrates to the two macrocyclic subunits of a macrotricyclic coreceptor molecule.

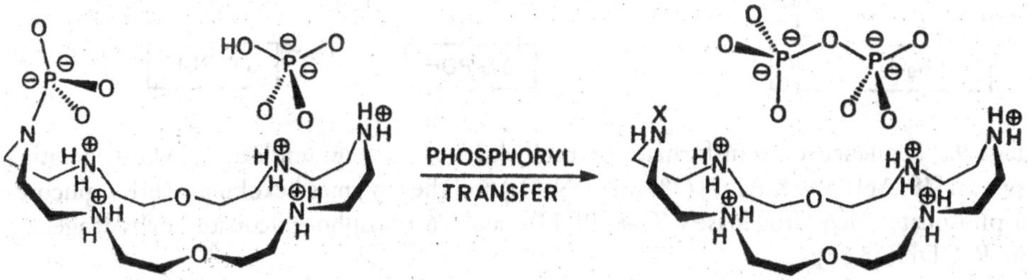

Fig. 8. Cocatalysis: pyrophosphate synthesis by phosphoryl transfer mediated by macrocycle **38** via the phosphorylated intermediate **81** [5.12].

This process was extended to the phosphorylation of various substrates, in particular to the synthesis of ATP from ADP in mixed solvent [5.62a] and in aqueous solution in the presence of Mg^{2+}, probably via formation of a ternary catalytic species **83** [5.62b]. The latter abiotic ATP generating system has been coupled to sets of ATP consuming enzymes resulting in the production of NADH by a combined artificial/natural enzymatic process (Fig. 9) [5.63].

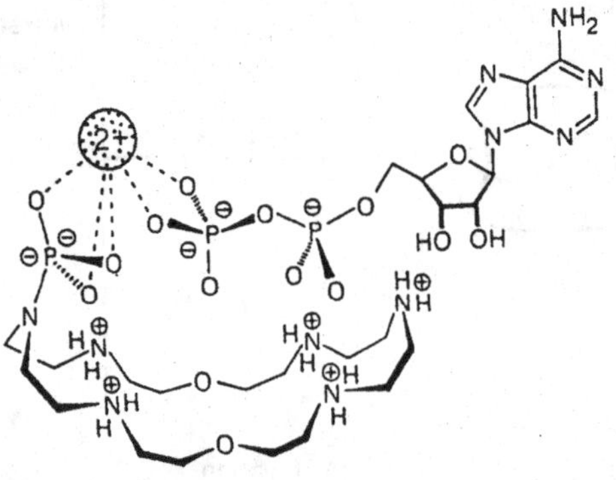

83

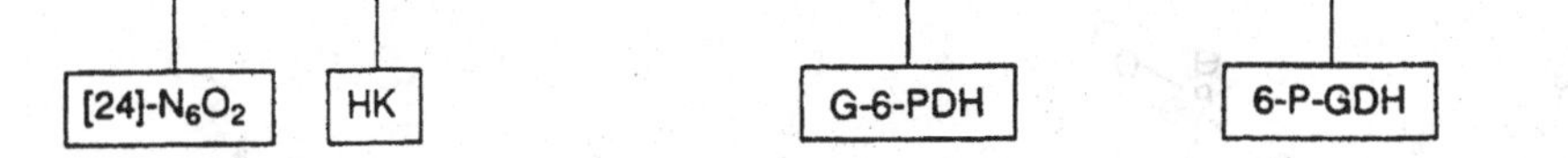

Fig. 9. Sequence of transformations catalysed by the supramolecular ATP-generating system [**38**, AcP, Mg^{2+}, ADP] (**38** = [24]-N_6O_2) and the enzymes hexokinase (HK), glucose-6-phosphate dehydrogenase (G–6–PDH) and 6-phospho-gluconate dehydrogenase (6–P–GDH) [5.32].

Templates possessing two hydrogen bonding subunits bind two substrates forming a ternary complex in which the substrates are positioned so as to facilitate bond formation between them [5.64a]. In a related way, the rate and stereoselectivity of a bimolecular Diels–Alder reaction are substantially increased by binding both the diene and the dienophile within the cavity of a tris-porphyrin macrocycle [5.64b].

A macrobicyclic thiazolium cyclophane **84** functions as a model of thiamine pyrophosphate-dependent ligases and effects benzoin condensations [5.38, 5.65a, A.11]. Acyl transfer is catalysed by formation of a ternary complex between a cyclophane receptor and two substrates [5.65b].

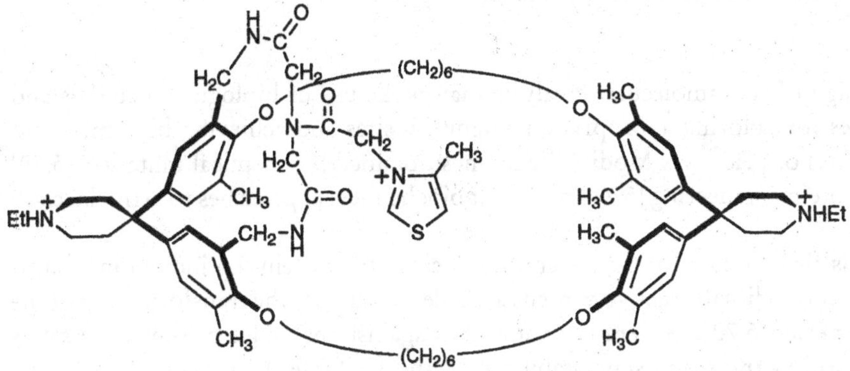

84

Difunctional binding and catalysis has been observed in functionalized cyclodextrins [5.66] and in a hydrogen bonding cleft [5.67]. Functionalized crown ethers have been used as reagent for peptide synthesis [5.68]. Self-replication processes also involve bond formation reactions; they will be discussed in Section 9.6.

Of special interest is the enhanced ligation of DNA (see **85**), effected by an imidazole-functionalized spermine binding in the minor groove of the double helix [5.69].

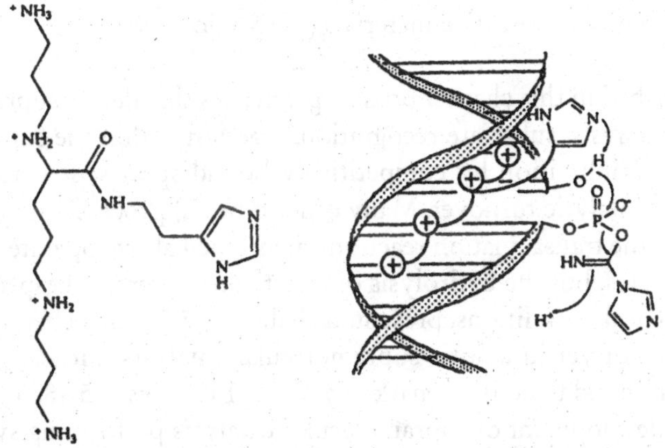

85

Bond-making processes such as those described above extend supramolecular reactivity to cocatalysis, mediating *synthetic reactions* within the supramolecular entities formed by coreceptor molecules.

5.6   Biomolecular and Abiotic Catalysis

The design of supramolecular catalysts may make use of biological materials and processes for tailoring appropriate recognition sites and achieving high rates and selectivities of reactions. Modified enzymes obtained by chemical mutation [5.70] or by protein engineering [5.71] represent biochemical approaches to artificial catalysts.

This is also the case for the generation of catalytic proteins by induction of antibodies. Antibodies to reactive haptens are able to facilitate the transformation of the bound species [5.72]. Generating antibodies against analogues of transition states should lead to transition state stabilization and facilitate the process [5.73]. Such *catalytic antibodies* or *abzymes* have been produced for a variety of reaction [5.74–5.76] and an active field of research has developed along such lines. It represents an approach to substrate specific, efficient and selective catalysis of supramolecular type, that is of much basic and applied interest. The strong affinity for the TS of the reaction of a given substrate leads it along the way and thus facilitates the process. In line with the remarks made at the beginning of this chapter, the antibodies should be generated not against a TSA itself but against a (TSA–X) isoster lacking the group(s) X, which belong(s) to the reactive function(s) of the protein that is (are) expected to perform the reaction. This requires designing a (TSA–X) species in which the face presenting the (–X) gap be ideally chosen so as to lead to the induction in the antibody of the desired reactive functional group at the correct position. An abiotic approach of much interest consists of the generation of TSA imprints on the surface or in the bulk of solid state materials (see Section 7.4).

The systems described in this chapter possess properties that define supramolecular reactivity and catalysis: substrate recognition, reaction within the supermolecule, rate acceleration, inhibition by competitively bound species, structural and chiral selectivity, and catalytic turnover. Many other types of processes may be imagined. In particular, the transacylation reactions mentioned above operate on activated esters as substrates, but the hydrolysis of unactivated esters and especially of amides under "biological" conditions, presents a challenge [5.77] that chemistry has met in enzymes but not yet in abiotic supramolecular catalysts. However, metal complexes have been found to activate markedly amide hydrolysis [5.48, 5.58a]. Of great interest is the development of supramolecular catalysts performing synthetic reactions that create new bonds rather than cleave them. By virtue of their multiple binding features, coreceptors open the way to the design of cocatalysts for ligation, metallocatalysis, and cofactor reactions, which act on two or more co-bound and spatially oriented substrates.

Supramolecular catalysts are by nature *abiotic* chemical reagents that may perform the same *overall* processes as enzymes, without following the detailed pathway by which the enzymes actually effect them or under conditions in which enzymes do not operate. Furthermore and most significantly, this chemistry may develop systems realizing processes that enzymes do not perform while displaying comparable high efficiencies and selectivities.

6 Transport Processes and Carrier Design

The organic chemistry of membrane transport processes and carrier molecules has been developed only rather recently, considering that their physico-chemical features and biological importance have long been recognized. The design and synthesis of receptor molecules binding selectively organic and inorganic substrates has made available a range of compounds that, if made membrane soluble, may become carrier molecules and induce selective transport by rendering membranes permeable to the bound species. Thus, transport represents one of the basic functional features of supramolecular species together with recognition and catalysis. [1.29, 5.12, 6.1]

The chemistry of transport systems has three main goals: to design transport effectors, to devise transport processes, and to investigate their applications in chemistry and in biology. Selective membrane permeability may be induced either by carrier molecules or by transmembrane channels (Fig. 10).

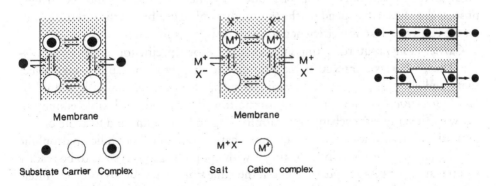

Fig. 10. Transport processes. Carrier mediated, via a neutral species (left), of an ion pair (centre). Channel mediated (right, top), gated channel (right, bottom).

6.1   Carrier-Mediated Transport

Carrier-mediated transport (or facilitated diffusion) consists of the transfer of a substrate across a membrane, facilitated by a carrier molecule located in the membrane. It is a cyclic process comprising four steps: (1) formation of the carrier–substrate complex at one interface; (2) diffusion of the complex through the membrane phase; (3) release of the substrate at the other interface; (4) back diffusion of the free carrier.

Because of its cyclic nature, this process presents analogies with molecular catalysis; it may be considered as *physical catalysis* operating a change in location, a *translocation*, on the substrate, like chemical catalysis operates a transformation into products. The carrier is the *transport catalyst* which strongly increases the rate of passage of the substrate with respect to free diffusion and shows enzyme-like features (saturation kinetics, competition and inhibition phenomena, etc.). The active species is the carrier–substrate supermolecule. The transport of substrate S_1 may be coupled to the flow of a second species S_2 in the same (*symport*) or opposite (*antiport*) direction.

Transport is a three-phase process, whereas homogeneous chemical and phase-transfer [2.87, 2.88] catalyses are single phase and two-phase respectively. Carrier design is the major feature of the organic chemistry of membrane transport since the carrier determines the nature of the substrate, the physico-chemical features (rate, selectivity) and the type of process (facilitated diffusion, coupling to gradients and flows of other species, active transport). Since they may in principle be modified at will, synthetic carriers offer the possibility to monitor the transport process via the structure of the ligand and to analyse the effect of various structural units on the thermodynamic and kinetic parameters that determine transport rates and selectivity.

The factors influencing selective transport may be divided into internal ones arising from the carrier, and external ones due to the medium. In a diffusion controlled process the rates will depend on the thermodynamic equilibria at the interfaces, i.e., on the relative extraction efficiency towards different substrates.

Carrier design requires taking into account factors specific for transport processes. Thus, whereas a molecular receptor should display high stability, high selectivity and slow exchange rate towards its substrate, a carrier molecule should be *highly selective*, but not bind its substrate too tightly, and be flexible enough to allow sufficiently fast exchange rates for loading and unloading and to avoid carrier saturation; in addition, the carrier must have a suitable lipophilic–hydrophilic balance (achieved eventually by fitting it with lipophilic groups) so as to be readily soluble in the membrane phase, while at the same time being able to reach the interface and enter into contact with the aqueous phase; the carrier should not be too bulky so as to diffuse rapidly; finally it may bear functional groups suitable for coupling of substrate flow with other processes (acid/base, redox) [1.7, 1.27, 6.1, 6.2].

The *external factors* comprise the nature of the membrane, the substrate concentration in the aqueous phase and any other external species that may participate in the process. They may strongly influence the transport rates via the phase distribution equilibria and diffusion rates. When a neutral ligand is employed to carry an ion pair by complexing either the cation or the anion, the coextracted uncomplexed counterion will affect the rate by modifying the phase distribution of the substrate. The case of a cationic complex and a counteranion is illustrated schematically in Figure 10 (centre).

Our initial work on the *transport of amino acids*, dipeptides, and acetylcholine through a liquid membrane employed simple lipophilic surfactant-type carriers [6.3]. Amino acids could be carried against their concentration gradient, pumped by protonation–deprotonation reactions in a pH gradient or by a gradient of inorganic salts. These studies were aimed at the physical organic chemistry of transport processes, exploring various situations of transport coupled to flows of protons, cations, or anions in concentration and pH gradients.

6.2 Cation Transport Processes — Cation Carriers

The question of carrier design was first addressed for the transport of inorganic cations. In fact, selective *alkali cation transport* was one of the initial objectives of our work on cryptates [1.26a, 6.4]. Natural acyclic and macrocyclic ligands (such as monensin, valinomycin, enniatin, nonactin, etc.) were found early on to act as selective ion carriers, *ionophores* and have been extensively studied, in particular in view of their antibiotic properties [1.21, 6.5]. The discovery of the cation binding properties of crown ethers and of cryptates led to active investigations of the ionophoretic properties of these synthetic compounds [2.3c, 6.1, 6.2, 6.4–6.13]. The first step resides in the ability of these substances to lipophilize cations by complexation and to extract them into an organic or membrane phase [6.14, 6.15].

Cryptands of type 7–9 and derivatives thereof carry alkali cations [6.4], even under conditions where natural or synthetic macrocycles are inefficient. The selectivities observed depend on the structure of the ligand, the nature of the cation and the type of cotransported counteranion. Designed structural changes allow the transformation of a cation receptor into a cation carrier [6.1, 6.4]. The results obtained with cryptands indicated that there was an optimal complex stability and phase-transfer equilibrium for highest transport rates. Combined with data for various other carriers and cations, they give a bell-shaped dependence of transport rates on extraction equilibrium (Fig. 11), with low rates for too small or too large

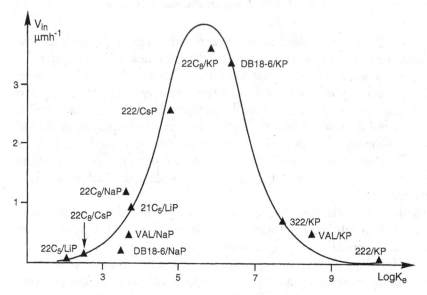

Fig. 11. Plot of initial cation transport rates for various carrier/alkali picrate pairs versus equilibrium extraction constants log K_e; the points are experimental data, the curve is calculated [6.1, 6.4, 6.17]. For analytical reasons the K_e values were determined in conditions different from those of the transport experiments; the carriers are cryptands (for [2.2.C$_5$] and [2.1.C$_5$] see structures in [6.1]); dibenzo-18-crown-6, DB18-6; and valinomycin, VAL; picrate, P.

extraction amounting respectively to very small and very high (carrier saturation) fraction of complexed carrier in the membrane; as expected, highest rates correspond to half-filled carriers [6.1, 6.4, 6.16].

Kinetic analyses allowed the experimental results to be related to the dependence of transport rates and selectivities on carrier properties [6.9, 6.17, 6.18]. Detailed studies of **8** and **9** in vesicles bore on the efficiency, the selectivity and the mechanism of the processes [6.19]. The rates of transport by proton ionizable macrocyclic carriers are pH dependent [6.12]. Diverse other ligands have been used as carriers, such as acyclic polyethers or calixarene derivatives [6.20, A.6].

Modulation of the complexation and transport properties by modifying structure and binding sites affords carriers of interest for a variety of applications, like control of levels of biologically active cations (see, for instance, [6.21] for the case of Li$^+$), selective transport of transition metal ions, removal of toxic heavy metal ions from biological fluids or from the environment, recovery of trace metals, separation procedures, etc. Macrocyclic polyethers can selectively transport organic primary ammonium cations, in particular physiologically active ones [6.22], and chiral carriers effect the resolution of racemic ammonium salts [6.23]. Crown ethers of suitable ring size facilitate the transport of the guanidinium cation [6.24].

In addition to specific carrier features, a number of external factors may also have marked effects on transport rates. The nature of the membrane phase (in particular for liquid or supported liquid [6.10b] membranes) influences the distribution equilibria as well as the stability and selectivity of the complex in the membrane and the cation exchange rates at the interfaces. The nature of the coextracted anion affects transport via a (cationic complex–anion) pair (Fig. 10) simply by modifying the amount of salt extracted into the membrane; this amount decreases with higher hydration energy and lower lipophilicity of the anion (for example, chloride compared with picrate). The concentration of salt in the aqueous phase will, of course, affect the amount extracted into the membrane and therefore the transport rates (for illustrations of these effects see for instance [6.1]).

6.3 Anion Transport Processes — Anion Carriers

Anion transport has been much less studied than cation transport. However, progress in the development of anion receptor molecules should provide a range of carriers for anionic species.

Simple lipophilic cations, like ammonium ions bearing long hydrocarbon chains, allow anion extraction into an organic phase and render liquid membranes permeable to anions by an anion exchange (antiport) process. Such carriers effect, for instance, selective transport of amino acid carboxylates [6.3] against inorganic anions like chloride.

Solubilization and transport of ion pairs by cation complexing agents, as discussed above from the point of view of cation transport, involves transport of the anion as external component of the (complexed cation–anion) pair. Lipophilic transition metal complexes [6.25a] or organometallic (such as tin [6.25b]) derivatives may serve as anion carriers by direct coordination of the anion to the metal cation and should provide a variety of selectivity features, depending on the specific interaction energy and geometry between a given anion and a given transition metal cation. The pair formed must, of course, be sufficiently kinetically labile. Anion binding to lipophilic porphyrin complexes is an interesting case of this type. Inorganic anions are carried by protonated cryptands [6.26] and oligopyrrole macrocycles [6.27].

Of special interest, because of its biological significance, is the transport of carboxylates and phosphates. In particular, the transport of nucleotides has been achieved [6.28–6.32] and may be made selective by introduction of additional base pairing interactions [6.31, 6.32]. The transport of ATP has significance with respect to bioenergetic processes. Antiviral chemotherapy may take advantage of enhancing the cellular uptake of modified nucleotides by carrier species.

Agents acting as carriers for polynucleotides and nucleic acid segments, and thus capable of mediating *gene transfer*, are of great value for biotechnology, genetic engineering and gene therapy [6.33, 6.34]. Recombinant viruses are efficient gene-transfer agents [6.34] but synthetic vectors (mixed, such as adenovirus–poly-lysine–DNA conjugates [6.35], or purely synthetic ones such as lipopolyamines [6.36]) are also able to induce marked transfection and represent very promising alternatives that would alleviate problems linked with using viral material. There is great basic and applied interest in both biology and medicine for the further development of artificial vectors capable of inducing efficient and stable gene–transfer. Lipophilic guanidinium species may be envisaged in view of the strong interactions between nucleic acids and the polyarginine peptides, protamines [6.37a]. Liposomes act as agents for DNA transfer [6.37b].

Liquid membranes of the water-in-oil emulsion type have been extensively investigated for their applications in separation and purification procedures [6.38]. They could also allow extraction of toxic species from biological fluids and regeneration of dialysates or ultrafiltrates, as required for artificial kidneys. The substrates would diffuse through the liquid membrane and be trapped in the dispersed aqueous phase of the emulsion. Thus, the selective elimination of phosphate ions in the presence of chloride was achieved using a bis-quaternary ammonium carrier dissolved in the membrane phase of an emulsion whose internal aqueous phase contained calcium chloride leading to phosphate–chloride exchange and internal precipitation of calcium phosphate [6.1].

Cation–anion cotransport was effected by an optically active macrotricyclic cryptand that carried simultaneously an alkali cation and a mandelate anion and displayed weak chiroselectivity [4.23a], as did also the transport of mandelate by an optically active acyclic ammonium cation [6.39]. Employing together a cation and an anion carrier should give rise to *synergetic transport* with double selection by facilitating the flow of both components of a salt (see the electron–cation symport below). Selective transport of amino acids is effected by a convergent dicarboxylic acid receptor [4.24b].

Neutral molecules are carried between two organic phases through a water layer by water-soluble receptors containing a lipophilic cavity [6.40, 6.41]. Urea and nucleosides are transported using a metallocarrier [6.42a] and complementary base-pairing agents [6.42b], respectively.

It is clear that numerous facilitated transport processes may still be set up, especially for anions, salts or neutral molecules, and that the active research in receptor chemistry will make available a variety of novel carrier molecules. Of special interest are those transport effectors, derived from coreceptors, that allow coupled transport (cotransport) to be performed.

6.4 Coupled Transport Processes

A major goal in transport chemistry is to design carriers and processes that involve the coupled flow of two (or more) species either in the same (symport) or in opposite (antiport) direction. Such parallel or antiparallel vectorial processes make it possible to set up a pumped system in which a species is carried in the potential created by physico–chemical gradients of electrons (redox gradient), protons (pH gradient), or other species (concentration gradient). Gradients may be generated by chemical reactions, as occurs in vectorial bioenergetics, for instance in the redox-linked proton pump represented by cytochrome c oxidase [6.43]. To this end, either two or more individual carriers for different substrates may be used simultaneously or the appropriate subunits may be introduced into a single species, a *cocarrier*.

6.4.1 Electron-Coupled Transport in a Redox Gradient

Substances undergoing redox reactions (such as quinone–hydroquinone, sulphide–disulphide, metal complexes, redox couples) may serve as electron carriers and allow the coupling of oxidation–reduction processes across membranes (see, for instance, [6.44–6.46]) to cation or anion transport.

Electron-cation symport has been realized in a double carrier process where the coupled, parallel transport of electrons and metal cations was mediated simultaneously by an electron carrier and by a selective cation carrier [6.47]. The transport of electrons by a nickel complex in a redox gradient was the electron pump for driving the selective transport of K^+ ions by a macrocyclic polyether (Fig. 12). The pro-

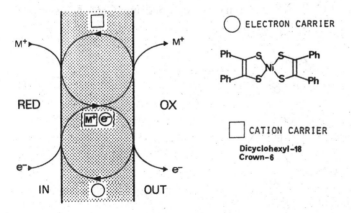

Fig. 12. Electron–cation coupled transport: a redox-driven electron–cation symport consisting of an electron carrier (nickel complex) and a selective cation carrier (macrocyclic polyether). RED, potassium dithionite; OX, $Na_3[Fe(CN)_6]$.

cess has the following features: active K^+ transport and coupled electron flow; two cooperating carriers acting synergetically; a redox pump; a selection process by the cation carrier; regulation by the cation–carrier pair. It represents a prototype for the design of other multicarrier coupled transport processes.

Quinone type carriers perform the cotransport of two protons and two electrons ($2e^-$, $2H^+$ symport) [6.48, 6.49] and take part in mitochondrial and photosynthetic electron transport. Cation receptor sites such as crown ethers or cryptands bearing a quinone [6.50a] or a ferrocene [6.50b] group (see also Section 8.3.1), bind and carry cations with redox coupling through switching between a low affinity state (quinone, ferricinium) and a high affinity state (reduced quinone, ferrocene).

Electron–anion antiport has been realized, for instance with redox active carriers such as ferrocene derivatives [6.51a] or alkylviologens [6.44–6.46, 6.51b] via ferricinium or reduced viologen species, respectively. The latter have been used extensively in light-driven systems and in studies on solar energy conversion [6.44–6.46].

6.4.2 Proton-Coupled Transport in a pH Gradient

Carriers bearing negatively charged groups may form a neutral complex by cation binding and effect cation exchange (antiport) across membranes; if one cation is a proton, a proton pump may be set up in a pH gradient. This has been realized for alkali cations with natural (for instance, monensin) or synthetic carboxylate-bearing ionophores [6.52].

A case of special interest is that of the transport of divalent ions such as calcium versus monovalent ones. The lipophilic carrier 86, containing a *single* cation receptor site and *two* ionizable carboxylic acid groups, was found to transport selectively Ca^{2+} in the dicarboxylate form and K^+ when monoionized, thus allowing pH control of the process. This striking change in transport features as a function of pH involves pH regulation of Ca^{2+}–K^+ selectivity in a competitive (Ca^{2+}, K^+) symport

$Z = CH_2C_6H_5$

86

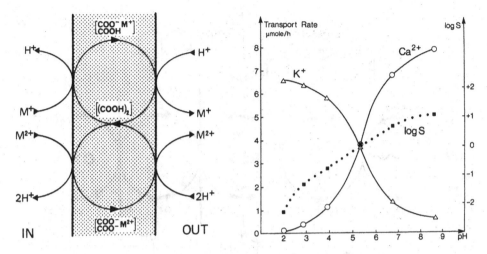

Fig. 13. (left) A competitive divalent/monovalent cation symport coupled to $M^{2+}/2H^+$ and M^+/H^+ antiport in a pH gradient by a macrocyclic carrier such as **86**; the state of the carrier is indicated as diprotonated, $[(CO_2H)_2]$, or complexed, $[(CO_2^-)_2, M^{2+}]$ and $[(CO_2H)(CO_2^-),$ $M^+]$; (right) Rates of K^+ and Ca^{2+} transport by the macrocyclic carrier molecule **86** through a liquid membrane as a function of pH in the starting IN phase; the dotted line represents the pH dependence of the transport selectivity log S.

coupled to $(Ca^{2+}, 2H^+)$ and (K^+, H^+) antiport in a pH gradient, which provides a proton pump (Fig. 13) [6.53].

This system demonstrates how carrier design allows transport processes to be endowed with regulation of rates and selectivity, as well as with coupling to energy sources for the transport of a species against its own concentration gradient. It may also be set up for effecting (Ca^{2+}, K^+) antiport, as has been realized in another dicarboxylic acid crown ether carrier [6.54]. pH regulation of calcium transport [6.55] by an acyclic amino acid derived ionophore has been described as well as proton coupled ion transport in a polymeric crown ether carboxylic acid [6.56]. Bis-crown ethers containing rings of different size carry Na^+ and K^+ in opposite directions [6.57].

6.4.3 Light-Coupled Transport Processes

Light-driven transport may be brought about by the photogeneration of a species that will induce the process or perturb it. *Light-induced electron transport* has been achieved through the proflavine-sensitized photogeneration of reduced methylviologen, MV^+, which transfers electrons to a quinone-type carrier contained in the membrane [6.49]. A $(2e^- + 2H^+)$ symport thus sets in under irradiation and stops when the light is switched off (Fig. 14).

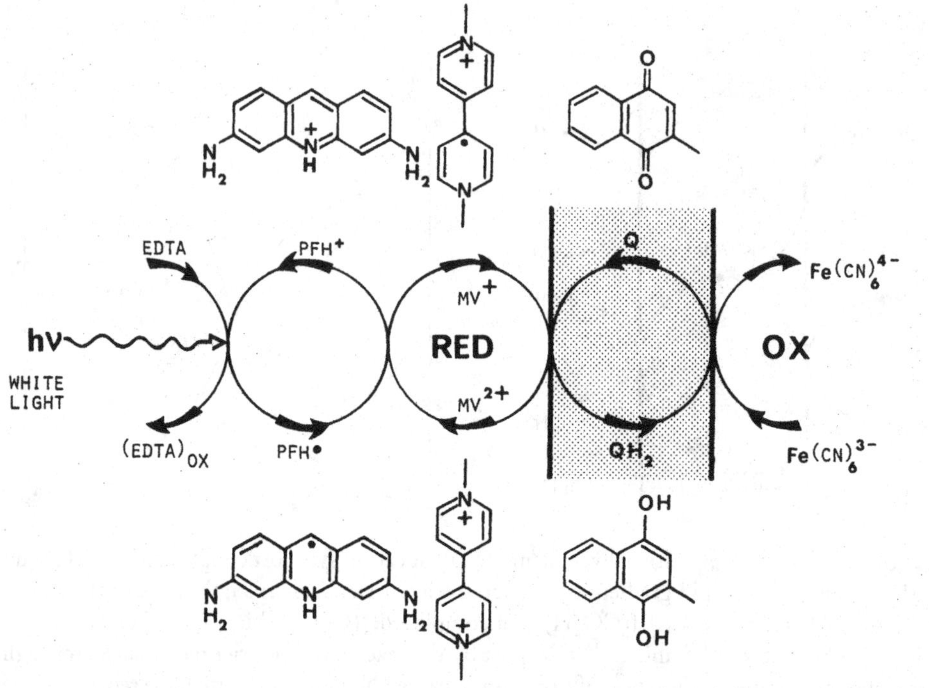

Fig. 14. Schematic representation of light-driven ($2e^- + 2H^+$) symport across a membrane via the quinone carrier molecule vitamin K_3 and its hydroquinone form; proflavine (PF)-sensitized photoreduction of methyl-viologen MV^{2+} in the RED phase, yields the reducing species MV^+, with simultaneous oxidative decomposition of EDTA used as electron donor; the OX phase contains ferricyanide as electron acceptor [6.49].

Such processes are of interest as components of artificial photosynthetic systems and as prototypes of photochemical cells and as batteries for energy storage. Various other photosensitized electron transfer reactions have been described, involving especially organized molecular assemblies [6.44–6.46, 6.58–6.60]. In particular, membranes permeable to electrons and protons may be of use for separating the oxidative and reductive half-cells producing respectively oxygen and hydrogen in photochemical water splitting cells. Our own work [6.61] towards the photodissociation of water evolved at least partially from our studies on electron transport [6.49]. Membrane systems may become part of artificial photosynthetic systems for solar energy conversion and storage.

Light-driven (electron, cation) symport occurs when combining this system with the (nickel complex, macrocycle) process described above [6.62]. Photocontrol of ion extraction and transport has been realized with macrocyclic or acyclic ligands (containing, for instance, azo or spirobenzopyran groups) that undergo a reversible structural change on irradiation, resulting in two forms having different ion affinities [6.52, 6.63–6.65].

The results obtained on coupled transport processes stress the role of cocarrier systems capable of transporting several substrates and driven by physical and chemical energy sources.

6.5 Transfer via Transmembrane Channels

Transmembrane channels represent a special type of multi-unit effector allowing the passage of ions or molecules through membranes by a flow or site-to-site hopping mechanism. They are the main effectors of biological ion transport. Natural and synthetic peptide channels (gramicidin A, alamethicin) allowing the transfer of cations have been studied [6.66–6.68].

Artificial *cation channels* could give fundamental information on the mechanism of cation flow and channel conduction [6.69, 6.70]. A solid-state model of cation transfer inside a channel is provided by the crystal structure of the KBr complex of **27c** ($Y = Y' = CH_3$), which contains stacks of macrocycles with cations located alternately inside and above a macrocyclic unit, like a frozen picture of cation propagation through the "channel" defined by the stack [6.71].

A polymeric stack of macrocycles has been synthesized [6.72] and a cyclodextrin-based model of a half-channel has been reported [6.73]. Channel-type conduction of Na^+ ions has been reported for a tris-macrocyclic ligand [6.74]. A derivative of the acyclic polyether ionophore monensin forms lithium channels in vesicles [6.75a], which may be sealed by diammonium salts [6.75b].

Cylindrical macrotricycles such as **52** (substrate removed) represent the basic unit of a cation channel based on stacks of linked macrocycles; they bind alkali cations [6.76], and cation jumping processes occur between the top and bottom rings, representing the elementary step of cation propagation [6.77]. Similar observations have been made for a bis-calixarene [6.78]. Theoretical studies give insight into the molecular dynamics of ion transport [6.79] and the energy profile in the gramicidin cation channel [6.80].

Electron channels, as transmembrane wires, represent the channel-type counterpart to the mobile electron carriers discussed above and will be considered in Section 8.3.2. *Anion channels* may also be envisaged.

Several types of studies directed towards the development of artificial ion channels and the understanding of ion motion in channels are being carried out. These effectors will be considered in Section 8.4; they deserve and will receive increased attention, in view also of their potential role as molecular ionic devices.

A further step in channel design must involve the introduction of (proton-, ion-, redox- or light-activated) *gates* and *control elements* for regulating opening and closing, rates, and selectivity. In this respect, one may note that ionizable groups on mobile carriers (such as **86**, see Section 6.3) correspond, for these effectors, to the gating mechanisms in channels. By the introduction of polytopic features, which may include selective binding subunits as well as gating components for flow regulation, the design of cocarriers and of artificial molecular channels should add a new dimension to the chemistry of the effectors and the mechanisms of transport processes.

Transport studies open ways to numerous developments of this chemically and biologically most important function.

From the *abiotic* point of view, they strive to explore the many facets of effector design, to identify the elementary steps and analyse the mechanisms involved, to set up coupling to chemical potentials, energy and signal transduction, transmembrane communication. They should lead to the development of selective carriers for use in pharmacology (drug delivery [6.81] or protection, etc.), in analytical chemistry and in separation science (analysis and separation of components of mixtures, optical resolution, recovery of minerals, recycling, etc.). They should also provide systems allowing temporary charge separation for use in artificial photosynthesis, solar energy conversion schemes and batteries for instance.

From the *biomimetic* point of view they offer models for biological transport mechanisms, in order to dissect the very complex membrane phenomena that regulate the relationship of living organisms with their environment, allowing them to retain their identity and to control their exchanges.

As regulation systems involving effectors, coupled transfers of charges and of mass, gates and pumps, transport processes extend towards the *chemistry of information storage and retrieval* at the molecular level, and are a major component in the design of molecular ionic devices (see Section 8.4). They thus open wide perspectives for the basic and applied developments of the functional features of supramolecular chemistry.

7 From Supermolecules to Supramolecular Assemblies

As pointed out in Chapter 1, supramolecular chemistry comprises two broad, partially overlapping areas covering on the one hand the oligomolecular *supermolecules* and, on the other, the *supramolecular assemblies*, extended polymolecular arrays presenting a more or less well-defined microscopic organisation and macroscopic features depending on their nature (layers, films, membranes, vesicles, micelles, microemulsions, gels, mesomorphic phases, solid state species, etc.).

Continuous progress is being made in the design of synthetic molecular assemblies, based on a growing understanding of the relations between the properties of the molecular components (structure, sites for intermolecular binding, polar–apolar domains, etc.), the characteristics of the processes that lead to their association and the supramolecular features of the resulting polymolecular entity.

Molecular organization and self-assembly into layers, membranes, vesicles etc., construction of multilayer films [7.1–7.5], generation of defined aggregate morphologies [4.74, 4.75, 7.6–7.8] etc., make it possible to build up specific supramolecular architectures. The polymerization of the molecular components has been a major step in increasing control over the structural properties of such assemblies [7.9–7.13].

Endowing these polymolecular entities with recognition units and reactive functional groups may lead to systems performing molecular recognition or supramolecular catalysis on external or internal surfaces of organic (molecular layers, membranes, vesicles, polymers, etc.) [7.1–7.13, A.41] or inorganic (zeolites, clays, sol–gel preparations, etc.) [7.14–7.20] materials.

7.1   Heterogeneous Molecular Recognition.
Supramolecular Solid Materials

The recognition processes discussed in Chapters 2–4 all occur in homogeneous solution. Heterogeneous molecular recognition may take place on surfaces or in the bulk of organized phases and solid materials.

This is the case on formation of solid state inclusion compounds, *clathrates*, with specific guest species that are compatible with the lattice defined by the packing features of the host molecules. Very extensive work has been performed on such entities and on the processes that lead to molecular recognition in the solid state depending on the type, the size, the shape of the guest molecule and its complementarity to the lattice component. Starting with the well-known clathrates of urea [1.32] all kinds of hosts have been studied [7.21, 7.22, A.32]. They form crystalline inclusion complexes either by chance or, more and more so, by design through a suitable choice of the host species. Isolation and separation of substances and even optical resolution by enantioselective clathrate formation may be achieved on crystallization of the solid state inclusion species [7.23–7.25]. The reactions of the partners in host–guest inclusion compounds are controlled by their arrangement [7.24c]. Zeolites yield selective inclusion complexes [7.26] and molecular guests form intercalation compounds with layered structures and related materials [7.27]. The extensive literature cannot be presented here; the references given provide entries into this area.

Molecular recognition may in principle be achieved by *imprinting* a specific shape and size selective mark on the surface or in the bulk of a material. It requires acquiring the ability to control the design of surfaces and solids [7.28], for instance by the use of a two-liquid phase medium [7.29a] or of liquid crystal templates [7.29b] or of crown ethers and cryptands [7.30] for the synthesis of zeolite molecular sieves, by intercalation to form pillared clays [7.31] or by generating hybrid organic–inorganic networks through sol–gel methods [7.32]. Inorganic–organic composite materials, obtained in particular by template mineralization of molecular microstructures, represent potential substrates for the introduction of shape selectivity and molecular recognition features [7.32c,d]. Layered structures [7.33a] and framework solids containing tunnels, cages and micropores [7.33b] have been obtained.

Imprinting into polymeric materials has been realized by either a covalent or a non-covalent approach. The former makes use of reversible covalent binding of the substrate to the monomer, which is then polymerized [7.34]. In the latter, suitably functionalized monomers are left to prearrange around the substrate through non-covalent interactions and the ensemble is then polymerized [7.35]. Removal of the imprint molecule from the polymer leaves recognition sites that are complementary to it in both geometry (size, shape) and functionality. Analogous imprinting may also be produced in sol–gel materials or in zeolites.

If surfaces and solids can serve as media for molecular recognition of adsorbed or included substrate molecules, conversely, molecular recognition may be used to engineer *supramolecular solid materials* [7.28] by directing the formation of surfaces or solids through specific interactions and framework species. A very fruitful and promising area may emerge from the combination of the chemistry of solid materials with supramolecular recognition processes.

7.2 From Endoreceptors to Exoreceptors. Molecular Recognition at Surfaces

The design of receptor molecules has mainly relied on macrocyclic or macropoly-cyclic architectures (see Chapters 2–4) and on rigid spacers or templates that allow binding sites to be positioned on the walls of molecular cavities or clefts in such a way that they converge towards the bound substrate. The latter is more or less completely surrounded by the receptor, forming an *inclusion* complex. This widely used principle of *convergence* defines a convergent or *endosupramolecular chemistry* with endoreceptors effecting endo-recognition. Biological analogies are found in the active sites of enzymes where a small substrate binds inside a cavity of a large protein molecule.

The opposite procedure consists in making use of an external *surface* with protuberances and depressions, rather than an internal cavity, as substrate receiving site. This amounts to the passage from a convergent to a divergent or *exosupramolecular chemistry* and from endo- to *exoreceptors*. Receptor–substrate binding then occurs by surface-to-surface interaction, which may be termed *affixion* and symbolized by // or by the mathematical symbol of intersection ∩ (if there is notable interpenetration of the surfaces) [1.37] giving [ρ//σ] and [ρ ∩ σ], respectively. *Exorecognition* with strong and selective binding requires a large enough contact area and a sufficient number of interactions as well as geometrical and site (electronic) complementarity between the surfaces of ρ and σ. Such a mode of binding finds biological analogies in protein–protein interactions, for instance at the antibody–antigen interface where the immunological recognition processes occur [7.36, 7.37].

Metallo-exoreceptors represent a type of exoreceptor that involves the arrangement of ligands bearing recognition sites around a central metal ion. The latter would provide both a specific disposition of the recognition units depending on the coordination geometry of the ion and additional strong electrostatic interactions due to the charge on the ion (Figure 15). Such species have been investigated by means of α,α'-bipyridine ligands bearing nucleosides, that yield "metallonucleate" complexes on binding to a metal ion [7.38a]. More recently, peptides (see Section

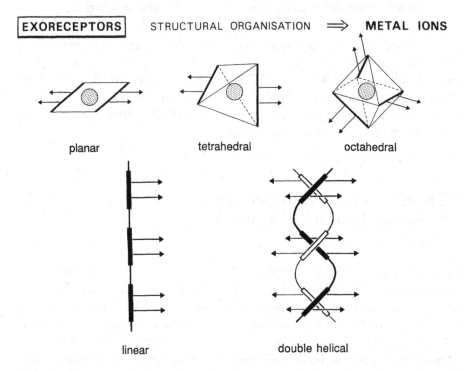

Fig. 15. Schematic representation of a planar, tetrahedral, octahedral and double-helical arrangement of external interaction sites (arrows) on ligands organized around metal ions of given coordination geometry in metallo-exo receptors; a linear strand is also shown.

9.3.1, [9.88, 9.89]) and sugar units [7.38b] have been organized by similar means; deoxyribonucleohelicates represent metallo-oligonucleosides arranged in a double-helical fashion on an inorganic framework (see Section 9.3.1, **144, 145,** [9.66]). Further studies are required to explore the potential of such an approach.

Molecular recognition at inorganic or organic surfaces and interfaces (monolayers, films, membranes, cell walls, organic or inorganic solids, etc.) involves outside directed interaction sites, i.e., exoreceptor function. It makes possible the control of crystal growth processes [7.39–7.42]. Suitably designed auxiliary molecules act as promoters or inhibitors of crystal nucleation inducing, for instance, the resolution of enantiomers or the crystallization of desired polymorphs [7.39–7.41].

Surface recognition takes place on molecular layers bearing recognition groups, allowing the selective binding of complementary substrates at the interface [7.43–7.45]. For instance, the binding of streptavidin on biotinylated lipid monolayers [7.43, 7.44, 7.46], of ATP [7.47a], of nucleotides [7.47b] and of amino acids [7.47c] on guanidinium bearing and on hydrogen bonding monolayers have been investigated. Amphiphilic complementary components form organized bilayer membranes through hydrogen bonding networks [7.48]. Recognition and chiral discrimination occurs in chiral monolayers [7.49]. Specific organisation and selective interactions at liquid interfaces [7.50] may be envisaged, in particular with respect to controlling phase transfer processes. Recognition at molecular layers is also the basis for the development of substrate selective sensors (see Section 8.4.5).

Insertion of recognition groups into vesicle membranes provides ways of performing vesicle targeting [7.51–7.53] (for instance for drug delivery) and of setting up models of cell–cell interactions [7.51, 7.54].

The fact that polymolecular assemblies define surfaces on which and through which processes can occur, again stresses the interest of designing exoreceptors operating at the interfaces, in addition to endoreceptors embedded in the bulk of the membranes. The designed use of molecular recognition events may lead to important developments in the control of the supramolecular architectures and of the physical properties of organized materials; thus, it could have a strong impact on materials science (see also Section 9.7).

7.3 Molecular and Supramolecular Morphogenesis

The design of exoreceptors requires a means of generating defined external molecular shapes. This may be achieved by the stepwise synthesis of molecular architectures via covalent bond formation (*covalent* or *molecular morphogenesis*). The polymerization of oriented systems in molecular assemblies may also serve this end [7.10].

Another approach to controlled molecular morphogenesis is provided by the generation of globular molecules such as the "starburst dendrimers" and the "arborols", based on highly branched structures formed via cascade processes and growing from a central core [7.55–7.60]. Most such molecular scaffolding hase been produced by repetitive processes; the use of sequences of different reactions is expected to give access to an even richer variety of *non-repetitive* branched architectures.

One may also assemble organic ligands on metal templates that impose a given coordination geometry, as noted in the preceding section. Inorganic tree-like species have been elaborated [7.61, 7.62]. Starting from a given core, it may be possible to produce a molecule or a metal complex of given size and external shape.

On the other hand, one may try to achieve the spontaneous generation of supramolecular shapes by self-assembly of molecular components (see Chapter 9). This represents a recognition-dependent *supramolecular morphogenesis*, an *automorphogenesis* based on non-covalent, intermolecular bonding interactions. It should open paths to the controlled assembly of organic or inorganic polymolecular architectures, patterns and surface topographies. On the polymolecular level, it may be related to the generation of shapes and structures in two-dimensional lipid mono-layers [7.6] and in aggregates of amphiphilic molecules [7.8].

Molecular and supramolecular morphogenesis is at its highest expression in the generation of biological structure in the course of the development of living organisms [7.63].

7.4   Supramolecular Heterogeneous Catalysis

All processes discussed in Chapter 5 represent homogeneous reactions and catalytic transformations. Heterogeneous supramolecular catalysis may be effected on organized organic or inorganic materials. It involves confering both substrate recognition features and reactive sites to external or internal surfaces of organic or inorganic materials such as polymolecular assemblies, polymers, zeolites, clays, sol–gel preparations, etc. [7.1–7.19]. This may be possible by generating a shape and size-selective footprint on the surface or in the bulk of a material, thus providing heterogeneous catalysts with substrate specificity based on molecular recognition.

Reactivity and catalysis in polymolecular assemblies and in macromolecules has been extensively studied, especially in the case of micelles and reverse micelles which may be considered as microreactors [7.64–7.68]. Amphiphilic molecules bearing suitable functional groups have been designed for effecting a variety of reactions in micellar conditions and surfactant systems have been investigated as enzyme models [7.64–7.68]. A range of structural selectivity and kinetic features (for instance, enantioselectivity [7.69]) have been obtained. Combining amphiphilic molecules bearing recognition groups (e.g., lipophilized macrocycles or H-bonding units) with others carrying reactive functions could lead to mixed assemblies capable of effecting a reaction on specific substrates in confined conditions, in closed vessels (micelles, liposomes, nano-size or larger capsules [7.70a]) or at interfaces [7.70b].

Heterogeneous catalysis on inorganic solid materials is of major importance in both basic and applied chemistry. Extending it into the supramolecular domain by incorporation of substrate recognition sites requires controlling the design of surfaces and of solids [7.28] and finding a means of introducing appropriate reactive sites at the substrate binding domain. Metal complexes may be encapsulated in the cavities of zeolites providing novel solid state metallo-reagents and inorganic biomimetic catalysts [7.71a]. Asymmetric induction has been obtained by grafting a NADH model on a silica surface [7.71b]. Of particular interest is the inclusion of enzymes into sol–gel materials [7.71c].

One may note that the analysis of solid state molecular geometries obtained from crystal structure data may reveal the changes occurring when two reagents approach each other, thus providing insight into chemical reaction paths [7.72].

Of special interest is the eventuality of stabilizing transition states by imprinting their features into cavities or adsorption sites using stable transition state analogs as templates. Studies towards such TSA "footprint" catalysis have been performed by generating TSA complementary sites as marks on the surface [7.73a] or as cavities in the bulk [7.73b] of silica gel. These imprinted catalytic sites showed pronounced substrate specificity [7.74a,b] (namely in the case of cavities [7.73b]) and chiral selectivity [7.74c,d].

Similar features may be achieved with organic polymeric materials. Template imprinting of complementary cavities containing appropriate functional groups yields models of enzyme active sites [7.34, 7.35, 7.75]. They perform the synthesis of amino acids with enantioselectivity [7.76] or the esterolysis of activated esters by TSA imprinting [7.77].

The area of molecular recognition and catalysis on solid material presents great potential. It requires the design of appropriate solids and surfaces by developing techniques for imprinting complementary sites for substrate selection and transition state stabilization, as well as for inserting reactive catalytic units. One may for instance consider generating chiral cavities in zeolites or in polymers and fitting them with reactive groups. The sol–gel techniques and the materials they yield deserve special attention in this respect due to their versatility [7.19].

The solid state, heterogeneous aspects of supramolecular catalysis should provide ground for a promising exchange and an increasing opening up of heterogeneous and homogeneous catalysis towards each other. They represent a particularly fruitful field of future investigation at the meeting point of supramolecular chemistry with the organic and inorganic solid state.

The incorporation of suitable photoactive, electroactive or ionoactive groups and components may yield *functional* supramolecular assemblies capable of performing operations such as energy-, electron- or ion transfer, information storage, and signal transduction [7.1–7.3, 7.43, 7.44, 7.78–7.90]. The combination of receptors, carriers and catalysts, handling electrons, ions and molecular substrates, with polymolecular organized assemblies, opens the way to the design of *molecular* and *supramolecular devices* and to the elaboration of chemical microreactors and artificial cells.

8 Molecular and Supramolecular Devices

Molecular devices have been defined as structurally organized and functionally integrated chemical systems; they are based on specific components arranged in a suitable manner, and may be built into supramolecular architectures [1.7, 1.9]. The *function* performed by a device results from the integration of the *elementary operations* executed by the components. One may speak of photonic, electronic or ionic devices depending on whether the components are respectively photoactive electroactive or ionoactive, i.e., whether they operate with (accept or donate) photons, electrons, or ions. This defines fields of *molecular* and *supramolecular photonics*, *electronics* and *ionics*.

Two basic types of components may be distinguished: *active components*, that perform a given operation (accept, donate, transfer) on photons, electrons, ions, etc.; *structural components*, that participate in the build-up of the supramolecular architecture and in the positioning of the active components, in particular through recognition processes; in addition, *ancillary components* may be introduced to modify or perturb the properties of the other two types of components. A basic feature is that the components and the devices that they constitute should perform their function(s) at the molecular and supramolecular levels as distinct from the bulk material. Incorporation of molecular devices into supramolecular architectures yields functional supermolecules or assemblies (such as layers, films, membranes, etc.).

Molecular and supramolecular devices are by definition formed from covalently and non-covalently linked components, respectively. One might envisage that covalently built devices made up of distinct but interacting components, retaining at least in part their identity as if they were bound together in a non-covalent fashion, could also belong to the supramolecular domain. Such a case has been argued for *supramolecular photochemistry* [A.10] and could be extended to other supramolecular functions.

This may be perceived as a significant stretching of the basic definition of supramolecular species. This definition is essential to charactere the identity of the field.

One may however consider that the integrated operation of its individual components confers to a molecular (covalently linked) device a flavour of supramolecular nature, since its function extends over the whole device in a sort of superstructure fashion. It then would appear justified to include in the domain of supramolecular chemistry systems that bear relation to supramolecular species on *functional* grounds, systems whose function is of supramolecular nature although they are clearly molecular in terms of structure and bonding. A further reason for not excluding such a possibility lies in the considerable enrichment that such an open view brings to the field!

8.1   Molecular Recognition, Information and Signals. Semiochemistry

Molecular recognition events represent the basis of information processing at the supramolecular level. They may give rise to changes in electronic, ionic, optical, and conformational properties and thus translate themselves into the generation of a signal. In endo receptors the binding (information) sites are oriented into a molecular concavity; exo receptors, with outward-directed sites, present "informed" surfaces on which recognition may occur (Section 7.2). By making use of the three-dimensional information storage/readout operating in molecular (endo- or exo-) recognition, in combination with substrate transformation and translocation, one may be able to design components for devices that would be capable of processing information and signals at the molecular and supramolecular levels.

Molecular recognition processes may play a role in several key steps: (1) the building up of the device from its components; (2) its incorporation into supramolecular arrays; (3) the selective operation on given species (e.g., ions); (4) the response to external physical or chemical stimuli (light, electrons, ions, molecules, etc.) that may regulate the operation of the device and switch it on or off; (5) the nature of the signals generated and of the signal conversion effected (photon–photon, photon–electron, electron–electron, electron–ion, ion–ion, etc.).

Recognition-dependent *informed* devices represent the molecular and supramolecular entities by which molecular recognition events may be transduced into processes and signals, through the design of suitable components responding to external stimuli and suitable for incorporation into the final superstructure. Recognition expressed via a chemical transformation, in the nature or in the rate of formation of the products, amounts to the generation of a specific *molecular signal*.

The resulting area of investigation has been termed *semiochemistry* (from the Greek σημεῖον: sign, signal), the chemistry of molecular signal generation, processing, transfer, conversion, and detection [1.9, 8.1]; it may be considered as the chemical sector of a general area of *semionics* covering all aspects of signal handling. Devices generating a signal and effectors carying it can be termed *semiogens* and *semiophores*, respectively, the elementary sign in a signal being itself a *semion*. An ensemble of information and signals conveys a specific *message*.

Signal transfer through a membrane may occur either by *translocation* of a species (carrier or channel facilitated) or by *transduction* via effector/receptor binding (see also Section 8.5.1).

Information storage, processing and signalization are, of course, basic features of living systems that bring them to their highest expression, from the genetic message to cellular communication and up to brain function.

8.2 Supramolecular Photochemistry. Molecular and Supramolecular Photonic Devices

The formation of supramolecular entities from photoactive components may be expected to pertub the ground-state and excited-state properties of the individual species, giving rise to novel properties that define a *supramolecular photochemistry* [8.2, A.10, A.20].

Thus, a number of processes may take place within supramolecular systems, modulated by the arrangement of the components: excitation energy migration, photoinduced charge separation by electron or proton transfer, perturbation of optical transitions and polarizabilities, modification of redox potentials in ground or excited states, photoregulation of binding properties, selective photochemical reactions, etc.

Supramolecular photochemistry, like catalysis, may involve three steps: binding of substrate and receptor, mediating a photochemical process (such as energy, electron or proton transfer), followed by either restoration of the initial state for a new cycle or by a chemical reaction (Figure 16).

The photophysical and photochemical features of supramolecular entities form a vast area of investigation into processes occurring at the level of intermolecular organization. They may depend on recognition events and then occur only if the correct selective binding of the complementary active components takes place, as illustrated for a two-component case in Figure 17.

In principle, supramolecular photonic devices require a complex organization and adaptation of the components in space, energy, and time, leading to the generation of photosignals by energy transfer (ET) or electron transfer (eT), substrate binding, chemical reactions, etc. Numerous types of devices may thus be imagined

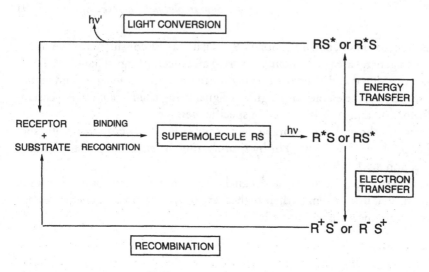

Fig. 16. Representation of photoinduced energy transfer and electron transfer processes involved in supramolecular photochemistry. Generation of R*S, RS*, R+S⁻, or R–S⁺ may be followed by a chemical reaction.

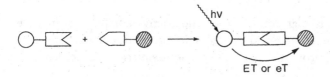

Fig. 17. Molecular recognition-dependent photochemical molecular devices. Photochemical processes such as energy transfer (ET) or photoinduced electron transfer (PeT) may be induced via association of two (or more) complementary units, each bearing a component of the device; the complementary units may be as small as heterocyclic bases or as large as antigen–antibody conjugates.

involving oriented energy migration, antenna effects, vectorial transfer of charge, conversion of light into chemical energy, optical signal generation, photoswitching, etc. (see for instance [8.3, A.10, A.20]).

8.2.1 Light Conversion and Energy Transfer Devices

Conversion of adsorbed light into emitted light of another wavelength occurs in any luminescent species. One would like, however, to perform separately the various steps of the overall process by means of different components that may be opti-

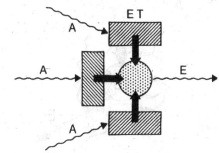

Fig. 18. Light-conversion photochemical molecular device, consisting of two components, a light collector (or antenna, light-absorbing groups A) and a light emitter E, and performing a three-step process: absorption (A), energy transfer (ET), and emission (E).

mized for a given purpose (wavelength of absorbed or emitted light; lifetime of emission, etc.). This is the case for a *light-conversion molecular device* consisting of two discrete components, a *light collector* (or *antenna*), formed by an array of strongly absorbing units and an *emitter*, thus allowing the separate optimization of absorption and emission. In order to function, intercomponent *energy transfer* must occur (as efficiently as possible). Thus, such a device operates in a three-step mode: absorption-energy transfer-emission (A–ET–E, Fig. 18).

This has been realized in *luminescent cryptates* of europium(III) and terbium(III) with macrobicyclic ligands [2.74, 2.75], such as that in structure **20**, which incorporate various heterobiaryl groups (2,2′-bipyridine, phenanthroline [2.71, 2.73, 8.4–8.7], bithiazole, biimidazole, bipyrimidine [8.8], biisoquinoline [8.5, 8.6], and their *N,N′*-dioxides [8.9-8.11]) serving as light-collecting antenna components. These complexes present a unique combination of features arising from both their cryptate structure and the nature of the emitting species: protection of the included ion from deactivation due to interaction with solvent (water) molecules; very high thermodynamic stability and kinetic inertness; multiple photosensitizing groups suitable for ET and displaying strong UV light absorption; characteristic long wavelength and long lifetime of the emission. They display a bright luminescence in aqueous solution whereas the free ions do not emit under the same conditions. Absorption of UV light by the organic antenna groups is followed by energy transfer to the included lanthanoid cation which then emits its characteristic visible radiation [2.74, 2.75, 8.2, 8.8–8.15]. The corresponding A–ET–E process is shown for the Eu(III)-cryptate in **87**.

Higher conversion efficiency and stronger luminescence are found for europium cryptates such as **88** containing N-oxide ligands [8.10, 8.11]. In the solid state, complex **87** exhibits especially efficient energy-conversion (about 60%) [8.16]. This is also the case for the corresponding Tb(III) cryptate up to about 100 K; at higher temperature, energy migration and back-transfer from the ion to the cryptand decreases the emission intensity [8.17]. Since the nature of the ligand groups strongly influences the photophysical properties, it may be possible to tune them to the desired purposes by suitable structural variations. Related macrocyclic complexes display

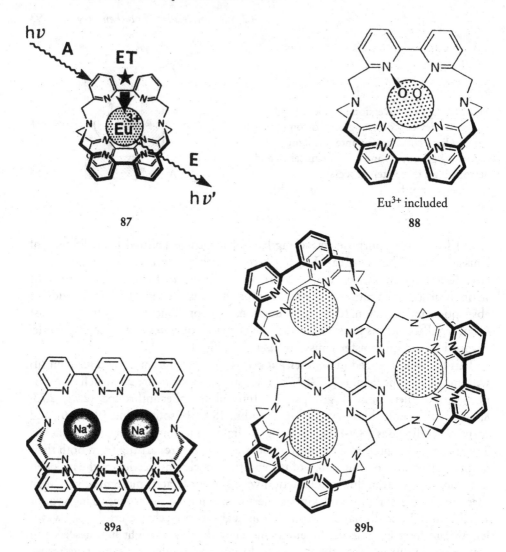

87

Eu³⁺ included

88

89a   89b

analogous luminescent properties [8.18]. Di- and trinuclear complexes, such as those derived from the sodium cryptates **89a** [8.19a] and **89b** [8.19b], would offer further developments for energy transfer studies.

Photoactive cryptates are of interest as novel luminescent materials (in solution or in the solid state) and as labels for biological applications subject to very stringent constraints (labelling of monoclonal antibodies, of oligonucleotides, of membrane components, cytofluorimetry, etc.) [8.20]. Linked, cascade systems performing very long range energy transfer (VLRET) may be imagined.

Of particular interest is the development of homogeneous *immunoassays* based on cryptate **87** as label. They make use of a second, long range energy transfer between a group **87** and an acceptor A (such as allophycocyanine) borne each by a specific antibody to different epitopes of the antigen as represented in **90** (antibodies bearing **87** and A: left and right; antigen: centre) [8.21a, b]; this process also results in signal amplification. Such immunoassays present great value for medical diagnostics [8.21c]. Europium or terbium complexes of crown ethers [8.22] and of a calixarene [8.23] display similar ET luminescence processes. Photochemical communication by ET or eT may take place between a luminescent metal complex borne by a β-cyclodextrin core and a guest molecule included in the receptor cavity [8.24].

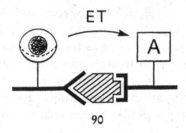

90

Polymetallic complexes presenting directional energy migration are of much significance for the design of photochemical molecular devices. Large arrays of multiple photoactive and redox-active building blocks (of ruthenium- or osmium tris(bipyridine)-type for instance) have been constructed for such purposes [A.10, 8.25–8.27].

Intramolecular transfer of excitation energy occurs in molecules containing several chromophores such as pyrene, anthracene or porphyrin groups connected by various linkers [8.28, 8.29]. Energy and photoinduced electron transfer processes (see Section 8.2.3) are found in bichromophoric supermolecules assembled through hydrogen bonding recognition via base pairing [8.30, 8.31].

Ultrafast energy hopping takes place in multichromophoric β-cyclodextrin derivatives [8.32] and triplet energy migration occurs in solid and liquid-crystalline phases of substituted phthalocyanines [8.33]. In the former case, very efficient ET from the multiple collector antenna leads to strong luminescence of an acceptor dye included in the cavity [8.32c]. Such results are of much interest for the design of inorganic or organic photochemical devices presenting directional energy migration, as was shown to take place in a photonic molecular wire built on a rigid linear array of porphyrin units [8.32d].

8.2.2 Photosensitive Molecular Receptors

Receptor molecules bearing photosensitive groups may display marked modifications in their photophysical properties on the binding of substrate species, leading to changes in their light absorption (e.g., colour generation) or emission features and allowing their detection by spectroscopic measurements [8.34–8.41] (see also Section 6.4.3). They represent molecular devices for *substrate-selective optical signal generation* and for *optical reading-out* of recognition processes. Such photo-chemosensors make possible the development of sensitive analytical methods for the detection of specific substrates [8.42].

Chromo- or lumino-ionophores of macrocyclic or macropolycyclic type respond to the binding of metal ions [8.34–8.41] and may be of much interest as analytical tools, for instance, for environmental applications or for the study of ionic changes in biological processes such as cell signalling [8.41]. Receptors such as **91–93** combine the strong and selective complexing ability of cryptands with the intense absorption or emission properties of photosensitive groups like azophenol [8.43–8.45], spirobenzopyran [6.65], anthracene [8.37, 8.46, 8.47] or coumarin [8.48, 8.49]. They display marked changes in absorption and fluorescence on complexation of alkali-metal cations or protonation, thus acting as fluorescent signalling systems (that may involve photoinduced electron transfer quenching [8.39, 8.40, 8.46]), for instance in a very selective colourimetric assay of lithium in serum [8.43b] (see also [8.50, 8.51]). Photoresponsive substances derived from other types of receptors like spherands [8.52] and calixarenes [8.53], present related properties.

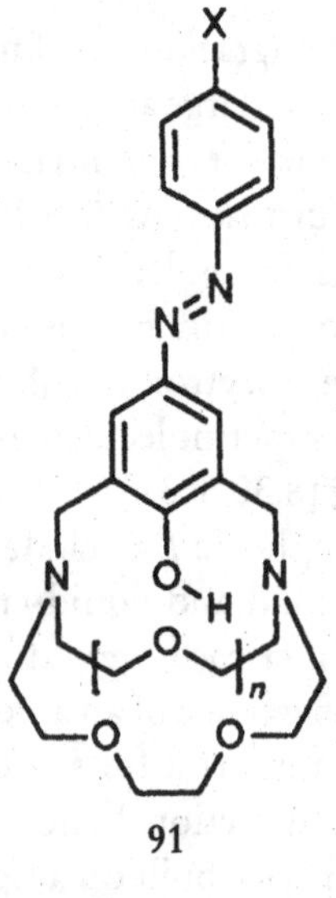

91

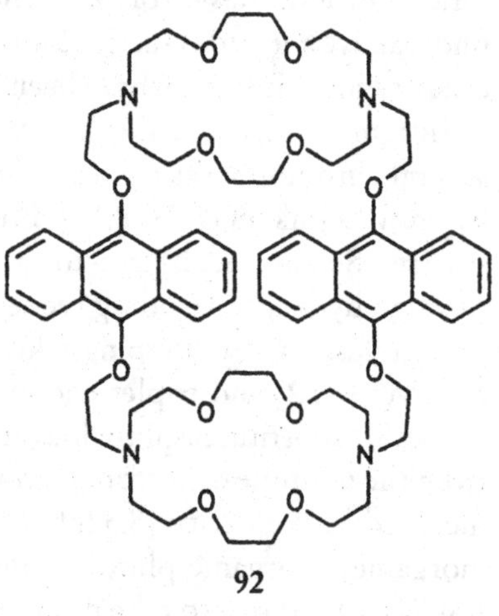

92

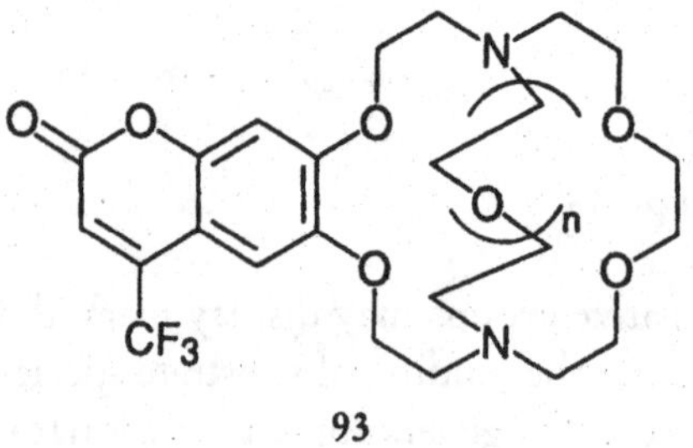

93

The macrotricyclic receptor **92** displays fluorescence detection of linear recognition of diammonium cations due to changes in monomer–excimer emission [8.54a] and the bis-naphthyl macrotricycle in **52** gives an extremely strong fluorescence enhancement on protonation, being thus a very sensitive proton sensor [8.54b]. Similar optical detection processes operate in the recognition of barbiturate by a luminescent bis-pyrene ligand [8.55a] and in the selective colouration of a spiropyran derivative on binding of guanosine derivatives [8.55b]. Fluorescent sensors may also be derived from cyclodextrins [8.56].

Photosensitive receptors for anionic substrates have been reported [4.27, 8.57]. Of special interest is the strong enhancement of the acridine fluorescence observed when ATP binds to a receptor molecule formed by a [24]-N_6O_2 macrocycle bearing a lateral acridine group, as shown in **82** [4.27]; guanosine triphosphate slightly decreases the emission. Thus, this receptor is a sensitive and selective ATP probe that generates a fluorescence signal on ATP binding. Such features may be of use for monitoring ATP levels and reactions involving ATP [8.58].

Photoresponsive receptor molecules present features that allow the development of molecular recognition dependent optrodes [8.42, 8.59, 8.60], optical sensors for the continuous monitoring of specific substrates in organisms, in the environment, etc.

Finally, one may note that the variations in the beautiful colours of flowers due to anthocyanins result from supramolecular effects, namely copigmentation, intramolecular sandwich-type stacking and intermolecular association [8.61].

8.2.3 Photoinduced Electron Transfer in Photoactive Devices

Light absorption, by markedly affecting the electronic properties of molecules and metal complexes, may induce intra- or intermolecular electron transfer processes leading to electron–hole separation.

The photogeneration of charge-separated states by photoinduced electron transfer (PeT) is of interest for initiating photocatalytic reactions (e.g., natural and artificial photosynthesis) and for the transfer of photosignals (e.g., through a membrane). It may be realized in a three component entity containing a photosensitizer PS linked to an electron donor D and an electron acceptor A, possessing suitable redox properties in their ground and excited (for PS) states so that irradiation of PS leads to electron transfer from D to A yielding the charge separated triad D^+–PS–A^- (Figure 19).

Numerous systems of this type, such as the carotenoid–porphyrin–quinone triad **94** [8.62a] (for PeT in a pentad see [8.62b]), have been extensively studied in many laboratories from the photochemical point of view and as models of natural photosynthetic centres [8.62–8.69, A.10, A.20], especially in order to achieve very fast charge separation [8.69] and slow recombination, for instance in multiporphyrin

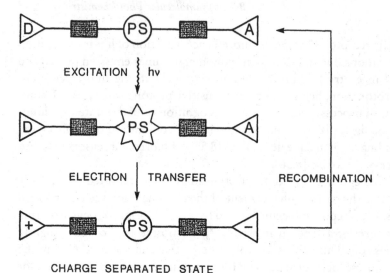

Fig. 19.   Schematic representation of a center for photoinduced charge separation consisting of a photosensitizer PS, a donor D and an acceptor A as well as insulating spacer groups [8.2].

94

complexes [8.69c]. In this respect, organized supramolecular assemblies have been used to separate the D, PS and A components and influence the rates of forward and reverse electron transfer processes so as to yield long lived charge-separated states [8.68].

The D, A or PS units may be metal coordination centres. Metal to ligand charge transfer (MLCT) in metal complexes (such as Ru(II) or Re(I)-diimine centres) has been extensively used for generating PeT processes [8.64–8.68, A.10, A.20]. Our own work has been concerned with photoinduced charge separation in macropoly-cyclic coreceptors containing both a photosensitive porphyrin group and binding sites for silver(I) ions as acceptor centres. Thus, complexation of silver ions by the lateral macrocycles, as shown in **95**, results in quenching of the singlet excited state of the Zn–porphyrin centre by an efficient intracomplex electron transfer, from the porphyrin to Ag(I), leading to charge separation and generating a porphyrinium cation of long lifetime [8.70a].

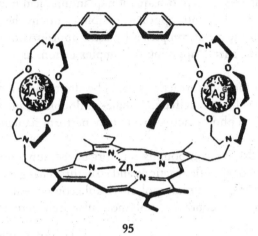

95

Long range thermal and PeT processes are of particular interest in both chemistry and biology (see also Section 8.3.2). Rigidly linked components allow the investigation of the geometrical factors influencing electron transfer [8.71a,b, A.10, A.20]. Electron hopping occurs in polynuclear assemblies of metal complexes bound, for instance, to a polymer support [8.71c]. Specific PeT effects take place in restricted reaction spaces, between groups on the surfaces of micelles, in dendrimers or bound to the DNA double helix [7.83].

Systems undergoing photoinduced electron-transfer processes represent components for *light-to-electron conversion devices*. Of special interest is the design of systems performing long-range electron transfer (see Section 8.3.2), in order, for instance, to generate transmembrane photoelectric signals or to serve as the basis of an electronic shift register memory at the molecular level [8.72].

8.2.4 Photoinduced Reactions in Supramolecular Species

Receptor molecules containing suitable light-sensitive groups may undergo photoinduced structural and conformational changes that affect their binding properties [6.63–6.65]. As a consequence, substrate uptake or release occurs, representing a way to generate protonic or ionic (or molecular) photosignals in *light-to-ion conversion devices*.

Receptor–substrate binding may affect the photochemical reactivity of either or of both species. The recognition process could thus modify the course of a reaction, affecting its rate or the nature of the products, and give rise to novel transformations.

The complexation of coordination compounds may make it possible to control their photochemical behaviour via the structure of the supramolecular species formed. For instance, the binding of cobalt(III) hexacyanide by macrocyclic polyammonium receptors markedly affects their photoaquation quantum yield in a structure-dependent manner [8.73–8.77]. It thus appears possible to orient the photosubstitution reactions of transition-metal complexes by using appropriate receptor molecules. Such effects may be general, applying to complex cations as well as to complex anions [2.114].

The course of a number of other photoreactions could be altered or oriented by selective binding. In particular, substances effecting single- or double-strand nucleic acid recognition and fitted with a photoreactive group may perform sequence specific photocleavage [4.56, 5.29, 5.30].

The incorporation of photoactive compounds into organized supramolecular assemblies allows the induction of specific reactions and structural changes such as, for instance, in the photocrosslinking of photopolymerizable components in lipid vesicles, a powerful approach to the control of polymolecular architectures [7.10, 8.78].

8.2.5 Non-Linear Optical Properties of Supramolecular Species

Organic and inorganic substances presenting a large electronic polarizability are likely to display marked optical non-linearities that depend first on molecular features [8.79–8.84], but novel effects should arise on going to the supramolecular level [8.85, 8.86].

Non-linear optical (NLO) properties are considered to result from both the intrinsic characteristics of the molecules and from their arrangement in a material (solid, powder, monolayer or multilayer film). An intermediate stage should also be taken into account, that of the supermolecules, where the association of two or more components may yield more or less pronounced NLO effects. Thus, three levels of non-linear optical properties may be distinguished corresponding to the molecule, the supermolecule and the material. The molecular level involves intramolecular effects and structures, whereas at the supramolecular level intermolecular interactions and architectures come into play together with collective effects in the material [8.85, 8.86]. Incorporation into supramolecular arrangements may perturb the electronic properties of individual molecules and impose a specific orientation, namely by recognition induced self-organization.

The design of molecules, supermolecules and materials presenting NLO activity involves molecular and supramolecular engineering. At the molecular level, a high polarizability, leading to large quadratic β and cubic γ hyperpolarizability coefficients, is sought. At the supramolecular level, it is necessary to achieve a high degree of organization, such as found in molecular layers, films, liquid crystals, the solid state, which may be generated by molecular recognition and inclusion complex formation. Both features are required for materials to display pronounced second order NLO effects; the structure must also be non-centrosymmetric, due either to the molecular components themselves or to their arrangement in the condensed phase. On the other hand, centrosymmetric species present third order but no second order NLO properties. In addition, bulk characteristics such as stability, preparation, processing and mechanical features will determine the practical usefulness of a given material (for general presentations see for instance [8.79–8.84]).

Of special interest are push–pull polyenes composed of a polyolefinic chain bearing a donor group on one end and an acceptor on the other (see **96**). Push–pull carotenoid derivatives (Figure 20) are highly polarizable conjugated polyenes that also represent polarized molecular wires.

They display intramolecular charge transfer [8.87] and marked second-harmonic generation. The determination of the hyperpolarizabilities of such compounds

96

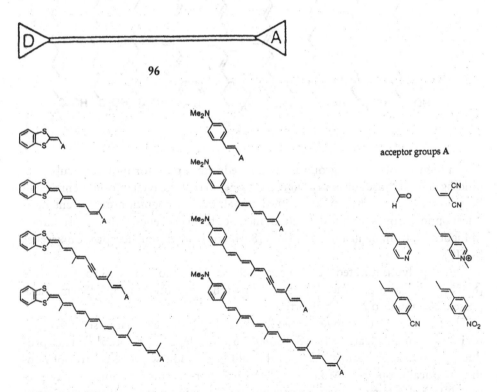

acceptor groups A

Fig. 20. Two series of donor–acceptor polyconjugated molecules bearing a benzodithia or a dimethylaminophenyl group as donor D and a variety of acceptor groups A.

yields exceptionally large values for the longer ones [8.88, 8.89] and provides insight into the dependence of quadratic and cubic hyperpolarizabilities on chain length and on substituents [8.83, 8.90, 8.91].

The incorporation of such compounds into mixed Langmuir–Blodgett films built from a fatty acid (see **97**) [8.92] or from an amphiphilic cyclodextrin [8.93] produces oriented arrangements presenting marked NLO features.

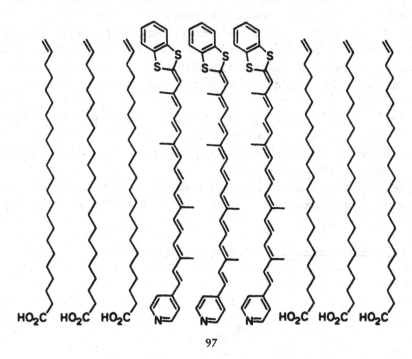

97

If a highly polarizable group is introduced into a receptor molecule, substrate binding should cause substantial perturbations, so that the recognition event would be converted into a non-linear optical signal. Such recognition-dependent non-linear optical probes may be derived for instance from polyenes such as those shown in Figure 20, from inclusion complexes [8.94a] or from donor–acceptor calixarenes [8.94b].

Other polyunsaturated substances such as oligothiophenes [8.95a] or dyes [8.95b] as well as organometallic or coordination compounds [8.85, 8.95c] may also present pronounced NLO properties.

The D–PS–A systems investigated for their ability to yield charge-separated states possess suitable features (see Section 8.2.3). Ion-dependent optical changes produced by indicator ligands [8.34–8.40] could lead to ion-selective control of non-linear optical properties. Molecular electron donor–acceptor complexes may present NLO effects since they possess polarized ground states and undergo intermolecular charge transfer on excitation. It is possible to more or less finely tune their polarization, polarizability, extent of charge transfer, absorption bands, etc., by many variations in basic structural types as well as in substituents.

It is clear that combining the design of NLO active molecules with the manipulation of selective intermolecular interactions may produce novel NLO materials that take into account supramolecular features.

8.2.6 Supramolecular Effects in Photochemical Hole Burning

Dye molecules incorporated in a crystalline or amorphous (polymer or glass) solid matrix are located in different environments and therefore present a distribution of absorption frequencies whose envelope results in a broad absorption band. When such dye doped media are irradiated with laser light under conditions where the guest species are frozen in their environment (very low temperature), given molecules are selectively affected so that very narrow and stable "holes" can be "burned" into the absorption band. Such photochemical hole burning in condensed materials is, because of its very high resolution, of great interest for high density optical storage [8.96–8.100].

The effects between dopant and matrix are clearly of supramolecular nature and each dye molecule with its surrounding matrix species may be considered to represent a supermolecule [8.98, 8.100]. Thus, hole burning is specifically a supramolecular photochemical process. As a consequence, the manipulation of intermolecular interactions [8.101] and recognition processes in the condensed phase could allow the designed generation of a range of environmental distributions, as well as induce specific changes in the absorption features of the dye molecules. Thus, taking into account explicitly the supramolecular effects opens ways to control the environmental features in hole burning experiments.

A number of far-reaching applications have been envisaged from ultra-narrow optical filters to high density optical storage (beyond 10^8 bits/cm^2) [8.96, 8.99], to holography and to optical "molecular computing" [8.98, 8.100, 8.102]. Optical data processing with bacteriorhodopsin displays very interesting features [8.99, 8.103]. Room temperature and single molecule hole burning represent exciting recent developments [8.99].

The photochemical and photophysical processes discussed above provide illustrations and incentives for further studies of photoeffects brought about by the formation of supramolecular species. Such investigations may lead to the development of photoactive molecular and supramolecular devices, based on photoinduced energy migration, electron transfer, substrate release, or chemical transformation. Coupling to recognition processes may allow the transduction of molecular information into photosignals. Thus molecular design, selective intermolecular bonding, and supramolecular architectures with photophysical, photochemical, and optical properties are brought together, building up recognition-dependent *supramolecular photonics*.

8.3   Molecular and Supramolecular Electronic Devices

Much attention has been given to the possibility of developing chemical systems capable of handling electrons in a way reminiscent of electronic components. Very extensive work has been performed on electroactive materials such as organic metals (e.g., conducting polymers and charge transfer salts) [8.104–8.106] or molecular semiconductors [8.107] that may lead to devices based on organic materials [8.108]. In such bulk compounds supramolecular effects occurring between the species forming the material play of course an important role.

On the other hand, the possibility has been envisaged to design electronic devices that would operate at the level of the molecule and the supermolecule. This defines a field of *molecular and supramolecular electronics* concerning the properties of single molecules, of oligomolecular associations and of polymolecular architectures such as Langmuir–Blodgett films [7.2, 7.3, 7.78, 7.79, 8.109–8.112, A.43] within and between which electron transfer processes [8.113] may occur. To realize such devices involves several steps. It is first necessary to imagine a component molecule that may possess the desired features, synthesize it, and study its properties. The second step is to incorporate it into supramolecular architectures, such as membranes or other organized structures, and to investigate whether the resulting entity possesses the required properties. The third step requires the connection of the basic unit to other components, in order to address it via relay molecules or with an external physical signal. Pursuing the design of molecular electronic circuitry may have many spin-offs along the way, in addition to raising novel questions about the handling of molecules. It is probably premature to try to define clearly the routes to the goal; indeed, the definition of the goal itself may change under the influence of progress made along a given path (for a view of the recent activities in the field see [8.114, 8.115]).

The possibility of designing devices such as molecular rectifiers [8.116], transistors, switches, and photodiodes has been envisaged and some of the required features are present in compounds such as metal complexes or D–PS–A systems that lead to photoinduced charge separation at the level of the isolated molecule ([8.62]; see Section 8.2.3). The advent of single electronics that concern processes involving a single electron in inorganic microstructures [8.117a] leads to the consideration of molecular monoelectronics [8.117b].

8.3.1 Supramolecular Electrochemistry

Molecular devices operating on electrons are basic elements for converting molecular recognition events into electronic signals. Redox properties may be affected by binding in a structure-dependent manner. Such perturbations by complexation lead to an electrochemical characterization of recognition. Electrochemical effects resulting from receptor–substrate binding define a domain of *supramolecular electrochemistry*.

Redox responsive receptor molecules are based on two types of component: substrate binding sites and electroactive groups. A number of substances have been studied in which the redox properties of groups such as metallocenes [6.50b, 8.118–8.121], quinones [6.50a, 8.122] or paraquat [8.123] are modified on substrate binding (see also Section 6.4.1). Conversely, redox interconversion of these groups allows the reversible switching between states of high and low affinity. The mutual effects of redox changes and binding strength in a receptor–substrate pair thus may make it possible to achieve electrocontrol of complexation and, conversely, to modify redox properties by binding.

The electrochemical properties of redox-active substrates are also affected on binding to a receptor molecule. This is the case for the complexation of metal hexacyanides by polyammonium macrocycles where the shift of the redox potential depends on the binding constants and the oxidation or reduction of the substrates leads to pronounced changes in stability (see Chapter 3) [3.21, 3.22].

As a consequence of these electrochemical effects, molecular receptors may be used to convert the chemical information present in binding and recognition processes into electronic signals [8.124–8.127]. They allow the development of substrate specific sensors, based on chemically sensitive thick or bilayer membranes [6.6, 8.125], and semiconductor devices [8.126, 8.127] such as field effect transistors (FET). Conversely, one may envisage performing electrochemically controlled, selective release of a given substrate from a material, for instance for drug delivery [8.128]. Ultramicroelectrodes [8.129] represent means for addressing small scale electrochemical changes.

Multinuclear metal complexes possess a rich electrochemistry [8.130a] and may serve as electron reservoirs. Organic redox systems containing multiple electrophores effect electron storage and transfer [8.130c]. Thus, dinuclear Cu(I) [4.7] and Ni(II) [8.131] complexes exchange two electrons in one step and four electrons in two steps, respectively. Attachment of several redox groups such as ferrocene [8.132, 8.133], viologen [8.134] or Ni(II) cyclam [8.135] to a core provides electron reservoirs that may be capable of powering multielectron catalytic reactions. Polyoxymetallates also take up a number of electrons (up to 32) [8.136]. Biological multiredox centres are found, for instance, in the electron transfer protein cytochrome c_3 that contains four heme groups [8.137], whose redox properties are affected by local solvation as in synthetic porphyrins [8.138].

Systems presenting many identical redox-active components bound to a central core or assembled by specific interactions represent approaches towards *molecular batteries*, i.e., molecular devices capable of performing the reversible exchange (storage and release) of many electrons at the same potential.

8.3.2 Electron Conducting Devices. Molecular Wires

Among the various devices and components performing molecular-scale electronic functions that may be imagined, a crucial one is a *molecular wire*, which might operate as a connector permitting electron flow to occur between the different elements of a molecular electronic system.

Three main classes of electron transfer processes may be considered: (1) *electron transport* by redox active molecules acting as mobile carriers through membranes; this may or may not also involve electron transfer from one carrier to another during an encounter; (2) *electron hopping* between suitable redox active groups attached to a backbone or assembled via non-covalent interactions; (3) *electron conduction* along a continuous conjugation path formed by π-bonds, which may also involve other electron transmitting groups such as strained σ-bonds, etc.; it represents an electron channel, and embodies specifically the idea of a *molecular wire*.

The components effecting these processes may be either organic or inorganic. The two extreme cases (1) carrier mediated and (3) channel-mediated electron transfer are represented in Figure 21. The first is coupled to the flow or the counterflow of a positively or a negatively charged species, respectively; it has been considered in Section 6.4. Numerous studies on electron transfer across vesicle bilayer membranes have been performed, employing a variety of effectors, especially in view of the relations to biological electron transfer [6.44–6.46, 8.139].

The design of a molecular wire should satisfy three criteria: (1) contain an electron-conducting chain; (2) possess terminal electroactive and polar groups for reversible electron exchange; (3) be long enough to span a typical molecular supporting element such as a monolayer or a bilayer membrane (Figure 21, right).

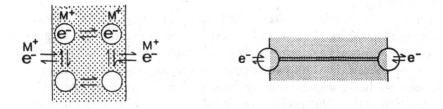

Fig. 21. Transmembrane electron transfer processes: carrier mediated via a redox carrier (left) or channel mediated via a molecular wire (right). Both processes may be coupled to light by introduction of photoactive groups in the carrier or in the wire.

Our first approach towards the design of molecular wires was based on the *caroviologens* (CV^{2+}), long, conjugated polyolefinic chains bearing pyridinium groups at each end, which combine the structural features of carotenoids with the redox properties of methylviologen. Such compounds (e.g.; **98**) were synthesized and shown to be incorporated in a transmembrane fashion into bilayer vesicles [8.140], as is also the case for other carotenoids bearing terminal hydrophylic groups [8.141]. The orientation of caroviologens in model lipid membranes was studied by linear dichroism measurements. Dichroic ratios were determined on caroviologens of different lengths solubilized in lyotropic lamellar phases of different thicknesses and macroscopically aligned between quartz plates. The orientation parameters obtained indicated transmembrane arrangement for caroviologens of length compatible with the bilayer thickness of the membrane [8.142].

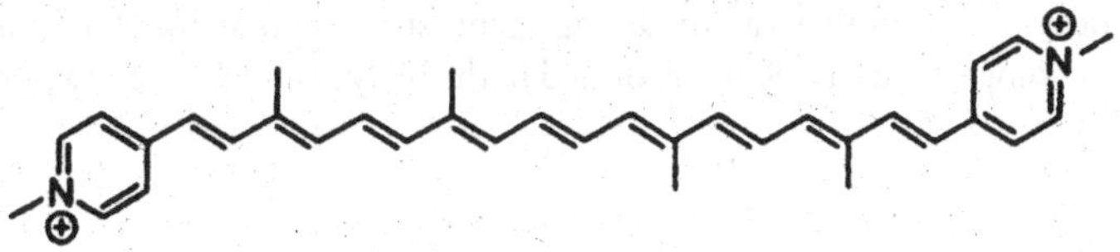

98

Electron-transfer experiments employing the CV^{2+}/dihexadecyl phosphate vesicle system did not give positive results, perhaps because of the high negative charge of the membrane surface, which may modify the redox potential of incorporated CV^{2+} and reduce interfacial electron transfer rates. However, on incorporation of zwitterionic caroviologens $CV^{2\pm}$ such as compound **99** into phospholipid vesicles, electron transfer occurred between an external reducing phase and an internal oxidizing phase (Figure 22). Using only a small amount of incorporated **99** (about 150 active molecules per vesicle) an acceleration factors of 4–8 over background was obtained. This clearly indicated that $CV^{2\pm}$ **99** did induce electron conduction [8.143]. Pulse radiolysis studies provided information about the reduced $CV^{\bullet+}$ [8.144a] and neutral CV species (see also [8.144b]) that are expected to be intermediates in the process.

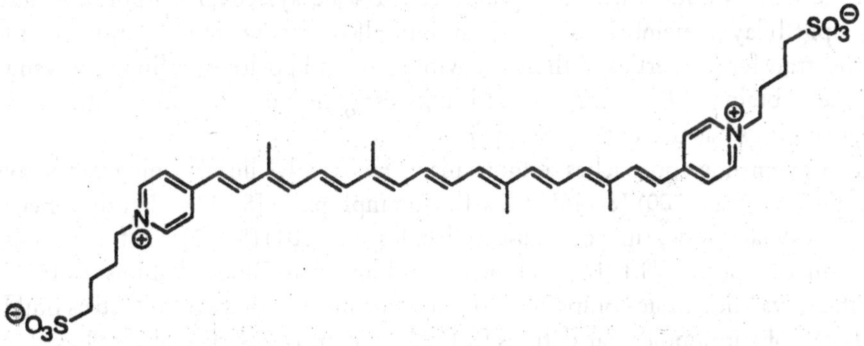

99

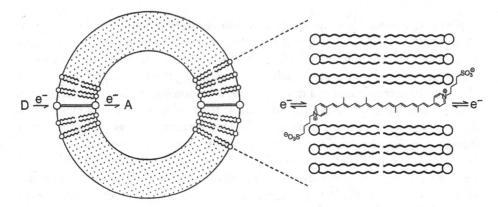

Fig. 22. Schematic representation of transmembrane electron transfer from a reducing agent (such as sodium dithionite) to an oxidizing agent (such as potassium ferricyanide) by a zwitterionic caroviologen such as **99** incorporated in the bilayer membrane of a phospholipid vesicle [8.143].

100

Thus, the caroviologen approach does produce *functional molecular wires* that effect electron conduction in a supramolecular-scale system. Incorporation into black lipid bilayer membranes (BLM) should allow further investigations of the electron-transfer properties of these caroviologens and positive preliminary results have been obtained [8.145a]. A theoretical investigation of electron conduction in molecular wires has been made [8.145b].

Other promising approaches to molecular wires involve linear conjugated arrays of porphyrins (e.g., **100**) [8.146], a bis-flavin amphiphile [8.147], oligothiophenes [8.148, 8.149] and long, rigid aromatic molecules (e.g., **101**) [8.150] that form persistent reduced species [8.151a] and oriented Langmuir–Blodgett films [8.151b]. Tetrathiafulvalenes, basic components of "organic metals", have been used as building-blocks of supramolecular entities [8.152]. One may note that species based on

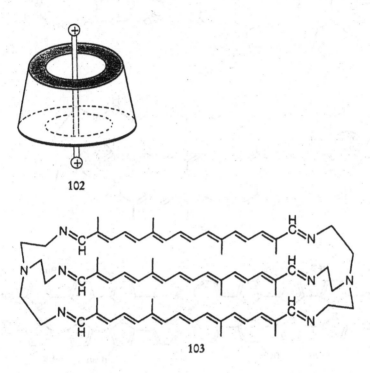

101

102

103

bicyclo[1.1.1]propellane ("staffanes") [8.153], bicyclo[2.2.2]octane [8.154] and car-
borane [8.155] components offer linear rigid frameworks for the oriented arrangement of redox active groups.

Since caroviologens are rather fragile compounds, they can be protected from the
environment by inclusion into polyanionic derivatives of β-cyclodextrin in a
rotaxane fashion 102 [8.156]. Also, in the design of molecular devices, it may be
desirable to introduce some extent of redundancy in order to reduce the risk of
device failure. This is the case in the tris-carotenoid macrobicycle 103 that represents
a sort of triple-threaded molecular "cable" whose crystal structure 104 has been
determined. It forms a dinuclear Cu(I) complex 105 in which the bound ions introduce a positive charge at each of the species, a feature of potential interest for transmembrane inclusion [8.157].

The development of electronic circuity will require the relative positioning of
components. This has been investigated, for instance, through orthogonal connection of two oligothiophenes via spiro junctions (e.g., 106) [8.158, 8.159]. Electron

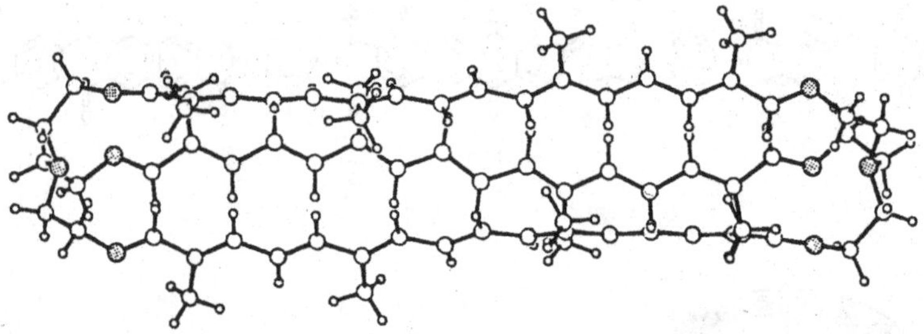

104

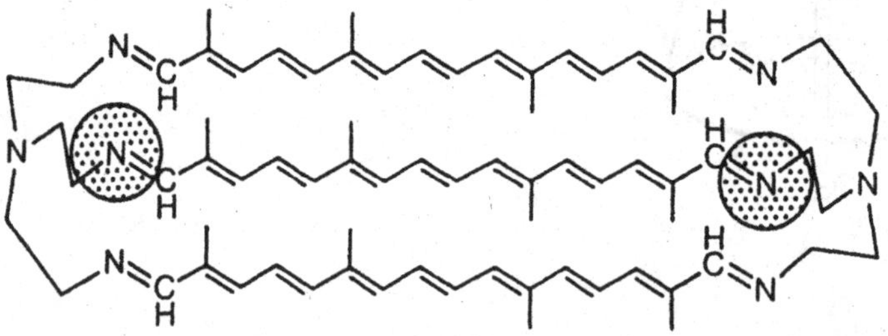

105

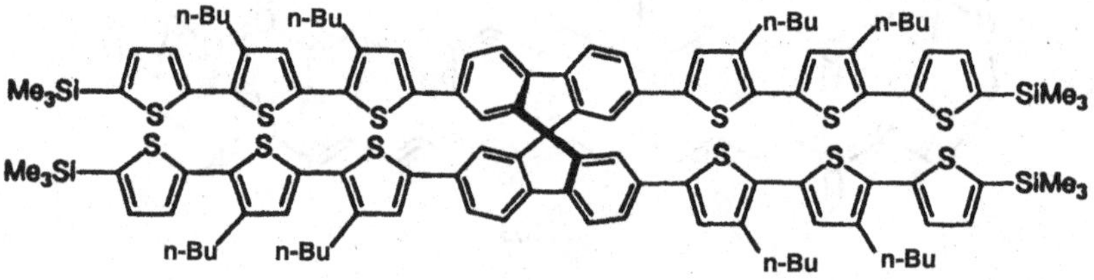

106

relaying between groups in substances containing many redox centres [8.160] may be used for electrically "wiring" enzyme redox centres to electrodes [8.161]. Miniaturization [8.129] and surface functionalization of electrodes [8.162] with suitably selected molecular units provide ways towards the electrical addressing of devices.

Electron transfer processes play a major role in biology and their mechanisms have been very actively investigated in order to gain information about the structural factors affecting transfer rates and directionality [8.163]. A series of appropriately derivatized cytochrome c protein molecules revealed that electron transfer rates correlate with the length of tunnelling pathways comprising covalent bonds and hydrogen bonds as well as through-space jumps [8.164]. The DNA double helix may serve as support for photoactive electron donor and acceptor groups [7.83b]. Such results also provide information of great value for the design of artificial electronic wires for oriented electron transfer over long distances in molecular and supramolecular systems. It depends on a number of features for which optimization conditions have been analysed to achieve ultrafast and highly efficient processes [8.165, 8.166].

8.3.3 Polarized Molecular Wires. Rectifying Devices

Conjugated polyolefinic chains bearing an electron-acceptor group on one end and a donor on the other end (e.g., **96**) represent polarized molecular wires that should display preferential *one-way electron transfer* and could act as a *rectifying compo-nent*. Such molecules, donor–acceptor carotenoids (Figure 20), were prepared and shown to possess very pronounced non-linear optical properties in powder form, in solution, or in Langmuir–Blodgett films (see Section 8.2.5). Oriented incorpora-tion into organized assemblies [8.92, 8.93] and electron-transfer experiments might reveal one-way conduction by polarized molecules of this or similar type.

Approaches to organic rectifiers have been made using D–σ–A species containing a donor and an acceptor group linked by an insulating covalent σ bridge [8.116]. A system based on a conjugated D–π–A unit incorporated in molecular layers has been reported [8.116c] as well as a bilayer rectifier at clay-modified electrodes [8.116d]. Theoretical analyses of electron transfer through such a single chain molecular wire have been given [8.116e]. Species of this type may contribute to the development of a molecular transistor.

8.3.4 Modified and Switchable Molecular Wires

Future developments in the design of molecular wires may concern three main fea-tures: (1) modifying the conduction component, replacing the polyolefinic structure by units such as connected or condensed oligothiophenes, oligopyrroles, aromatic groups or metal coordination centres, for instance (see Section 8.3.2); (2) varying the redox active end groups, which may also serve to anchor the device to a surface or into a membrane; (3) arranging the wire units into organized arrays or circuits through link-ing to a structural framework or by self-assembly based on recognition processes [8.167].

Attaching metal centres (such as ferrocene groups, pyridine or 2,2′-bipyridine complexes) at the end of a polyconjugated chain would provide molecular wires combining the rich electrochemical and photochemical activities of metal complexes with the long-range conjugation properties of the chain, thus leading towards systems capable of performing electro- or photoinduced long range electron trans-fer (LReT). Coupling molecular wires with photoactive groups is expected to yield photoresponsive electron channels. Species of the D–PS–A type (see Section 8.2.3) may act as photoactivated molecular wires in bilayers [8.168] capable of generating LReT, charge separation and signal transfer.

Carotenoid chains bearing metal complexes as terminal groups, *metallocarotenates*, such as **107** and **108**, belong to this class [8.169]. Thus despite a large metal–metal separation, **108** presents electronic coupling between the terminal subunits both in the Ru(II) Ru(III) mixed valence complex and in the triplet state of the bis-Ru(II) complex [8.170]. Bis-Ru(III) (NH$_3$)$_5$ complexes of bis-pyridine polyenes also display features resulting from the properties of the metal centres [8.171]. The bridging ligand controls the extent of electronic communication between terminal metals [8.172]. A ferrocene unit introduced into donor-acceptor type conjugated systems, as in **109**, may serve as donor centre or redox switch in films deposited on electrodes [8.173]. Substances of such types, which incorporate electro- and photo-

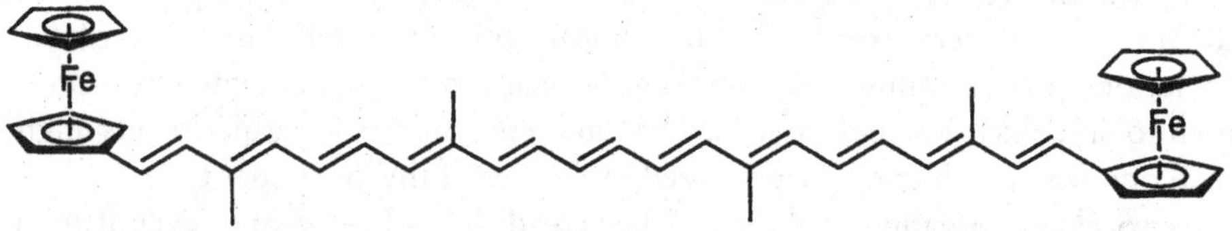

107

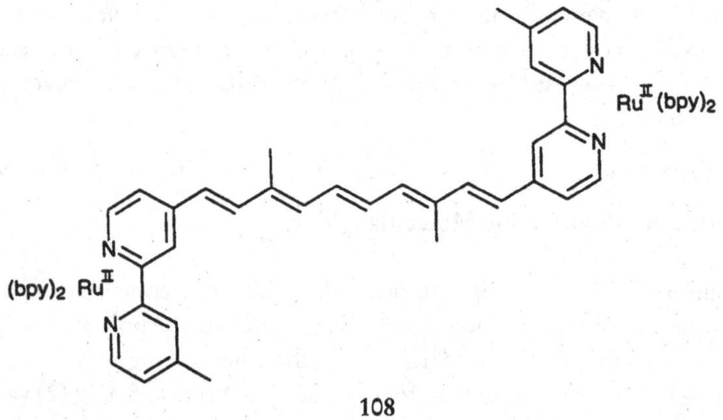

108

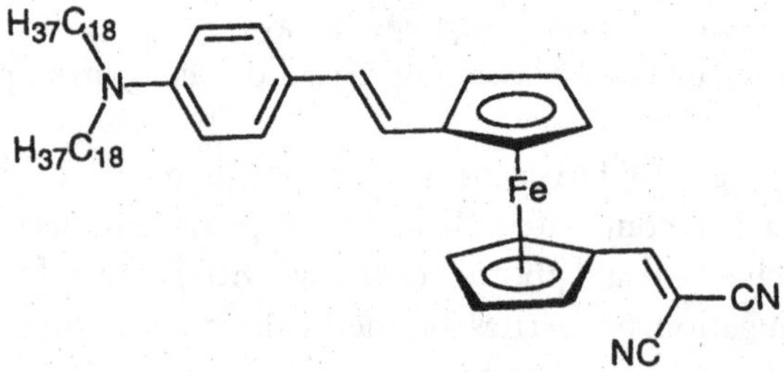

109

sensitive switches at the end of or inside the conjugated chain, represent switchable molecular wires (see Section 8.5.2) responding to external stimuli.

8.3.5 Molecular Magnetic Devices

Molecular magnetism [8.174] is by essence of supramolecular nature since it results from the collective features of components bearing free spins and on their arrangement in organized assemblies or solid lattices. The engineering of molecular magnetic systems requires the search for high-spin components of organic (such as free radicals, carbenes or charge-transfer salts), organometallic [8.174–8.177] or inorganic (metal coordination centres [8.178, 8.179]) nature and their arrangement in suitable supramolecular architectures so as to induce spin coupling and alignment. Of special interest is the search for molecular ferromagnetism towards which significant progress has been made [8.174–8.176, 8.179].

It is clear that the design of molecular free-spin components and the supramolecular control of spin coupling in three dimensions represents a major area of investigation towards magnetic devices and novel magnetic materials that may be of great interest for information storage and processing.

8.4 Molecular and Supramolecular Ionic Devices

The numerous receptor, reagent, and carrier molecules capable of handling inorganic and organic ions are potential components of molecular and supramolecular *ionic devices* that would function via highly selective recognition, reaction, and transport processes with coupling to external factors and regulation. Such components and the devices that they may build up form the basis of a field of *molecular ionics*, the field of systems operating with positively or negatively charged ionic species as support for signal and information storage, processing, and transfer. In view of the size and mass of ions, ionic devices may be expected to perform more slowly than electronic devices. However, ions have a very high information content by virtue of their multiple molecular (charge, size, shape, structure) and supramolecular (binding geometry, strength and selectivity) features.

Molecular ionics offers a promising field of research that may already draw from a vast amount of knowledge and data on ion binding and transport by natural and synthetic receptors and carriers. The molecular recognition events involved may be directly related to information and signal processing by ions, as is the case in biology [8.180, 8.181]. Indeed, biological signals and communication events are based on ionic and molecular species (sodium, potassium, calcium, chloride, acetylcholine ions, etc.). While the interest of components operating with photons and electrons is well recognized, one may make a special case for the development of molecular ionics as a counterpart to molecular photonics and electronics, and a wide open playground for the design of ion- or molecule-based devices [8.182].

Selective ion receptors represent basic units for ionic transmitters or detectors; selective ion carriers correspond to ionic transducers. These units may be fitted with triggers and switches sensitive to external physical (light, electricity, heat, pressure) or chemical (other binding species, regulating sites) stimuli for connection and activation.

Binding, transport and triggering may be performed by separate species, each having a specific function, as in multiple carrier transport systems (see Section 6.4). This allows a variety of combinations between photo- or electroactive components and different receptors or carriers. Light- and redox-sensitive groups incorporated into receptors and carriers affect binding and transport properties [6.50, 6.52, 6.63–6.65]. Coreceptors and cocarriers provide means for regulation via cofactors, co-bound species that modulate the interaction with the substrate. Thus, a simple ionizable group, such as a carboxylic acid function, represents a *proton switch* and leads to proton-gated receptors and carriers responding to pH changes, as seen for instance in the regulation of transport selectivity by **86** [6.53] (see also Section 8.5). Analytical methods based on optical or electrical effects induced by ion binding and transport have been developed (see Sections 8.2.2, 8.3.1 and 8.4.5).

Ion transfer takes place through mobile carriers or ion channels [6.69] (see Chapter 6; Fig. 10); a rotating shuttle process may also be considered. Artificial transmembrane channels have been much less explored than carriers, probably because of the inherently larger molecular structures involved, despite the fact that biological transport is thought to occur mainly via channel-type species [8.180, 8.181]. Various routes to the design of ion channels and ion-responsive membrane systems have been investigated; some of them have already been mentioned above (Section 6.5). One may imagine a variety of ion channel types depending on the nature of the ion binding sites, on the way in which they are arranged, on the molecular type and on the overall structural features. Some are represented schematically in Figure 23.

Approaches to artificial ion channels have, for instance, made use of macrocyclic units [6.72, 6.74] (see also below), of peptide [8.183–8.185] and cyclic peptide [8.186] components, of non-peptidic polymers [8.187] and of various amphiphilic molecules [6.11, 8.188, 8.189]. The properties of such molecules incorporated in bilayer membranes may be studied by techniques such as ion conductance [6.69], patch-clamp [8.190] or NMR [8.191, 8.192] measurements. However, the nature of the superstructure formed and the mechanism of ion passage (carrier, channel, pore, defect) are difficult to determine and often remain a matter of conjecture.

In addition to attempts at constructing ion channels on the basis of multiconnected polyether macrocycles [8.193], we have investigated mainly two approaches

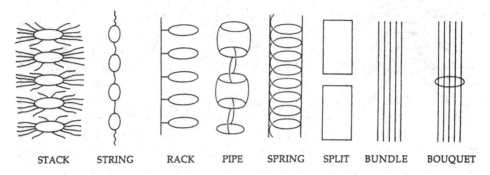

Fig. 23. Schematic representation of different possible types of ion channel structures (from left to right): stack, string, rack, pipe of macrocycles, helical strand, two half-channel units, self-aggregated and macrocycle core bearing bundles of chains.

towards the design of molecular structures that may function as ion channels and lead to ion-responsive membrane systems: tubular mesophases and "bouquet molecules". Self-assembling structures will be considered below (Section 9.4).

8.4.1 Tubular Mesophases

A stack of macrocyclic rings defines a molecular tube through which ions might flow, as illustrated by a solid-state model of such an ion channel [6.71]. Since in discotic liquid crystals, the spontaneous axial stacking of molecular discs gives columnar mesophases, it appeared that if such phases could be obtained with ring elements displaying an internal void, their arrangement in an axial stack would yield a hollow column, that is, a molecular tube 110. Macrocyclic polyamines bearing suitable side chains, such as 111 and 112, indeed give tubular mesophases that contain stacks of macrocycles [8.194, 8.195]. Photocyclization of cinnamoyl derivatives of 111 yields covalently linked columnar aggregates [8.195]. Although the macrocyclic elements used at present are not well suited for ion binding, further elaboration could lead to ion channels that would be phase (temperature) dependent and could therefore also present non-linear ion transfer behaviour.

Channel-like architectures are formed in the mesophase given by complexes of long chain crown ether derivatives [8.196a,b] and long-chain calixarene derivatives display columnar liquid-crystalline arrangements [8.196c]. Self-assembled tubular structures based on cyclic peptide components have been described [8.186].

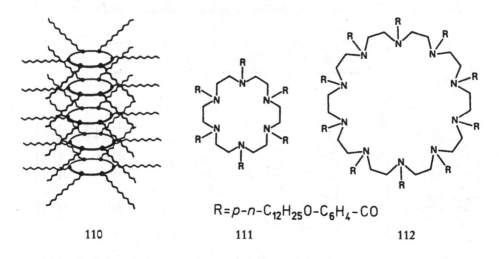

$$R = p\text{-}n\text{-}C_{12}H_{25}O\text{-}C_6H_4\text{-}CO$$

110   111   112

Ionic conductivity has been established with flexible polymers of poly(ethylene oxide) type [8.197]. The more rigid components of tubular mesophases or solid state macrocyclic structures may allow fast ion propagation if a smooth profile of the interaction potential for ion migration along the stack of macrocycles can be established. This may require incorporating the counteranion into the framework and employing ion binding sites that form loose rather than strong complexes. The latter would be expected to cause local trapping of the ions in a potential well. Molecular modelling should provide helpful information about the design of such structures. Similar considerations apply to the potential ion channels that could be formed by the stacking of polyether macrocycles borne by phthalocyanine [8.198–8.201], porphyrin [4.68], peptide [8.202] or polymeric [6.72] backbones. Cooperative ion binding has been found to occur in such multisite species [8.201].

8.4.2 Ion-Responsive Monolayers

Deposition of macrocycles bearing lipophilic side chains at the air–water interface could yield ion-responsive molecular films.

Indeed, macrocycles 111 and 112 form such monolayers [8.195, 8.203]. Comparison of measured surface areas with molecular models indicated that the macrocycles were lying flat on the water surface while the aliphatic chains were tilted upright in the compressed film, as shown in Figure 24 [8.203]. Monolayers containing binding units may interact selectively with ions or other substances present in the subphase [8.204].

Various developments may be envisaged, in particular towards monolayers effecting ion recognition at an interface (of interest, for instance, for selective detection and sensors, see Section 8.4.5) or ion-binding, electro- or photoactive films.

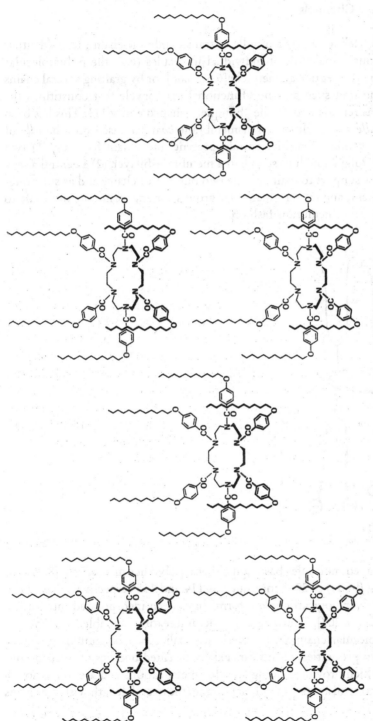

Fig. 24. Schematic representation of the arrangement of macrocycle 111 in a compressed film at the air–water interface; the aliphatic chains are represented in their fully extended form, but some curling may occur.

8.4.3 "Bouquet"-Type Molecules and the "Chundle" Approach to Molecular Channels

A channel may be delineated by a bundle of transmembrane chains, formed either by the spontaneous association of individual molecules (e.g., the polymolecular channel formed by the peptide alamethicin [6.67, 6.68]) or by grafting several chains onto a supporting unit such as a polyfunctional macrocycle that constitutes the organizing core, as represented by the "bouquet"-shaped entity **113**. This has been termed the *chundle* approach to a transmembrane *ch*annel based on a b*undle* of chains. Suitable *"bouquet"* entities should present three main features: (1) two bundles of chains long enough to span half a membrane bilayer, (2) a central annulus serving both as support to maintain the two bundles of chains and as substrate-selection component, and (3) terminal polar groups for anchoring the molecule to each interface in a transmembrane fashion.

113

"Bouquet"-shaped molecules based on either a polyether macrocycle [8.205] or a β-cyclodextrin (β-CD) [8.206] core, such as **114** and **115** respectively, have been synthesized and their polycarboxylate forms have been incorporated into vesicle bilayer membranes [8.207]. They present structural features suitable for studies of chundle-type molecular channels: (1) the functionalized cyclic annulus possesses substrate selection properties; (2) it bears axially oriented bundles of oxygen-containing chains, which provide binding sites for metal cations and are long enough for the molecule to span a typical lipid membrane (the overall length with the chains

114   **115**

in an extended state may be estimated to lie in the 45–50 Å range); (3) the terminal carboxylate groups allow for anchorage at water–membrane interfaces and trans-membrane orientation, as well as for eventual coupling to proton gradients.

When incorporated into the phospholipid membrane of liposomes, compounds **114** and **115** increased the permeability to Na^+ and Li^+ ions, and ion transfer proceeded by a cation–cation antiport mechanism (Figure 25) [8.208].

Variations in the bouquet molecule including changing from poly(oxyethylene) to polymethylene chains and changing from a crown ether annulus to a cyclodextrin (CD) annulus did not have strong effects on ion–transport activities. A carrier mechanism could be excluded for the compounds studied since ion-transport activity was undiminished in membranes in the gel state [8.208]. However, the results gave no firm indication about the exact mechanism of the process, whether it involved passage through the interior of the molecules or occurred through a pore induced by aggregation of several molecules, as is the case for the pore-forming polyene antibiotics (e.g., amphotericin, candicidin, nystatin, etc.) [8.209]. Models of the latter have been investigated as channel forming structures [8.210].

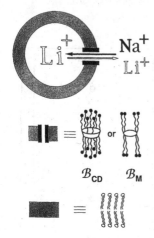

Fig. 25.   Schematic representation of the transport experiment in which aqueous NaCl solution is added to a preparation of liposomes containing bouquet molecules in the membrane in aqueous LiCl solution, creating opposing gradients in Na^+ and Li^+ ion concentrations. The entry of Na^+ ions, initially found only in the external volume, into liposomes is followed by ^{23}Na NMR spectroscopy; the exit of Li^+ may also be followed by ^7Li NMR spectroscopy.

The "bouquet"-type compounds **114** and **115** and related molecules thus form functional ion transfer entities. Further studies are required to ascertain the mechanism of transport and the effects of parameters such as membrane composition, nature of alkali metal ions and concentration of bouquet molecules on the transport rate in order to better describe the mode of action of these compounds. Ion transport by "bouquet"-type molecules related to **114** and bearing lateral macrocyclic units has been investigated [8.188].

A very large molecule such as **115** bearing seven chains on each side of the rigid β-CD core of about 6 Å diameter and 400 $Å^3$ internal volume is of interest in itself [8.206]. By its size, its internal space, and its chains, it represents a sort of *molecular vesicle* formed not by self-assembled amphiphilic chains but by chains covalently grafted on a central core. Because it is dissymmetric, this β-CD chundle could act as an oriented transmembrane device. Its transfer selectivities might have analogies with those of the channel of the acetylcholine receptor. Physical, chemical, membrane-incorporation, and substrate-transfer properties of this and other β-CD derivatives will provide basic information about the handling and the behaviour of large functional molecular species. Whatever the efficiency of the present or other chundle molecules as ion channels, they nevertheless allow the testing of approaches to channel design and to the handling of selective ionic conduction.

Incorporation of molecular ionic devices into supporting structures (such as membranes, layers, and films) allows the exploration of their functional properties and of their ability to produce ionic signals. Regulation of flow, switching, and coupling to other components will require further structural modifications. Cooperative effects and regulation of ion binding may be produced with ligands containing two or more sites for the same or different substrates [8.182, 8.211, 8.212], whereas photoresponsive ligands provide switching capacity (see Section 8.5).

8.4.4 Molecular Protonics

The proton, being the smallest ion and the elementary particle counterpart of the electron, deserves special attention. The designed handling of protons defines a field of *molecular protonics,* whose development and features may parallel those of molecular electronics. Despite the much higher mass of the proton compared with that of the electron, the rates of proton transfer can be very fast especially if tunnelling processes are involved. On the other hand, the extremely wide range of proton transfer rates may be considered an asset for the design of protonic devices. Molecular protonics thus represents a particularly interesting special case of molecular ionics. Proton transfer and gradients are of basic importance in bioenergetics, where they drive transport processes and ATP synthesis, and in the chemiosmotic scheme [8.213].

Numerous processes may be linked to proton transfer and protonation/deprotonation reactions (for general descriptions of proton transfer see, for instance, [8.214–8.217]). Proton-triggered yes/no or +/– switching is contained in the ability of polyamine receptors and carriers to bind and transport cations when unprotonated and anions when protonated; also, zwitterions such as amino acids may change from bound to unbound or vice versa, when they undergo charge inversion as a function of pH.

Linear arrays of protonatable or hydrogen bonded sites may allow the directed long range transfer of protons, thus functioning as proton-conducting channel, i.e., as *proton wire*. Relevant systems would be linear polyamines or polyphenolic condensed aromatic units [8.218], self-assembled hydrogen bonded heterocyclic ribbons such as **116** (see Section 9.4.4) or polyelectrolyte membranes [8.219] in which collective proton motion may take place and lead to proton conductivity.

116

Multiple cyclic proton transfers occur in hydrogen bonded arrays of heterocyclic units [8.220, 8.221] or inside rings such as porphyrins [8.222]. Macrocyclic polyamines present various protonation patterns [3.13a] that could be of interest as information units. Data storage in a molecular "memory" by hole burning makes use of proton transfer processes for instance, in porphyrins and hydrogen bonded dye structures [8.96–8.99, 8.223]. Proton transfer mediates charge transfer [8.224] and phase change [8.225] in H-bonded solid molecular materials.

Photoinduced proton transfer may be generated through the large variation in acidity or basicity of functional groups in the excited states of specific structures [8.226] and lead to photoinduced pH jumps [8.227, 8.228]. Changes of optical properties by tautomerisation in the excited state [8.229] occur, for instance, in the fluorescent states of bipyridyl diols [8.230a] and form the basis of a proton transfer laser process [8.230b].

Of much significance is the realization of long-lived photogenerated tautomeric states and long-range proton transfer (LRpT) processes. The latter could lead to proton transfer charge separated states to be put in parallel with the extensively studied charge separation by photoinduced electron transfer (see Section 8.2.3). A number of systems present photochromism on the basis of photoinduced proton transfer [8.229].

In photosynthesis, light absorption results in charge separation by photoinduced electron transfer. Realizing a proton-based analogue would be of much interest. The first requirement is to devise systems presenting long-lived photoinduced tautomeric states, compared with the very short lifetimes of the acid–base features of excited states (for instance, in 2-naphthol) [8.226]. Significant progress towards such a goal has been achieved by means of derivatives of the photochromotropic dinitrobenzyl–pyridine **117**, which turns deep blue in the solid state on exposure to light and reverts slowly to the colourless form in the dark [8.231, 8.232]. This colour change results from a photoinduced proton transfer generating the tautomeric form **118**. Irradiation of the phenanthroline analogue **119** has been found to give a tautomer

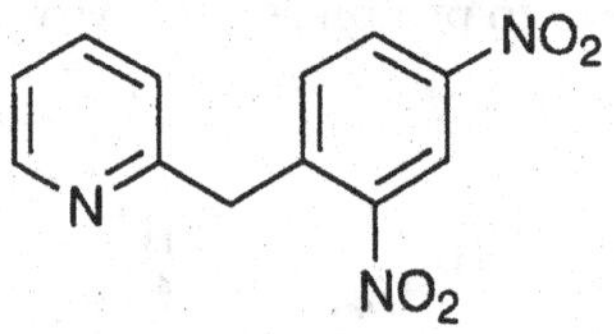

117

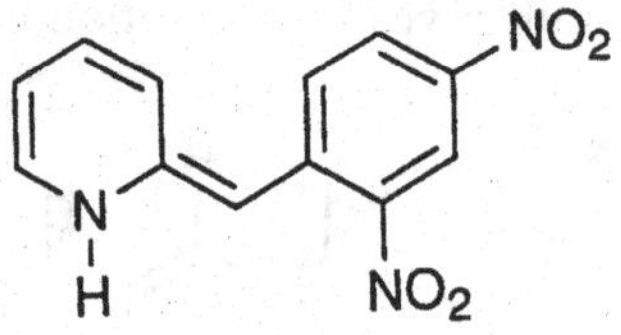

118

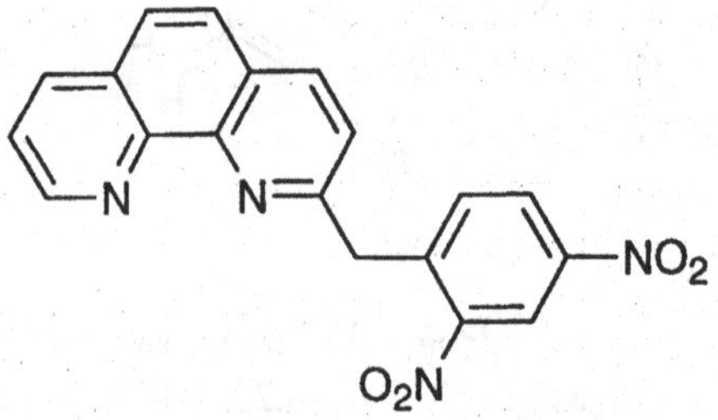

119

of much longer lifetime, leading to the observation of the colour change even in solution [8.233]. Long-lived proton transfer states might also be achievable by proton capture, for instance, as a proton cryptate [2.29] (Section 2.3).

Thermally or photochemically induced proton transfers represent bistable switching processes and are of interest for information storage. A lateral transfer of information on the surface of biological membranes is thought to occur by fast proton conduction through protonic networks [8.234].

Photoisomerization of azophenols causes changes in pK_a [8.235a]. Reversible pH changes are also generated by the light-induced release of hydroxide [8.235b]. Assistance to LRpT may be provided by conformational reorientation of intermediate groups acting in a sort of proton crane mode [8.236a]. Electrochromism due to proton transfer results from electric field dependence of a tautomeric equilibrium [8.236b].

Molecular protonic devices acting as proton wires, rectifiers, charge separation components, etc., represent highly significant goals in the development of information handling systems based on ionic species. They deserve active synthetic and physico–chemical studies.

8.4.5 Ion and Molecule Sensors

An important area of application of supramolecular chemistry lies in the development of selective chemical sensors [8.237, 8.238] for analysing the chemical composition of a medium, the environment or an organism (parts of or the whole) through the determination of the concentrations of the substances of interest. Such sensors require highly specific membranes performing molecular recognition by means of the incorporation of selective receptors (see also Section 7.2). Substrate binding can be transduced into an electrical or optical signal via the changes it causes in a physical property of the active layer such as conductivity, capacitance, mass, light absorption or emission etc. Such substrate selective sensors are realized in chemically sensitive field effect transistors (CHEMFET) [8.126] and in light-addressable potentiometric sensors (LAP) [8.127] (see also Sections 8.2.2 and 8.3.1).

Multisensor devices may be based on an array of sensing units that present different sensitivities for each component of a collection of substrates to be determined [8.238], resulting in a specific signature for each compound. With a sufficient number of signals the composite response may be deconvoluted and the nature and concentrations of the substances determined using techniques such as multifactorial system analysis aided by neural network procedures.

The development of efficient and broad range sensing methods has important applications in areas such as environmental control, quality checking or in the chemical characterization of an organism, for instance for medical diagnostics.

8.5   Switching Devices and Signals. Semiochemistry

8.5.1 Switching Devices. Signals and Information

If the operation of the devices considered above on their specific substrate (photon, electron, ion) is triggered by external optical, electrical or chemical stimuli, their features may be switched between two (or more) states presenting different characteristics. Such switching devices are therefore formed by two main components: a trigger, the switching unit, activated by an external stimulus and a substrate, the switched species; they should operate with efficiency, reversibility and resistance to fatigue.

The substrate carries information defined by its molecular recognition features and constitutes a signal that may be modulated by the switching process. The stimulus to which the switch responds and the substrate on which it operates may both be either of photonic, electronic, ionic, magnetic, thermal or mechanical nature. Thus, a whole set of switches based on (substrate–stimulus) pairs, is defined by the combination of the various types: opto–photo, opto–electro, electro–photo, etc.

It is clear that the design of switched devices is intimately linked to the chemistry of signal generation, processing, transfer, conversion and detection, i.e., to *semiochemistry*. The device is the effector generating a signal, the *semiogen*, and the substrate is the carrier of the signal, the *semiophore* (see Section 8.1).

Switching also implies *molecular* and *supramolecular bistability* since it resides in the reversible interconversion of a molecular species or supramolecular system between two thermally stable states by sweeping a given external stimulus or field. Bistability in isolated molecules or supermolecules is, for instance, found in optical systems such as photochromic [8.229] or thermochromic substances or devices, in electron transfer or magnetic processes [8.239], in the internal transfer of a bound substrate between the two binding sites of a ditopic receptor (see Section 4.1; see also Fig. 33) [6.77]. Bistability of polymolecular systems is of a supramolecular nature as in a phase transition or a spin transition, both of which involve an assembly of interacting species.

Multistability is present in systems that are interconvertible between more than two states such as polynuclear–multivalence metal complexes or polytopic receptors where one or more substrates may occupy and transfer between several different binding sites.

Molecular hysteresis results when the interconversion between the states does not follow the same pathway on sweeping the external stimulus or field in one or the opposite directions [8.240a, 8.241]. A hysteresis loop has, for instance, been described for the electrochemical interconversion of redox states in a dinuclear metal complex; it involves the switching between two different binding sites [8.240b]. Spin transition systems are of interest with respect to information storage and processing [8.241b]; a spin transition in a polymeric iron complex presents a thermal hysteresis at room temperature [8.241c].

A hysteresis loop can formally be drawn for the interconversion of a photochromic substance between two states A and B characterized by two well separated absorption bands as shown in Figure 26; sweeping the frequency up from v_0 to v_A converts the system from state A to B when v_0 reaches the absorption band of A; the system remains in state B if v_A goes back to v_0; sweeping v_0 to v_B converts the system from state B to A, where it remains when v goes back to v_0. Such state interconversion curves are also characterized by the non-linearity of the response with respect to scanning the triggering stimulus.

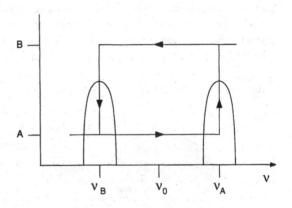

Fig. 26. Formal photohysteresis loop for the photochemical interconversion of a photochromic substance between two states A and B possessing two well separated active absorption bands at frequencies v_A and v_B (see text).

Signal conversion may be brought about when a device responds to a given stimulus by generating a signal of another nature. Specific cases have already been mentioned above (Sections 8.2.3 and 8.2.4). Substrate binding to a suitable receptor possessing a photoactive, a redox active, a proton exchanging site or a binding site for another substrate may lead to modifications of the electrochemical, acid–base or binding properties, inducing the generation of a signal based on the release or the uptake of a photon, an electron, a proton, an ion or a molecule, respectively (Figure 27).

Logic devices may be designed on the basis of functional receptors that respond differently to different substrates. A ditopic receptor responding differently when either or both of two different substrates are bound represents an "AND" gate. When the response is the same for two different substrates an "OR" function results. When the binding is of sequential or cascade type (see Chapter 4), the binding of a second substrate requiring the prior binding of a first, an IF operation is obtained, which also implies a regulation capability. The input–output may be optical, electronic or ionic. When the input is ionic (metal ions or molecular substrates) and the output optical, photoionic logic devices are obtained. This is the case for receptors responding to substrate binding by a change in light absorption or emis-

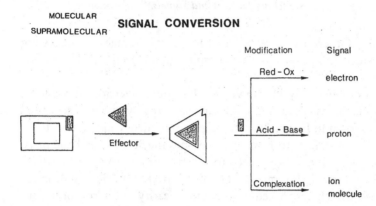

Fig. 27. Schematic illustration of signal conversion processes; binding of an effector to a given site of a receptor in the resting state is accompanied by a change in redox, acid–base or complexation activity at another site of the receptor and leads to the release of an electron, a proton, an ion or a molecule.

TRANSFER AND READING OF INFORMATION

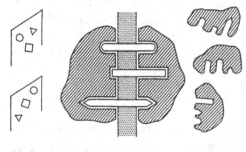

3D Structure + Kinetics (time) = 4D Process

Fig. 28. Schematic representation of information and signal transfer by transduction across a membrane by means of molecular transmembrane rods; (left): two possible in plane configurations with cross sections of the rods; (centre): binding of the transmitter and receiver receptors to the rods respectively on the left and on the right of the membrane; (right) other non-complementary receptors.

sion properties, like a change in fluorescence (see Section 8.2.2) as in the AND gate based on the simultaneous binding of protons and sodium ions [8.242] (see also [8.112e]).

Transfer by transduction and reading of information and signals across a barrier such as a membrane could be performed following the principle illustrated in Figure 28.

A few transmembrane molecular rods having different lengths and terminal shapes and moving freely by lateral diffusion may bind to a given receptor–transmitter on one side (left) of the membrane in a configuration in the membrane plane and in a displacement perpendicular to the membrane that will depend on the disposition and depth of the binding sites on the transmitter. This will impose on the other side of the membrane a given topography for the set of rods that can be read out selectively by binding of the complementary receptor–receiver. Such three dimensional information transfer and reading should be highly efficient, selective and economical. Indeed with only a few molecular rods an extremely large amount of information may be expressed since the set of rods can adopt a very high number of topographies defined by in-plane configurations and out-of-plane displacements. If, in addition, kinetic effects of association and dissociation rates come into play, a 4D process results. The rods could be bolamphiphile molecules of different shapes and nature.

The time modulation of transmitter signals, through pulsations of different durations and shapes may lead to frequency and amplitude *coding* of signalling events, the detection requiring molecular receivers suitably tuned to frequency patterns and thresholds. Such processes occur in intercellular communication by cell to cell signalling [8.243].

Signal amplification is expected to take place when the triggering stimulus yields a non-linear response as a function of the intensity (concentration) of the stimulus (stimulating species). This holds in systems displaying cooperativity and self-organization, as in a phase change, for example. The chiral twisting of a mesomorphic phase to become cholesteric on addition of a small amount of chiral component is such a case. Stimulus induced perturbation of organized supramolecular assemblies, such as optical switching of membrane self-assembly (see Section 8.5.5) or recognition-induced formation of liquid crystalline polymers (see Section 9.4.2), may be considered as an amplification of a microscopic signal to the macroscopic scale. Positive cooperativity is found in a polytopic ligand undergoing aggregation on ion binding [8.201]. Further discussion of this and related questions will be given in Chapter 9.

8.5.2 Photoswitching Devices

Most switching devices studied make use of light for interconverting a molecule between two different states. Both forward and backward processes may be photoinduced or one of them, usually the reverse transformation, may be thermal.

Opto-photoswitching occurs in bistable photochromic substances [8.229] that are converted into a coloured form on irradiation and revert to the initial colourless or less coloured state either thermally or on irradiation with light of another frequency. Such substances and the materials they may form, for instance by combination with polymeric matrices, are of great interest for optical data storage, in particular of reversible nature [8.244, 8.245]. They should display (1) high quantum yields for both forward and reverse processes; (2) well separated absorption bands in the two states; (3) high thermal stability; (4) low fatigue on multiple cycling; (5) read-out capability without erasing; (6) rapid response; (7) sensitivity to wavelengths of commonly used light sources such as red diode and green YAG lasers.

A number of photochromic systems have been extensively investigated that undergo cis–trans isomerization (indigos, azo compounds) cleavage (spiropyrans), electrocyclic processes (fulgides, 1,2-diarylethenes) [8.229, 8.244, 8.245]. For instance, cis–trans isomerization of a thio–indigo derivative allows the reading of pyrene excimer or monomer fluorescence [8.246]. The 1,2-dithienylethene system presents particularly attractive interconversion properties by photoreversible cyclitation [8.245].

Electro-photoswitching devices possess a dual mode operation resulting from different electrochemical properties in the two photo-interconvertible states. They may display both photochromism and electrochromism. They allow the photomodulation of electrochemical properties as was shown to occur in photochromic dihydroazulene derivatives [8.247]. Separate photochemical and electrochemical interconversion has been realized in systems containing a photoactive azo [8.248, 8.249] or dihydroazulene [8.250] unit appended with an anthraquinone moiety.

Bis-pyridinium polyenes, the caroviologens [8.140], may be considered as prototypes of molecular wires, allowing electron flow to take place, [8.143] and push–pull polyenes present very pronounced non-linear optical properties [8.88–8.90] (see Section 8.2.5). The combination of these components with an externally triggered switching process could lead to electro-photo or opto-photo switches. This may, in principle, be achieved by inserting into the polyenic path a bistable "on/off" unit that establishes or interrupts the electronic conjugation by reversible interconversion between a closed and an open state, thus modulating electron conduction on the one hand and NLO properties on the other (Figure 29). Photomodulation of electron conduction has been realized in a compound that undergoes reversible interconversion between unconjugated 120a and conjugated 120b states by irradiation in the UV and in the visible, respectively; 120b is easily reduced electrochemically, whereas 120a is not. Thus, the pair 120a /120b indeed represents a prototype of a switched molecular wire [8.251].

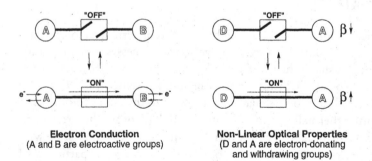

Electron Conduction
(A and B are electroactive groups)

Non-Linear Optical Properties
(D and A are electron-donating and withdrawing groups)

Fig. 29.   Optical switching of electronic conduction (left) and non-linear optical properties (right) respectively in redox and push-pull polyene molecular wires; β is the hyperpolarizability.

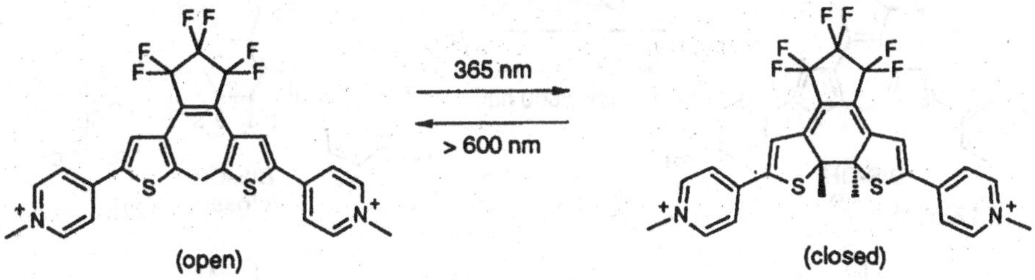

(open)
120a

365 nm
> 600 nm

(closed)
120b

(open)
121a

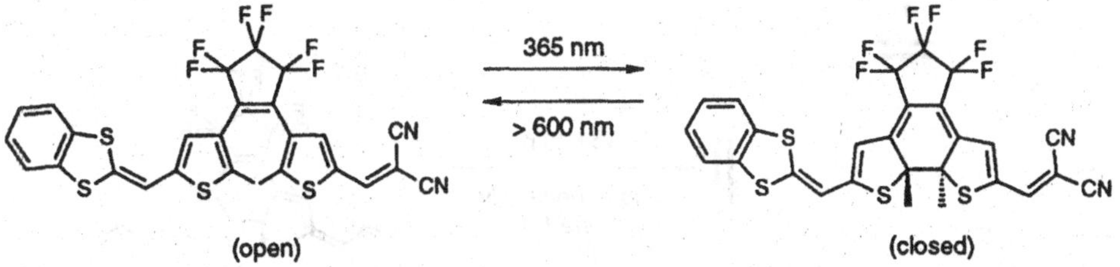

365 nm
> 600 nm

(closed)
121b

Similarly, **121b** possesses markedly enhanced optical non-linearity compared with **121a**, thus displaying a photoswitching of NLO properties.

When the dithienylethene system bears two phenol groups, as in **122a**, then a *dual-mode optical-electrical switching* process is obtained combining photochromism and electrochromism. Compound **122a** is photocyclized to **122b**, which can be electrochemically oxidized to the extended quinone **123**; the latter is photostable; reduction of **123** back to **122b** allows the photochemical opening to **122a** with visible light; all three compounds have different absorption bands.

The special features of the three species are, in principle, well suited as basis of an optical memory system with non-destructive write–read–erase capacity (see also [8.245]). After photochemical writing with UV light (**122a** → **122b**), the information may be "safeguarded" by an electrical process (**122b** → **123**). It may then be read many times and finally, after electroreductive unlocking (**123** → **122b**), the information may be erased with visible light (**122b** → **122a**). This cyclic process represents an EDRAW (Erase Direct Read After Write) process as compared to the usual WORM (Write Once Read Many) process [8.252].

It is clear that compounds based on the same principles as the present ones and bearing various functional features (metal complexes, ionizable groups, energy and electron transfer units) may be conceived, thus allowing for the reversible switching of a variety of physicochemical properties. They provide entries, at the macroscopic level, to materials for reversible information storage. At the microscopic level they lead to molecular signalling processes and represent optically and electrically addressable molecular switching devices.

Photochemical switching of the electrical properties of conductive Langmuir–Blodgett films results from photoinduced conformational changes produced by an azobenzene unit [8.253].

Electro-photoswitching devices represent means of generating photoinduced electronic signals. Whereas controlled back and forth switching between two thermally stable states is usually sought, fast switching is of interest for the design of molecular electronic components operating on very short time scales. Long-lived photoinduced charge separation is required for artificial photosynthetic investigations. Conversely, vectorial short-lived charge separation is desired for fast transfer and switching of electronic signals. The numerous studies concerning photoinduced electron transfer in metal complexes and in organic donor–acceptor systems are of much value in this respect, even if it was often long-lived processes that were sought (see Section 8.2.3) [8.62–8.68, A.10, A.20]. Picosecond optical switching in a bis-porphyrin donor–acceptor–donor molecule has been described [8.254]. D–PS–A and D–bridge–A triads display similar properties (see Section 8.2.3) [8.255]. Systems presenting very long-range photoinduced charge separation allow in principle the oriented transfer of electronic signals over long distances. For instance, insertion into membranes represents a means of generating photoinduced transmembrane signals. Such switching systems operate much faster than photochromic species and contain no moving parts.

Photoswitching of ionic and mechanical processes, iono–photo and mechano–photo switching, will be considered below.

8.5.3 Electroswitching Devices

Photo-electroswitching takes place in substances where an electrochemical change induces more or less pronounced variations in light absorption or emission properties. This occurs namely for metal complexes subjected to redox interconversion between their various oxidation states. A two-component device combining a luminescent centre and an electroactive unit may function as a photo–electroswitch in which the emission properties are modulated by redox interconversion via energy or electron transfer quenching (Figure 30).

Electrochemical switching of the emission of a metal complex occurs in a system in which a luminescent (tris-bipyridine)ruthenium(II) centre is linked to a quinone unit. Interconversion of the redox couple **124–125** allows the reversible switching

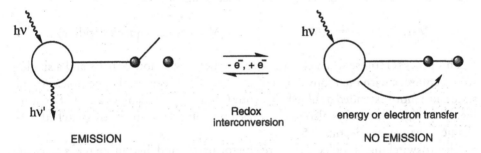

Fig. 30. Schematic representation of a photo–electroswitch where the emission properties of a photosensitive centre are modulated by the electrochemical interconversion of a redox centre inducing luminescence quenching by energy or electron transfer.

between the luminescent hydroquinone **125** and the quinone **124** in which the luminescence is quenched by very fast electron transfer from the metal centre to the organic acceptor unit.

The quinone–hydroquinone redox couple built into complexes **124** and **125** fulfils the requirements for the design of a bistable electro–photoswitch: both the oxidized and reduced forms are isolable and stable; the reduced form **125** is luminescent, whereas the oxidized form **124** is quenched; the electrochemical interconversion of the two species is reversible [8.256].

Of much interest are the electroluminescence properties of conjugated polymers that allow the development of electrically switched light-emitting devices [8.257]. Electric field activated bistable molecules are expected to switch at a critical field strength [8.258].

Electroswitching of structure takes place when a redox change induces a reversible structural or conformational process in a molecule, such as an electrochemically activated intramolecular rearrangement [8.259]. On the supramolecular level it consists of the electroinduced interconversion between two states of different superstructure. A case in point is the reversible interconversion of a double-helical dinuclear Cu(I) complex $M_2L_2^{2+}$ [8.260] and of a mononuclear Cu(II) complex ML^{2+} in a sequential electrochemical–chemical process [8.261]:

$$M_2L_2^{2+} \xrightleftharpoons[+2e]{-2e} 2ML^{2+} \qquad (M = Cu; \ L = quaterpyridine)$$

This process represents an electroswitching between a double helix and a single-strand complex with interconversion of the ligand between a rather compact strand state and a more extended double helical one (Figure 31). It displays a sort of breathing of the ligand and thus also constitutes a prototype for an electroswitched mechanical change (see Section 8.5.5).

Electroswitching of molecular aggregate formation and disruption takes place on interconversion of a lipid bearing a redox headgroup between a neutral and a charged state, as for instance in hydrophobic ferrocene–ferricinium derivatives [8.262, 8.263].

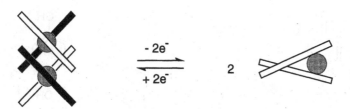

Fig. 31. Electroswitching of a double helix–single strand interconversion between a double helical dinuclear Cu(I) complex and a single strand Cu(II) complex in a sequential electrochemical–chemical process.

8.5.4 Switching of Ionic and Molecular Processes

Iono–photoswitching, the photochemical switching of ion binding and transport processes has already been considered in Sections 6.4.3 and 8.2.4. The modulation of ion binding strength by light provides a means for generating photoinduced ionic signals in principle in a reversible and pulsed fashion.

The nature of the ion(s) and the frequency and amplitude coding of *ionic pulses* may carry and process information, as in biological ionic processes [8.243]. The affinity and selectivity changes in reported bistable photoactivated receptors for ions and molecules are in most cases still quite small, but proper design should provide a range of substances for the photogeneration of ionic and molecular signals.

The cis–trans photoisomerization of complexing agents containing a thioindigo unit leads to changes in ion binding and transport properties [8.264–8.266]. Other systems could be based on the photoelectrocyclic isomerization of 1,2-diarylethylenes such as in **120a–120b**.

Complexes of photocleavable chelating ligands release irreversibly-bound divalent cations (calcium, magnesium) on irradiation [8.267, 8.268]. Photocleavable macrobicyclic cryptands [8.269] generate selective *photoionic signals*. On irradiation of the cryptate **126** cleavage of one of the bridges of the macrobicycle occurs thus releasing the bound cation. The process is very efficient due to the high binding constants of the cryptates and the large change in affinity that occurs on cleavage (from 10^4–10^5 M^{-1} to less than about 10 M^{-1}), thus leading to a large concentration jump (Figure 32).

126

In principle, any cation cryptate (see Section 2.3) can be treated in this way, thus allowing the initation of ion transport and cellular ion regulation processes by a light pulse [8.48b, 8.270, 8.271]. The same type of approach can be applied to anions and molecular substrates by incorporating the photocleaving unit into the corresponding receptor molecule (macrobicyclic polyamine for anions [8.272], cyclophane receptor for hydrophobic molecules, for instance). Photosensitive protecting groups allow the photogeneration of a free species like ATP, nucleotides, etc.

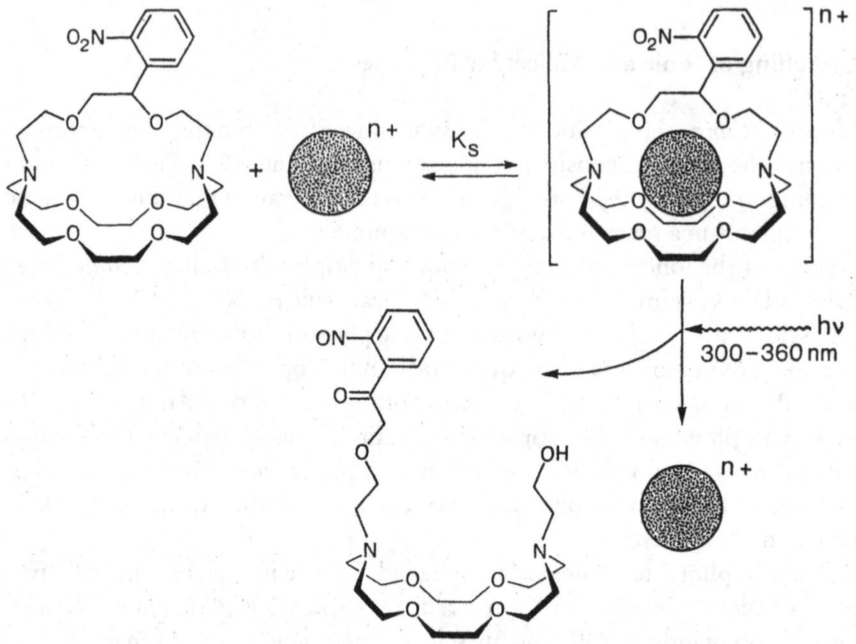

Fig. 32. Generation of photoionic signals by photocleavage of a macrobicyclic cation cryptate; the cation can be any one of those forming a cryptate.

[8.273]. Such photorelease of various substrates is of much interest for the study of both semiochemical processes and biological regulation and communication.

Photoprotonic signals can be generated by light induced proton release, this might happen for the quaternary N–R$^+$ derivatives of **117**, for instance [8.233]. The photoproduction of proton gradients across membranes could serve as a light-powered proton pump for inducing vectorial processes such as the transport of protons [8.233, 8.234] or H$^+$–ATPase model reactions [8.274].

Ion switches may be based on ionic jumps inside polytopic ligands between different sites presenting different optical or electrochemical properties. *Iono–electroswitching*, for instance, is produced by redox interconversion of a given cation between two different sites providing suitable binding for two different oxidation states in a ditopic receptor. This could occur in heteroditopic receptors, such as that in **47** or a macrotricycle of type **46** with two different rings (N$_2$S$_2$ and N$_2$O$_3$, for example) (see Figure 33). Such a complex should possess different optical properties in each site so that electroswitching would interconvert two species in which the different locations of the ion would correspond to different optical signals. Redox triggered iron translocation takes place in triple-stranded helical complexes [8.275].

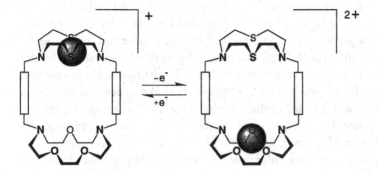

Fig. 33. A potential iono- or mechano–electroswitch based on a heterotopic macrotricyclic ligand; the electrochemical interconversion of a copper ion could switch the location of the ion from the [12]-N_2S_2 ring in the Cu(I) state (left) to the [15]-N_2O_3 subunit in the Cu(II) state (right).

Electroionic signals are generated by electrochemical interconversion of a selective receptor molecule containing a redox-active group (metallocene, quinone, etc.) between states of low and high affinity for a given substrate (see Sections 6.4.1 and 8.3.1).

Iono–ionoswitching, i.e. ionic switching of ionic processes, takes place when the binding of a given species interferes with that of another. Ditopic ligands may present allosteric effects leading to ionic modulation of affinities and thus to ion uptake or release when the binding of a first species affects that of the next [8.211, 8.276]. Proton transfer causes proton-induced switching of metal–metal interaction in dinuclear complexes [8.277a] and substrate binding to the curcurbituril receptor is modulated by pH in a bimodal fashion [8.277b].

8.5.5 Mechanical Switching Processes

Positional changes of atoms in a molecule or supermolecule correspond on the molecular scale to mechanical processes at the macroscopic level. One may therefore imagine the engineering of molecular "machines" that would be thermally, photochemically or electrochemically activated [1.7, 1.9, 8.3, 8.109, 8.278]. Mechanical switching processes consist of the reversible conversion of a bistable (or multistable) entity between two (or more) structurally or conformationally different states. Hindered internal rotation, configurational changes (for instance, *cis–trans* isomerization in azobenzene derivatives), intercomponent reorientations in supramolecular species (see Section 4.5) embody mechanical aspects of molecular behaviour.

Although they are random and not oriented, the motions of a substrate inside a cyclodextrin cavity (see Section 4.5) have molecular ball-bearing flavour. Correlated internal rotations between bulky groups presenting conformational gearing has the character of cogwheel processes at the molecular level [8.279a] and a molecular brake activated by metal ion binding has been described [8.279b]. Mechanical movement is produced in a protein by the proline switch that is operated by the peptidyl–prolyl *cis–trans* isomerases [8.279c]. Mechanical bistability is represented by the orientational isomers of guests in carceplexes [8.279d]. Mechanical analogies are also displayed by threaded and interlocked systems such as rotaxanes and catenanes [8.280–8.284] (see also Section 9.6.2), which are by construction mechanical objects of supramolecular nature. This is the case for the thermally activated motions of one ring around the other in a [2]-catenane [8.285a] and the shift of a ring along positions around a large ring in a "molecular train" fashion [8.285b]. Similar shifting of a ring between binding positions in a rotaxane represents a molecular shuttle or a sort of (supra) "molecular abacus" [8.286, 8.287]. Formally such motions are also related to ion jumping between sites in polytopic receptors (see above). These processes are all thermally activated. A further step will be to go from random back-and-forth displacements to oriented motions.

Mechano–photoswitching takes place in photoswitchable [2]-catenanes incorporating an azobenzene unit [8.288] and in a photoactivated molecular shuttle [8.289a]. The latter might be developed into a light-driven version of a supramolecular abacus as schematically represented in **127**. Such developments point to thermally or photochemically activated *mechanical switching devices* for information storage and processing.

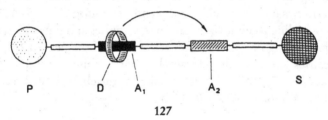

P   D   A₁   A₂   S

127

Mechano–electroswitching has been mentioned above (see section 8.5.3) (Figure 31). It takes place in the electrochemically (or chemically) triggered positional switching of the macrocycle along the threaded chain in a rotaxane [8.289b]. It may also be considered to occur when bound cations can be brought to shift from one site to another in a polytopic receptor by redox triggering due to the dependence of coordination geometry and binding preference on the oxidation state (see iono–electroswitching above).

Mechano–ionoswitching takes place, for instance, in polytopic ligands presenting ion binding induced allosteric changes [8.201, 8.276] or when an ion binds to a catenand to form a catenate [8.281, 8.282]. Modulation of the binding of a second substrate may be brought about.

Large structural changes occur on binding of metal cations to poly(ethylene oxide) [8.197], in the cooperative transition from a globular neutral form to an extended coil displayed by hydrophobic polyamines on protonation [8.290] or in stimuli-responsive self-organized systems such as liquid crystals, micelles, vesicles monolayers [8.291] and gels [8.292] (see also [8.262, 8.263]). Triggering supramolecular assemblies by optical, electrical or chemical stimuli may result in a large scale change of the organization, thus amounting to signal amplification to and expression on the macroscopic level. Such effects are of interest for the development of chemo–mechanical actuators for the conversion of light, electrical or chemical energy into mechanical energy and motion, a major goal being, as already pointed out, to achieve controlled oriented motion. Biological chemo–mechanical energy transduction occurs in the actomyosin molecular motor on muscle action [8.293] and more generally in the production of motion in living organisms from reversible chemical changes [8.294].

In general, optically, electrically or chemically triggered switches would seem to be preferable to mechanically activated ones, as are photo-, electro- and chemo devices with respect to mechano devices and electronic or photonic computing with respect to mechanical computing. The ultimate in (nano)mechanical manipulation of a molecular device is represented by the realization of a bistable switch based on the motion of a single atom by means of the scanning tunnelling microscope [8.295] (see also Section 9.9).

There is clearly a vast domain of investigation into the design and properties of molecular and supramolecular devices as basis for the development of chemical information processing and signalling as well as for the exploration of their relationships with related biological phenomena. One may envisage connecting artificial devices and biological entities, such as field effect transistors and neural cells [8.296], as well as the use of biological molecules for the engineering of devices, as in chip construction by DNA technology [8.297].

A highly relevant question is whether the goal should be to operate at the level of the single component or not. In order to minimize the risk of device failure the incorporation of some redundancy or repair procedures appears unavoidable, as in biological processes, so that it will be necessary to develop suitable addressing and detection procedures, error-checking and self-healing mechanisms for multidevice arrays.

Methods for analysing the response of complex multicomponent–multifunctional systems will be needed. Global responses may be submitted to deconvolution procedures and multicomponent analysis [8.298], making use for instance of pattern recognition [8.238, 8.299], neural network [8.238, 8.300] and fuzzy logic [8.301] approaches.

The formation of photonic, electronic, ionic switching devices from molecular components and their incorporation into well-defined organized assemblies represents the next step towards the development of circuitry and functional materials at the nanometric scale. This involves supramolecular engineering in the polymolecular state, at interfaces or in the bulk, for instance in functionalized Langmuir–Blodgett films (see Chapter 7, Section 9.7). The controlled build-up of such architectures requires the ability to direct *self-assembly* and *self-organization* processes through explicit instructed procedures.

9 Self-Processes — Programmed Supramolecular Systems

9.1 Self-Assembly. Self-Organization

In line with the key role played by molecular information in the field, the most recent developments of supramolecular chemistry concern the explicit application of molecular recognition as a means of controlling the evolution of supramolecular species and devices as they build up from their components and operate through *self-processes*. It is through our work on the spontaneous generation of inorganic double helices, the helicates, which was perceived as a self-assembly process [9.1], that we were led to implement the concepts of *self-assembly* and *self-organization* in supramolecular chemistry [1.7, 1.9]. As was then noted "Such control of self-organisation at the molecular level is a field of major interest in molecular design and engineering, which may be expected to become a subject of increasing activity" [1.7]. This has indeed been the case, judging from the increasing activity devoted to the field in recent years by a growing number of investigators [1.9, 8.283, 9.2, 9.3]. Self-assembly and self-organization have mainly been studied in biology and in physics [9.4–9.8]. Supramolecular chemistry provides ways and means for chemical science to explore this area and apply its power of design and control.

Supramolecular chemistry has relied on more or less rigidly organized, synthetically built up molecular receptors for effecting molecular recognition, catalysis, and transport processes and for setting up molecular devices. The use of macrocyclic and macropolycyclic structures was dictated by the need to achieve better control over the geometry and rigidity of molecular receptors. Such preorganization (prepositioning) was pursued from the start in the design of crown ethers, cryptands, spherands, etc (see Section 2.3). Beyond *pre*organisation, based on covalent bonding, lies the design of systems undergoing *self*-organisation, that is, systems capable of *spontaneously* generating a well-defined (functional) supramolecular architecture from its components under a given set of conditions.

The terms of "self-assembly" and "self-organization" may cover different notions [9.4–9.8] and have often been used loosely and interchangeably together with other words and phrases of imprecise or multiple meaning [9.8b]. One may wonder whether the term "self-organizing" should be applied to "structures that represent the only physically allowed pattern of behaviour" [9.8c] and consider using the term "self-assembly" for structure generating chemical systems at equilibrium, reserving "self-organization" to dynamic multistable systems [9.8].

On the other hand, in order to provide two levels of description that would take into account both the current investigations and the corresponding "working" terminology, one may propose to distinguish the two notions as discussed hereafter. Both classes of phenomena thus designated have their own characteristics but also overlap partially and adjustments may have to be made in the future as the field progresses.

Self-assembly is the broader term. It can be taken to designate the evolution towards spatial confinement through spontaneous connection of a few/many components, resulting in the formation of discrete/extended entities at either the molecular, covalent or the supramolecular, non-covalent level.

Molecular self-assembly will not be considered here (see [9.2]); in fact, it is a special type of synthetic procedure where several reactions between several reagents occur in one experimental operation to yield the final covalent structure; it is subject to control by the intramolecular conformational features of intermediates and by the stereochemistry of the reaction(s); the efficient assembly of a covalent structure may require that the connecting reaction(s) be reversible so as to allow searching for the final structure. Examples are found in the generation of macropolycyclic structures by multiple (amine–aldehyde) condensations (see Section 4.1) or of porphyrinogens, porphyrins and phthalocyanins (see also in [9.13a].

A limiting case of covalent self-assembly may be considered to be the formation of the curved carbon structures of the spheroidal fullerenes C_{60}, C_{70}, etc. [9.9a] and of carbon nanotubes [9.9b] in the high-temperature regime of carbon vapour that permits a certain structural fluidity.

Supramolecular self-assembly concerns the spontaneous association of either a few or many components resulting in the generation of either discrete oligomolecular supermolecules or of extended polymolecular assemblies such as molecular layers, films, membranes, etc. The formation of supermolecules results from the recognition-directed spontaneous association of a well-defined and limited number of molecular components under the intermolecular control of the non-covalent interactions that hold them together. The name *tecton* (from τέκτων: builder) has been proposed for designating the components that undergo self-assembly [9.10].

Self-organization could be considered as a set intersecting self-assembly, ordered self-assembly, that would (1) contain the systems presenting a spontaneous emergence of order in either space or time or both; (2) cover spatial (structural) and temporal (dynamic) order of both equilibrium structures and of non-equilibrium, dissipative structures, incorporating non-linear chemical processes, energy flow and the arrow of time; (3) concern only the non-covalent, supramolecular level; (4) be

multicomponent and result in the formation of polymolecular assemblies presenting supramolecular organization and long-range order (with or without regularity
and periodicity) by virtue of specific interactions operating either through recognition events between the molecular components or in a dynamic process. The higher
the degree of confinement or order (1D, 2D, 3D, 4D) of self-assembled entities, the
more they may be considered as organized (molecular layers, membranes, micelles,
colloids, liquid crystals [9.11], molecular crystals). Self-organization thus involves
interaction (between parts) and integration (of the interactions) leading to collective
behaviour, such as is found in a phase change or in the generation of spatial or temporal waves. The phenomenological description of the macroscopic features of self-
organizing systems must eventually find an etiological explanation at the microscopic, molecular, and supramolecular levels.

Self-assembly and self-organization of a supramolecular architecture are both
multistep processes implying information and instructed components of one or
several types. They may follow a sequence and a hierarchy of assembly steps. They
require *reversibility* of the connecting events, i.e., kinetic lability and rather weak
bonding (compared with covalent bonds), in order to allow the full exploration of
the energy hypersurface of the system. They may also amount to a (non-covalent)
replication at the supramolecular level. *Self-replication* procedures make use of
recognition events for selecting and positioning the reacting molecules (see Section
9.6).

Self-organization and self-assembly are also to be distinguished from mere *templating*. Templating [2.23, 9.12, 9.13] is a synthetically most efficient procedure
involving the use of temporary or permanent "helper" species, of organic or inorganic nature, for the stepwise assembly of molecular or supramolecular structures
of high complexity (see Section 9.6.2). However, if self-assembly may and often
does involve templating, templating itself is not, strictly speaking, self-assembly but
may be considered as the unit step in self-assembly processes that comprise several
(many) steps occuring spontaneously in a single operation.

There is thus a progression in terms of supramolecular structure and order generation through molecular information and programming: from templating to self-
assembly and then to self-organization, each stage representing a significant conceptual and operational step from the previous one. Depending on the nature of the
species involved, these processes may be either organic, inorganic, or mixed organic–inorganic, when both metal ion coordination and other interactions come into
play (see Sections 9.3 and 9.4).

Self-assembly and self-organization require molecular components containing
two or more interaction sites and thus capable of establishing multiple connections.
Self-complementary components associate with themselves undergoing *homoas-
sembly*, whereas complementary components (pleromers) associate with one another giving *heteroassembly*. *Allostery*, *cooperativity* and *regulation*, three directly
interrelated features, result when occupation of a given site leads to a change in the

binding features of the other site(s), making binding either easier or more difficult (positive or negative cooperativity).

Allosteric effects play a major role in biology, for instance in the conformational changes induced by the binding of an effector and regulating the activity of an enzyme [9.14a]; they have also been studied in synthetic receptors [8.201, 8.211, 9.14b, c]. Similarly, cooperativity, a thermodynamically well defined process [9.15a], is displayed by a number of biological species as well as by abiotic ones [8.70a, 8.201, 9.15b–d] (see also Section 9.3.1, [9.64, 9.65]), in particular in organized media such as polymer solutions or gels [7.8cd, 8.290, 8.292].

Self-organization displays positive cooperativity, implying that a given step in the assembly sets the stage for and facilitates the following one, for instance through an allosteric conformational change; it is self-assembly driven to completion.

Positive cooperativity is a basic characteristic of *molecular amplification devices*, since once initiated, the subsequent steps of the assembly are facilitated. It represents a non-linear process and confers features of an *error filter*; only the correct input will in principle lead to the cooperative structure generation.

Self-assembly may occur with *self-recognition*, mixtures of components yielding defined superstructures without interference or crossover (see Section 9.5). A special case is *self-resolution* which results in the formation of homochiral supramolecular entities from racemic components (see Sections 9.4.2, 9.4.4 and 9.7).

The kinetic lability gives to recognition induced self-assembling systems the ability to undergo *annealing* and *self-healing* (self-repair) of defects. In contrast, covalently linked, non-labile species cannot heal spontaneously and defects are permanent. Self-processes under equilibrium control are subject to *"Boltzmann defects"*; they follow a Boltzmann distribution of defects. In general, self-organizing supramolecular systems may undergo self-maintenance [9.8a], error-checking and correction by exchanging incorrect components for the correct ones. They also display *adaptation* by being able to react to environmental changes and to adjust to novel conditions.

Self-assembly and self-organization may result in a range of collective properties such as, in addition to phase changes already mentioned, directed light propagation, electronic conduction, magnetism, correlated protonic motions, etc.

Of special interest is the generation of short range order in molecular liquids resulting from a certain degree of organization or orientation as indicated for liquid quinoline by dynamic NMR studies [9.16a] or by direct spin coupling in liquids subjected to intense magnetic fields [9.16b]. Hydrophobic effects [4.30] lead to structure formation in aqueous solution [9.16c], in line with the remarkable water clathrates of quaternary ammonium ions [9.16d,e]. Self-organization in liquids represents a particularly intriguing area of investigation into the spatial and temporal behaviour of condensed matter.

Numerous biological supramolecular structures result from self-assembly and living systems offer the ultimate in self-organization [9.4–9.8]. They provide illustration and inspiration for the persevering investigation of self-processes in supra-molecular chemistry. Typical examples are protein folding, the spontaneous formation of the double helix of nucleic acids, of the viral protein coat [9.5, 9.17] and of various multiprotein quaternary architectures (such as actin filaments, microtubules, ribosomes, bacteriophage T₄ [9.5], multisubunit enzymes, etc.). The build-up of highly complex biological supramolecular structures may not occur spontaneously from the dissassembled components but may require helper proteins (the chaperones [9.18]) and follow an initiation–growth–termination sequence (see, for instance in [9.5]). This points to higher levels of structure (and function) generation through *sequential* and *effector-regulated* self-assembly involving temporal conditions, feedback and a hierarchy of events.

One may note that conversely to the development of self-assembling systems, the search for *assembly inhibitors* also presents intriguing perspectives. On the chemical side, they may allow the control and regulation of self-assembly itself. On the biological side the design of such inhibitors is of interest for interfering with the association of the components of multiprotein complexes; in particular, assembly inhibition (or conversely, assembly induction or stabilisation) could represent a fruitful approach to drug design; for instance, the inhibition of insulin aggregation may facilitate its absorption.

9.2 Programmed Supramolecular Systems

Recognition-directed self-processes may be considered to represent molecular information handling procedures. The information necessary for a process to take place must be stored in the components and the algorithms (the "Aufbau" rules) that it follows operate via selective molecular interactions. Systems presenting such features have been termed *programmed supramolecular systems* [1.9], a designation introduced in order to draw an analogy with and to provoke an analysis in terms of information science attributes such as storage capacity, set of instructions, programme, reading process, algorithm and input–output features. It also stresses a distinct character of intentionality through explicit design, steering and predictability.

Molecular programming involves the incorporation into molecular components of instructions for the generation of the desired supramolecular architecture. Depending on the design of the interaction patterns between the components, more or less loose or strict programming of the output entity will be achieved. The programme is molecular, the information being contained in the covalent framework; its operation through non-covalent recognition algorithms is supramolecular. The processing of molecular information via molecular recognition events implies the passage from the molecular to the supramolecular level.

Three levels of information input may be distinguished in self-assembly: (1) molecular *recognition* for the *selective binding* of complementary components; (2) *orientation* in order to allow *growth* through binding of the components in the correct relative disposition; (3) *termination* of the process, requiring a built in feature, a *stop signal*, that specifies the end point and signifies that the process has reached completion; this may be a closure relation generating a closed structure; termination control presents a particular challenge. In addition, temporal information may be involved if the progressive build-up of the final superstructure occurs through a *defined sequence* of molecular instructions and algorithms, a given component or recognition event coming into play at a well-defined stage in the overall process. In this case, the generation of a given intermediate structure will depend on the previous one and sets the stage for the next one; cascade systems (see Chapter 4) are of this type. Such sequential self-assembly represents the next step in the design of artificial systems presenting higher levels of complexity.

Information storage and processing as well as signal generation may make use of polytopic receptors in which the progressive occupation of the different binding subunits, eventually following a given sequence, leads to the generation of specific structural and functional patterns. The dynamics of exchange between different sites add a further regulation feature. Cases in point could be the proteins calmodulin [9.19] and cytochrome c_3 [8.137] containing respectively four calcium binding sites and four iron-porphyrin sites for electron exchange that may exist in different ion occupation or redox states. Polynuclear self-assembled inorganic species such as the cylindrical and grid structures considered below (Sections 9.3.2 and 9.3.3) represent prototypes of entities that may possess a variety of inscribed optical, electronic or ionic patterns.

Molecular programming points to a *molecular information science*, molecular *informatics* and *semiochemistry* (see Section 8.5) involving information and signal processing and communication at the molecular and supramolecular levels (see also Section 2.1). The design of molecular information dependent, instructed and functional self-organizing systems reveals new horizons in supramolecular chemistry.

9.3   Self-Assembly of Inorganic Architectures

Inorganic self-assembly and self-organization involve the spontaneous generation of well-defined metallo-supramolecular architectures from organic ligands and metal ions. The latter serve both as cement holding the ligands together and as centre orienting them in a given direction. In the process, full use is being made of the structural and coordination features of both types of components, which in addition convey redox, photochemical or chemical functionality, depending on their nature.

The formation of any complex species from an organic ligand and a metal ion is in principle an assembly process, that occurs spontaneously. The emphasis here lies in the design of the ligands and the choice of the metal ions in order to produce defined architectures in a controlled fashion from multiple subunits. Metal ions have properties of special interest as components of supramolecular systems and linkers for self-assembly. They provide (1) a set of coordination geometries, (2) a range of binding strengths, from weak to very strong, and of formation and dissociation kinetics, from labile to inert, and (3) a variety of photochemical, electrochemical and reactional properties. In addition, and most significantly, they allow the reversible assembly–disassembly of supramolecular architectures and represent switchable interaction sites, for instance, by electrochemical interconversion between oxidation states of different coordination geometries.

In view of the fact that metal ion complexation presents a very wide range of binding strengths, inorganic self-assembly may be considered to span the range from non-covalent to approaching covalent type self-assembly. Since there is a continuum of interaction strength, a clear-cut categorization is not feasible and the problem recalls the situation addressed by fuzzy sets [8.301]. No such distinction will therefore be made for the present purposes, although it is realized that some complexes considered may display thermodynamic and kinetic stabilities akin to covalent structures.

A great diversity of polynuclear metal clusters has been obtained by self-assembly with increasing ability to direct the process. They may or may not involve bridging species. They have various geometries, are of bioinorganic significance or of interest for materials science, representing approaches towards "supramolecular metals", for instance. They contain from a few to very many metal centres, as in huge clusters of 70 or 146 copper centres [9.20a,b], of 309 or 561 palladium atoms that display metallic behaviour [9.20c] or catalytic activity [9.20d], or in transition metal colloids [9.21] (for a selection of recent work see [4.11, 4.12, 8.136a, 9.20–9.29]).

Numerous metallomacrocycles self-assemble from their components giving species possessing various structures, for instance: triangular shape [9.30, 9.31] containing a cavity [9.32], which may include a guest molecule [9.33]; square [9.34, 9.35] or star like [9.36–9.39] shapes; wheel-shaped or toroidal hexameric [9.40], octameric [9.41] or decameric [9.42, 9.43] structures; square [9.44, 9.45] rectangular [9.46, 9.47] or bent [9.48] boxes into which substrate molecules may bind [9.49, 9.50]; adamantanoid shape [9.51, 9.52] with cation inclusion [9.51c], formally related to that of the spheroidal macrotricycle **21**; catenane type [9.53]. Coordination species of dendrimer or arborol nature have been constructed [7.61, 8.27, 9.54].

Many other types of metallomacropolycyclic structures may be imagined and will no doubt be obtained (for a recent overview on inorganic supramolecular chemistry see [A.40]). Choosing the nature of the building blocks makes possible a molecular engineering approach to inorganic or organometallic solid state materials (see

Section 9.7). Our work involved first the self-assembly of double-helical metal complexes. This particularly interesting case of self-assembly of a coordination compound represented a first step in our studies on the design of inorganic programmed systems.

9.3.1 Self-Assembly of Double-Helical and Triple-Helical Metal Complexes: The Helicates

Self-assembly may occur in repetitive chain ligands, acyclic coreceptors containing several identical binding subunits arranged linearly, if substrate binding to a given site is connected with binding to the other sites so that full site occupation is only compatible with a defined final architecture.

Double Helicates. This is the case for the ligands **128–131** containing two to five 2,2'-bipyridine (bipy) groups [9.55]. In the presence of Cu(I) ions, spontaneous assembly takes place giving double-stranded helicates, the di- to pentahelicates **132–135**, in which two ligand strands are wrapped around each other in a double-helical fashion with Cu(I) ions holding them together [9.1, 9.56, 9.57]. Their *double helix* structure rests on the determination of the crystal structures of **132** [9.58] and **133** [9.1] (X = H) and on spectroscopic data. It results from the tetrahedral-like coordination imposed by each Cu(bipy)$_2$$^+$ site and from the design of the ligands, which disfavours binding to only a single strand. These two features make up, respectively,

	X = H	X = COOR or CONR$_2$
n = 0	128	137
n = 1	129	138
n = 2	130	139
n = 3	131	140

132   133   134   135

the recognition process (the algorithm) and the molecular steric "programme" that leads to preferential formation of the double-helical structures. In other words, helicate formation amounts to a tetrahedral reading of the molecular information stored in the bipy strands.

One may draw an analogy between nucleic acids and helicates, with on one side the polynucleotide strands and their interaction through hydrogen bonding and on the other side the oligobipyridine strands and their binding together via metal ion coordination.

Only metal ions presenting the appropriate coordination features with bipy will form double helicates. This has been shown also to be the case for Ag(I), which yields the tri-, tetra-, and pentahelicates, as confirmed by determination of the crystal structure of the silver trihelicate 136 [9.59]. Similarly, the analogous oligobipyridines containing two or three bipy units separated by a CH_2CH_2 group give the corresponding Cu(I) helicates [9.60, 9.61]. Closely related ligands have been

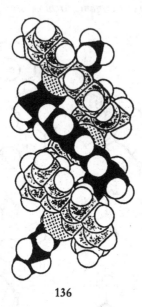

136

used to generate an allosteric ion complexation site on Cu(I) binding [9.62] as well as complexes of hairpin type in order to investigate a possible duplex-hairpin structural interconversion [9.63].

The formation of double-helicates may be expected to occur only in a given range of coordination and structural parameters. Of particular interest is the finding that it takes place with *positive cooperativity*, as was shown by a detailed analysis of the complexation of Cu(I) [9.64] and Ag(I) [9.65] by substituted tris-bipy ligands of type **138**. It is therefore driven to completion and may in this respect be considered as a self-organization process along the lines discussed above. Binding of the first ion prepares the stage for the second one (see also [8.70a]); cooperative protonation is present in a cryptate containing a water molecule as effector (see **25**, Section 2.4). These results provide a physico–chemical basis for the remarkable ability of these species to assemble spontaneously with high efficiency and selectivity.

A major feature of the spontaneous generation of a double-helical structure is that it can be employed as an *organization framework*, allowing the arrangement in space in a double-helical fashion of substituents attached to the bipy units. To this end, the oligo-bipy strands **137-140**, containing substituents in the 4,4′ position on the bipy groups, were synthesized [9.55].

Especially attractive was the possibility to connect nucleosides, as has been realized, for instance, with the hexathymidine **141** and with the elongated and alternating strands **142** and **143**. These compounds represent artificial oligonucleosides, which may interact with natural polynucleotides or nucleic acids. On treatment with Cu(I), **142** and **143** gave the double-helical complexes **144** and **145**, respectively, inside–out analogues of double-stranded nucleic acids, which may be termed *deoxy-*

141

142

143

ribonucleohelicates (DNH) [9.66]. Since these species are positively charged and present the hydrogen-bonding bases on the periphery, they bind to nucleic acids with formation of mixed natural–artificial species [9.67]. Such interactions may present selectivities resulting both from the overall shape of the DNHs and from the thymidine groups. The DNH molecules **144** and **145**, which bear outward-oriented recognition sites, represent polytopic variants of exoreceptors of the metallo-nucleate type (see Section 7.2).

144: n = 0
145: n = 1

Of course, it is possible to organize numerous other units by means of the helicate framework, thus bringing together groups capable of conferring, for instance,
energy-transfer, electron-transfer, or specific ion-binding features.

A point of major interest concerns the *chirality* of the double-helical complexes.
The helicates obtained from the achiral oligo-bipy chains **128–131** are a racemic
mixture of left- and right-handed double helices (Figure 34). High helicity induction was found when the optically active tris-bipy strand **146a** was used, giving
probably the right handed double-helical species **146b** on copper(I) binding [9.68].
One may note that the dihelicate **132** has features reminiscent of the two homochiral
components obtained in the "Coupe du Roi" (Figure 35) [1.9, 9.69].

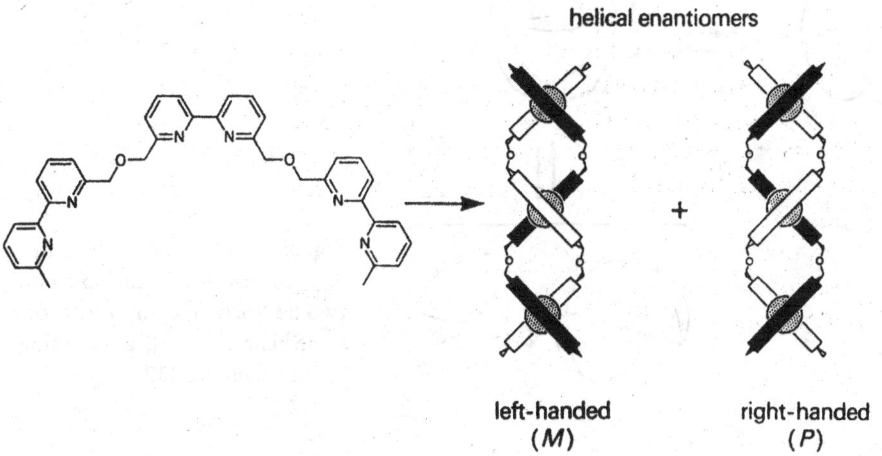

helical enantiomers

left-handed   +   right-handed
(*M*)   (*P*)

Fig. 34. Formation of enantiomeric double-stranded helicates from two tris(bipyridine) strands and three tetrahedrally coordinated metal ions (Cu(I), Ag(I); dotted circles) [9.68].

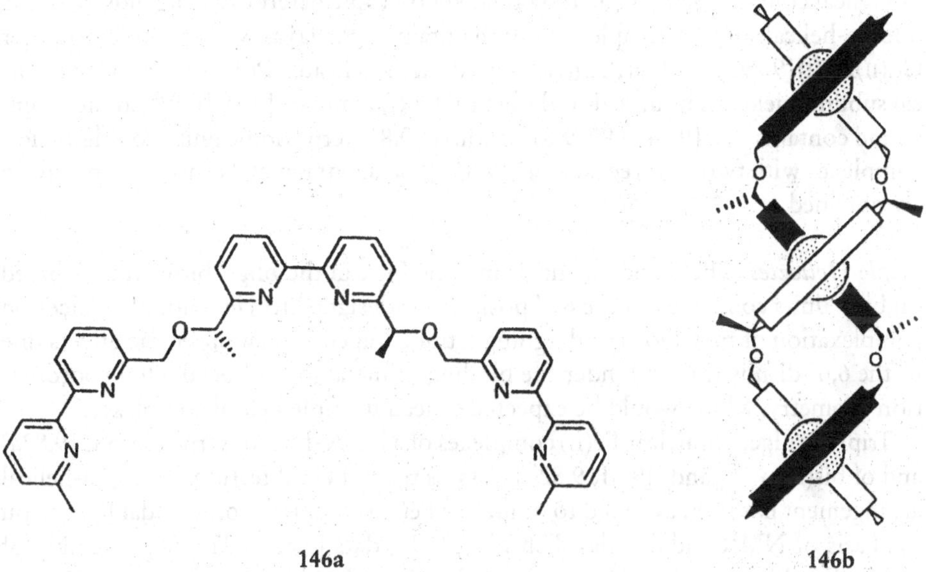

146a   146b

   A number of helical and double-helical complexes have been obtained with poly-pyridine ligands, which bind various metal ions yielding helical and double-helical [9.70–9.74] complexes that present interesting redox and metal–metal interaction properties [9.75]. The graphs of the double-helicates represent braids based on two threads and several crossings [9.1, 9.76], that may serve as templates for the synthe-

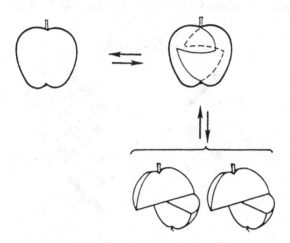

Fig. 35. Left: The "Coupe du Roi" or how to cut an apple into two homochiral components of same helicity [9.69], as an analogy to the dihelicate **132**.

sis of multiply interlocked or knotted structures presenting an increasing number of crossings [8.282]. Thus, suitable connection of the termini of a dinuclear or of a trinuclear Cu(I) double helicate yielded respectively a trefoil knot [8.282, 9.77a] and a doubly interlocked catenane [9.77b].

Helical complexes are formed by related polytopic heterocyclic ligands [9.78]. A double-helical sodium complex has been obtained [9.79a] as well as heterodinuclear Co(II)Ag(I) [9.79b] and Cu(I)Ag(I) [9.79c] double-helicates. Polyassociation of a helical subunit yields an infinite double-helical Cu(I) complex [9.79d]. When the ligand strand contains two [9.74, 9.80, 9.81] or three [9.81] terpyridine units double-helical complexes with two or three (see **147** [9.81b]) octahedral metal centres, respectively are obtained.

Triple Helicates. The steric information contained in the oligo-bipy strands based on bipy units connected in the 6,6' positions is designed to yield double helices on complexation of metal ions undergoing tetrahedral coordination. Steric effects due to the 6,6'-disubstitution hinder the binding of metal ions of octahedral coordination geometry, which would be expected to lead to triple helical complexes.

Triply bridged dinuclear Fe(II) complexes of a bis-N-hydroxy-pyridinone [9.82a] and of bis(bipy) ligands [9.61, 9.82b] possess triple-helical features. A triple-helical arrangement has been assigned to dinuclear Fe(III) complexes of tripodal ligands on the basis of NMR and circular dichroism (CD) data [9.82c]. The self-assembly of well-defined triple-helical dinuclear cobalt(II) [9.83] and lanthanide(III) [9.84] complexes has been achieved.

Using oligo-bipy ligands, the formation of triple helices becomes possible through a modification of the steric instruction by shifting the 6,6'-disubstitution, as in **128–131**, to a 5,5'-disubstitution, which should remove the hindrance towards binding of three bipy units to an octahedral metal centre. Indeed on reaction with

148

147   149   150

nickel(II) ions the tris-bipy ligand **148** gives the trinuclear triple helical complex **149**. Its crystal structure **150** shows a helical pitch much larger than that of the copper(I) and silver(I) double helicates [9.85]. An additional remarkable feature is that **149** crystallizes with partial spontaneous resolution. These results represent a successful application of molecular programming to the spontaneous but directed generation of desired supramolecular architectures [9.86].

In the self-assembly of helicates, in addition to the nature and the coordination geometry of the metal ions involved, three main features that bear structural information determining the nature and shape of the helical species formed may be distinguished: (1) the precise structure of the *binding site*; it specifies in particular the ability to coordinate metal ions with a given geometry as well as the number of strands bound; (2) the *spacer* separating the binding sites; it must favour inter- over intrastrand binding, and it influences the tightness of the helix (the pitch) through the rotation of one metal centre with respect to the next; (3) most importantly the *configuration of the coordination centres* determines whether the structure is indeed of helical nature; helicity implies that all metal centres have the same screw sense.

The programme containing these information features is molecular and linear, its operation is supramolecular and takes place through the reading of the information according to the coordination algorithm of the metal ion.

Further developments involve the investigation of the mechanism of formation of double- and triple helicates and of the effect of variations in ligand structure on their features, the determination of their physico–chemical (thermodynamic, kinetic, electrochemical, photochemical) properties, the exploration of the coordination chemistry of the ligand strands. For instance, it may be possible to obtain quadruple helical complexes with ions of high coordination number such as the lanthanides and linear ligands containing bipy or terpy units. Using cubic metal ions would also be of interest.

One may note the formation of a Mn_4Pd_4 cluster containing an orthogonal arrangement of helical units [9.87]. The potential of using metal coordination as template for the organization of large structures is shown in the metal ion-assisted self-assembly of polypeptides into triple-helix [9.88] and quadruple-helix [9.89] metalloprotein bundles. Organic templates have also been used to assemble model proteins [9.90]. Beyond these systems lies the investigation of processes involving several ligands and metal ions, that is, multicomponent self-assembly (see below).

Molecular helicity is a fascinating property displayed by chemical and biological macromolecular structures, such as the α-helix of polypeptides, the helical conformation of polymers and the double helix of nucleic acids, as well as by several types of small molecules and solid-state supramolecular networks [9.91, 9.92]. The helicates represent a novel class of helical structures providing a means to study the intricate features of the wrapping and unwrapping processes that take place on formation and dissociation of multiple helical entities, to devise helically organized large supramolecular architectures and to serve as bioinorganic model substances.

9.3.2 Multicomponent Self-Assembly

The self-assembly of the helicates and related structures involves just one type of ligand and metal ion. Further progress in the understanding and the control of the self-assembly and self-organization of inorganic (as well as of purely organic) superstructures requires the conception of systems capable of spontaneously generating well-defined architectures from a larger set of components, comprising at least two types of ligands and/or metal ions. The design and choice of these components must fulfil criteria at all three levels of molecular programming and information input that determine the output of the desired final species: recognition, orientation and termination.

Multiligand multimetal self-assembly has been achieved by means of a flat ligand containing three chelating subunits, hexaphenylhexaaza-triphenylene **151** and bipy derived ligands. One molecule of **151** and three 6,6′-dimethyl bipy units yield the circular complex **152** on binding of three Cu(I) ions [9.93], whereas with the tripodal ligand **153** a capped complex **154** is obtained [9.94]. When the ditopic bis-bipy component **155** is employed, two units of **151** and three of **155** bind six Cu(I) ions to form the cylindrical complex **156a** (Figure 36), whose crystal structure **156b** has

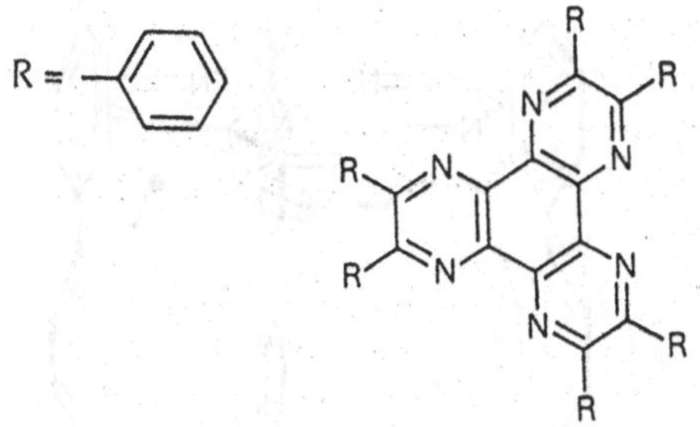

151

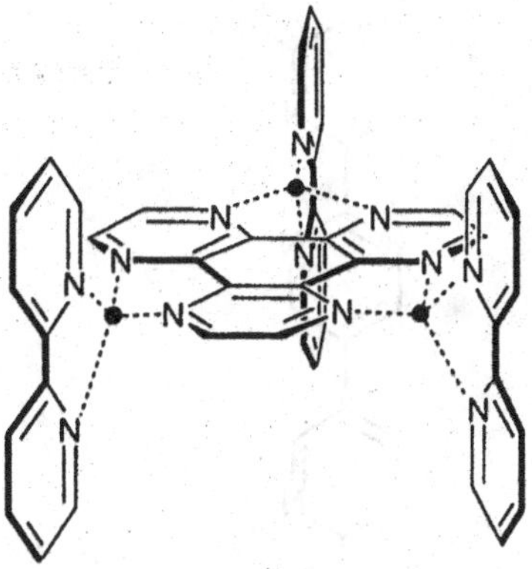

152

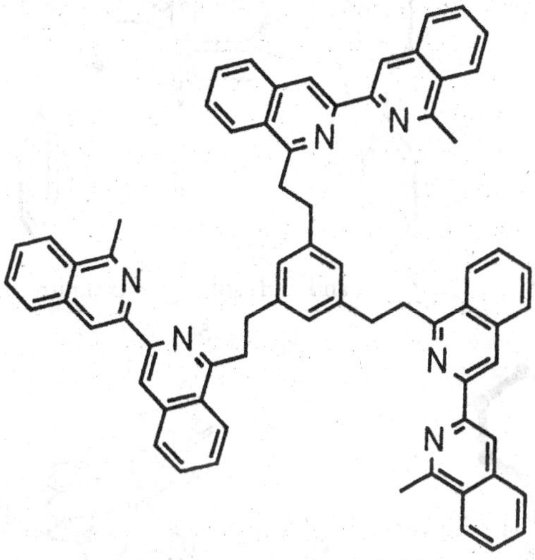

153

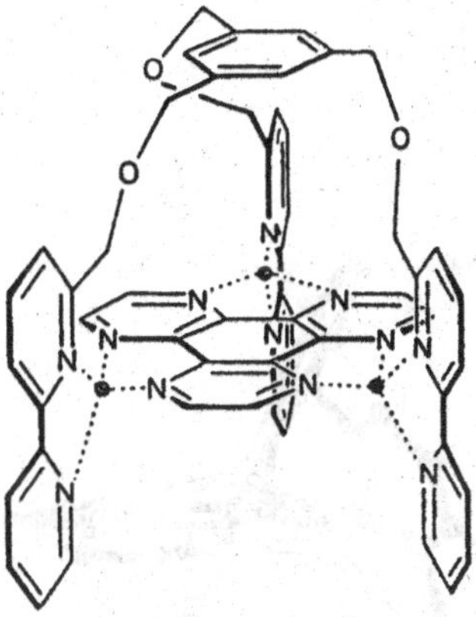

154

been determined [9.93]. This amounts to the self-assembly of five ligands of two different types and six metal ions, altogether eleven particles, to give the closed, cage-like entity **156** in one stroke!

Electrospray mass spectrometry (ESMS) is a powerful tool for gaining insight into the self-assembly process itself by allowing the identification of the species present in various solutions of such complicated mixtures of self-assembling ligands and metal ions leading up to the final architecture [9.94, 9.95].

Extension of the cage forming process to linear, rigid components containing three or four bipy subunits allows the generation of the triple-decker **157** or the quadruple-decker **158** cylindrical complexes with silver ions by self-assembly processes that involve a total of 15 and 19 particles [9.96], respectively.

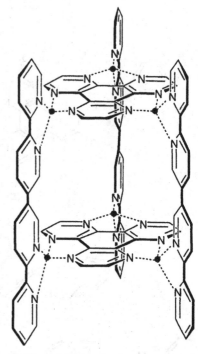

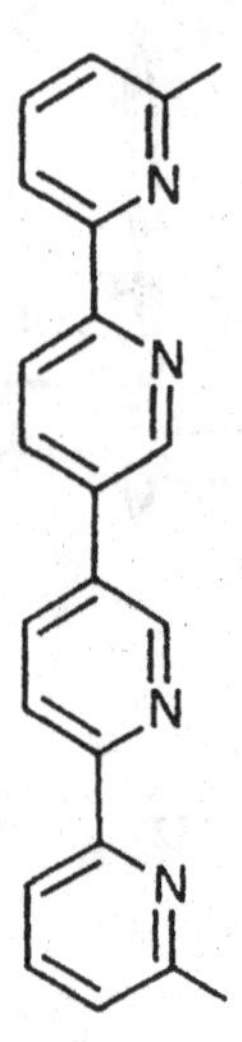

155

C$_6$H$_5$ and CH$_3$ substituents omitted

156a

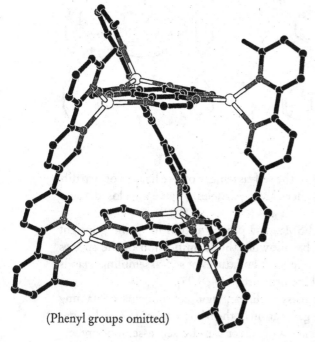

(Phenyl groups omitted)

156b

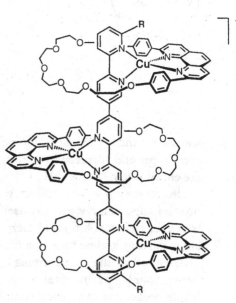

159

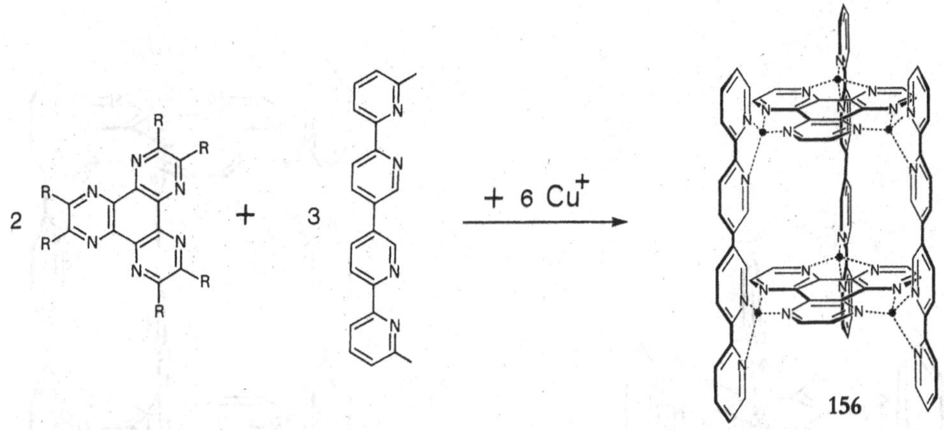

Fig. 36. Self-assembly of a multicomponent cylindrical complex **156a**; Me and Phe substituents omitted for **156a**; R = Phe.

The factors that induce these notable self-assembly processes are both structural (ligand nature, metal ion coordination, steric) and thermodynamic (site occupancy, entropy). They will be considered in more detail below (Section 9.5).

The formation of the large multicomponent superstructures **156**, **157** and **158** demonstrates the remarkable self-assembly of closed inorganic architectures by the spontaneous and correct association of a large number of particles consisting of two types of ligand and one type of metal ion. The operation of this instructed supramolecular system follows the three stages of molecular programming that determine the spontaneous generation of the final discrete structures.

9.3.3 Supramolecular Arrays of Metal Ions. Racks, Ladders, Grids

Various other inorganic superstructures may be imagined and selected as goal for self-assembly.

A *rack*-like arrangement schematically represented by **160**, would be formed by the complexation of several metal ions to rigid, linear sequences of binding sites. Thus, binding of bipy or phen units to **155** and to its extended tri- and tetratopic analogues present in **157** and in **158** via metal ions of tetrahedral coordination such as Cu(I) and Ag(I) should yield rack structures of type **160**. With the bipy or phen unit included in a macrocycle, inorganic rotaxane-type structures (see also Section 9.6) such as **159** have been obtained [9.97a]. A linear sequence of connected terpy sites such as **161** yields racks with individual terpy units and metal ions of octahedral coordination [9.97b].

Ladder superstructures **162** may result from the complexation of linear ligands of the type of **155** and bispyrimidine derivatives **163** or extended units **164** with

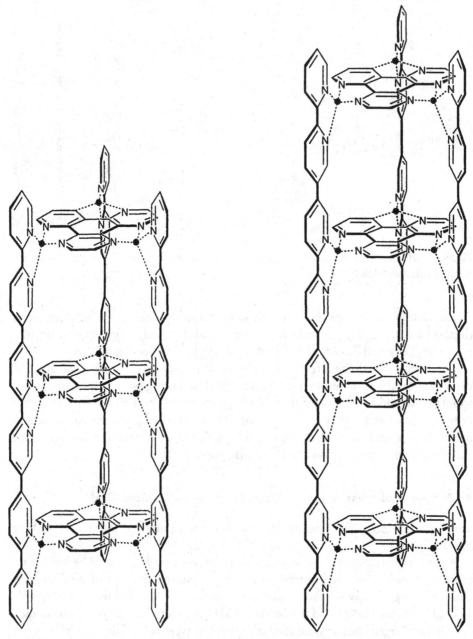

C$_6$H$_5$ and CH$_3$ substituents omitted

157   158

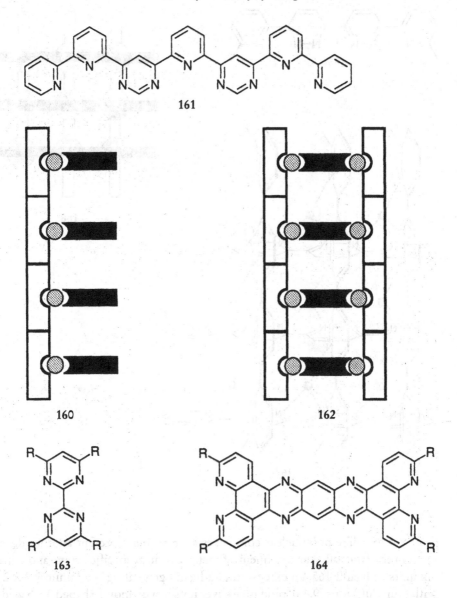

161

160

162

163

164

tetrahedral metal centres [9.97, 9.98]. Ligand **161** could lead to similar structures with octahedral metal centres and bis-terpy components.

Arrays of porphyrins are accessible through the assembly of porphyrin cores bearing binding groups (such as meso-pyridine units) by means of metal ions of suitable coordination geometry. Square arrangements [9.99a] or more extended grid-type arrays can be envisaged (see also Section 9.7). Such organized porphyrin structures should present interesting photochemical and electrochemical properties [9.99b].

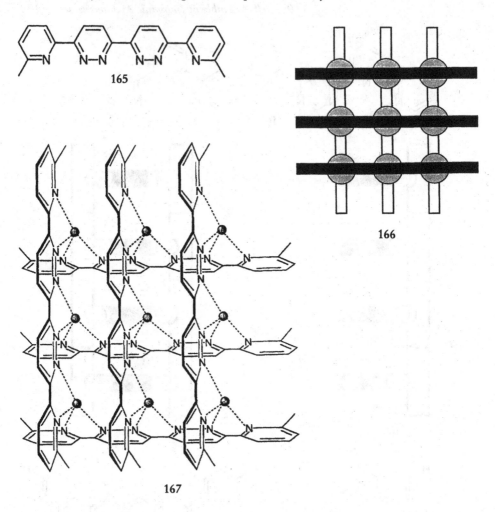

165

166

167

$M \times n$ *grid*-like polynuclear complexes such as **166** represent another development in the controlled arrangement of metal ions into specific arrays and patterns. Six units of ligand **165**, an extension of a ligand generating the minimal 2×2 grid with four Cu(I) ions [9.45], bind nine silver ions in a diamond-shaped 3×3 grid-like array [3.3]G **167**, whose crystal structure has been determined [9.100]. Grids based on octahedral metal centres may be obtained with ligands of type **161** [9.97c]. Such two dimensional superstructures (and their extensions into three dimensions) provide entries into arrays displaying novel photochemical and redox properties; they may represent architectures for information storage in and retrieval from inscribed patterns that could be adressable by light or electrically and recall formally the hardware units in electronic computers (see also Section 10.1).

9.4 Self-Assembly of Organic Supramolecular Structures

The self-assembly of organic supramolecular species makes use of interactions other than metal ion coordination, such as electrostatic, hydrogen bonding, van der Waals, stacking or donor–acceptor effects as found in proteins, nucleic acids, liquid crystals and molecular complexes. The designed use of these forces for the directed self-assembly of a given structure requires knowledge of their strength and of their dependence on distances and on angles.

The spontaneous generation of organized structures depends on the design of molecular components capable of self-assembling into supramolecular entities presenting the desired architectural and functional features. The nature of the species obtained will be determined by the information stored in the components. Thus, the self-assembly process may be directed by molecular recognition between two or more complementary subunits so as to form a given supramolecular architecture. If these molecular units incorporate specific optical, electrical, magnetic, binding, etc., properties their ordering may induce a range of novel features. Depending on the subunits involved the association may lead either to supermolecules or to organized assemblies, such as membranes, molecular layers and films, mesophases, polymeric species or solid state lattices.

9.4.1 Self-Assembly by Hydrogen-Bonding. Janus Molecules

The self-assembly of supramolecular structures via molecular recognition between complementary hydrogen bonding components has developed into a central theme for constructing well-defined arrangements of molecules in solution, in liquid crystals and in the solid state. The structural and energetic features of hydrogen-bonding interactions have been extensively studied (for a few recent references see [2.124, 2.135, 9.101–9.106]). The choice of molecular shape and size and especially the arrangement of hydrogen–bonding donor and acceptor sites are crucial to the correct tessellation of a supramolecular array.

The build-up of two-dimensional and three-dimensional architectures requires the presence in the molecular components of two or more hydrogen bonding subunits whose disposition determines the final supramolecular architecture. When two subunits are incorporated into a single group, it will then possess two recognition faces and may be termed a *Janus* molecule, a double faced H-bonding recognition unit. If the faces are the same or different the unit is homotopic or heterotopic; when they are complementary, the unit is self-complementary and may be termed *plerotopic* (see Section 2.1). Each recognition site may be monodentate, bidentate or tridentate depending on whether it contains one, two or three hydrogen bond donor

(D) or acceptor (A) centres. Of course, the stability and selectivity of the associations depend on the nature and position of the centres [9.106].

The bases of nucleic acids are well known heterocyclic groups that interact through hydrogen bonding, usually forming base pairs through Watson–Crick type association. Different modes of pairing are also possible, such as Hoogsteen type, and pairing specificity may be influenced by other structural factors [9.107]. These interactions lead to the self-assembly of the natural nucleic acid double helix and of the double-stranded structures of the synthetic hexose nucleic acids [9.107]. Combining the pairing modes allows the formation of base triplets and triplex nucleic acid structures [5.27, 5.30].

Supramolecular macrocycles based on a cyclic array of hydrogen-bonded components (cyclic oligomers or cyclamers) are formed for instance in the case of the tetramers of guanine, oligoguanosine nucleotides and derivatives [9.108–9.111]; they have been structurally characterized [9.112, 9.113] and have important biological implications [9.114]. Folic acid yields a cyclic tetramer [9.115] and a hexameric ring including a molecule of benzene in a nicely complementary fashion is formed by 1,3-cyclohexadione in the solid state [9.116a]; carboxylic acid dimerization yields hexameric rings in trimesic acid [9.116b] and iso-phthalic acid derivatives [9.116c].

Considering tridentate sites, 2,6-diaminopyridine and uracil are complementary single site groups (see Sections 9.4.2 and 9.4.3 below). On the other hand barbituric acid (BA) and 2,4,6-triaminopyrimidine (TAP) or -triazine (TAT) are Janus molecules containing two identical recognition sites while **168** is plerotopic by virtue of its two complementary sites (see Section 9.4.4).

The choice of the array of D and A centres allows the more or less strict programming of the resulting supramolecular architecture. Other factors may also need to be taken into account, such as steric effects and solvation. Two bidentate sites of α-pyridone type have been used to generate either cyclic structures or a polymeric motif [9.117, 9.118]. Tridentate groups of BA and TAP types may associate either as a linear or as a cyclic structure (Figure 37); indeed ribbon or tape [9.119, 9.120] and "crinkled" tape [9.121] structures as well as a supramolecular macrocycle, a "rosette", [9.121] have been obtained. On the other hand a strict and univocal program is contained in **168**, which on self-association can *only* lead to a ribbon structure [9.122]. Other heteroassembling tridentate groups may be designed.

The three-ring heterocyclic Janus molecules **169–171** contain uracil, cytosine- and guanine-type tridentate H-bonding sites [9.123]. These compounds may be termed homotopic according to the designation of polytopic receptors (Chapter 4). They are also enantiotopic when considering chirality in two dimensions [9.124]. Substances **170** and **171** contain a matching pair of faces bearing acceptor–donor hydrogen-bonding recognition units that are reminiscent of the cytosine-guanine pair, in which the disposition of hydrogen bond donors and acceptors avoids repulsive secondary electrostatic interactions [9.106] and encodes a program for self-assembly; their

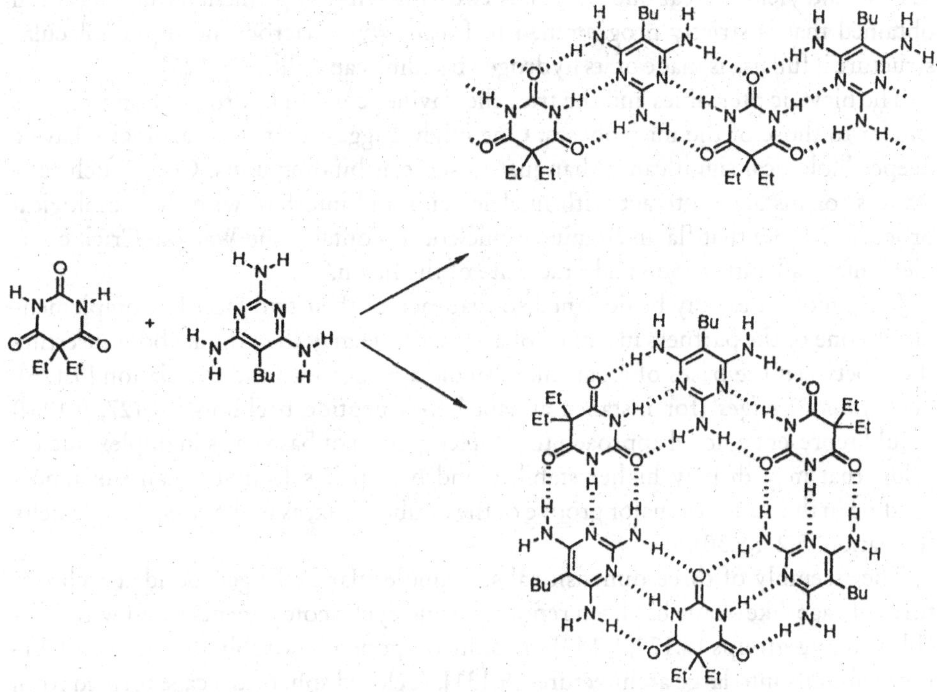

Fig. 37. Self-assembly of either a supramolecular ribbon (top) or a supramolecular macro-cycle (bottom) from barbituric acid and 2,4,6-triaminopyrimidine units [9.122].

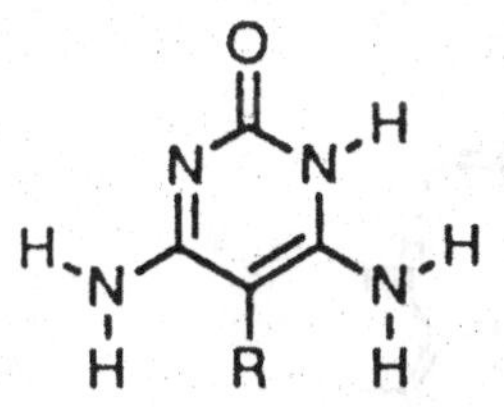

168

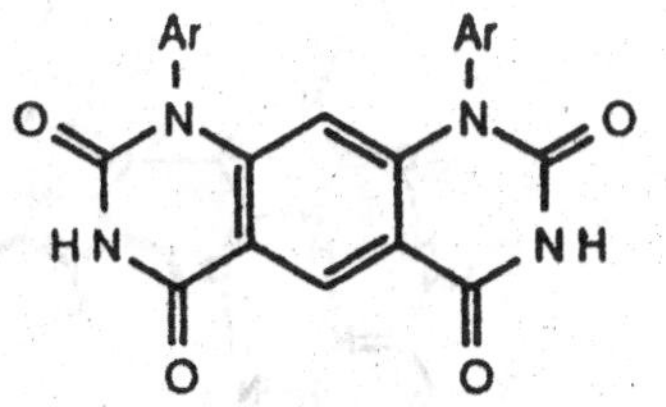

169

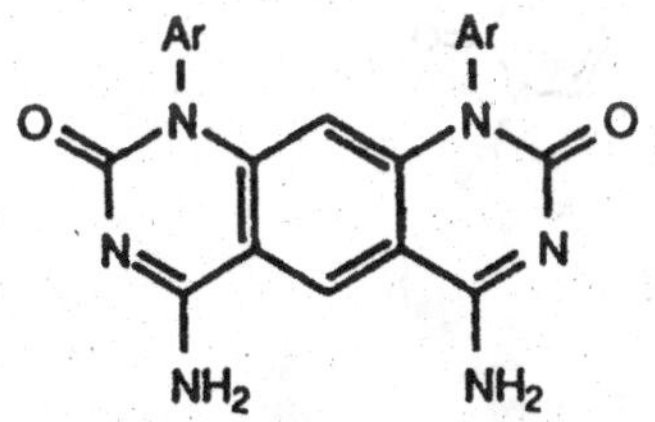

170

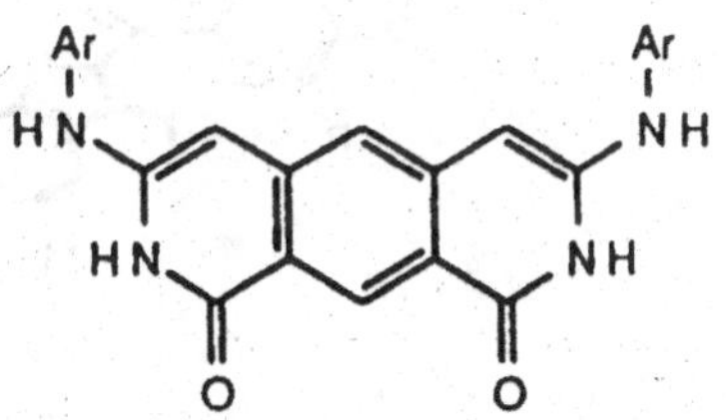

171

heteroassembly should give a cyclic structure, whereas **169** combined with TAP or TAT would yield a linear one [9.123]. A two-ring self-complementary unit has been obtained that is strictly programmed to form *only* a macrocyclic supramolecular structure if full use is made of its hydrogen bonding capabilities [9.125].

The bicyclic pteridines and the tricyclic flavines contain hydrogen-bonding sites similar to those of the nucleobases. One might suggest that these analogies have a deeper biological significance than just to serve as binding units. Could such substances for instance interact with nucleic acids and interfere with their biological processes? Note that flavin adenine dinucleotide contains the Watson–Crick complementary adenine group and uracil site of the flavin.

Janus molecules may be designed so that each of their two faces be complementary to one of the partners in a nucleobase pair. One may then think about inserting these between the bases of a pair like driving a wedge into the association [9.126]. Such *Janus wedges* (for instance attached to a peptide backbone [9.127, 9.128]) would represent a novel approach to the recognition of base pairs in duplex nucleic acids, that may display higher stability and base pair selectivity than substances binding in the minor or major groove of the double helix, as in the case of triple helix formation [5.27, 5.30].

The assembly of three-dimensional supramolecular hydrogen-bonded architectures of cage-like shape has been reported using cyclic components fitted with suitable binding groups [9.129, 9.130] or multicomponent assembly of BA- and TAT-type subunits into large architectures [9.131]. A closed spheroidal cage formed from two self-complementary components [9.132a] encapsulates guest species [9.132b, c]. Tris-pyridone species give three-dimensional networks with large cavities [9.10]. The self-assembly of a porphyrin (or its Zn(II) complex) bearing two uracil groups by means of two TAP units yields a supramolecular bis-porphyrin cage **172** into

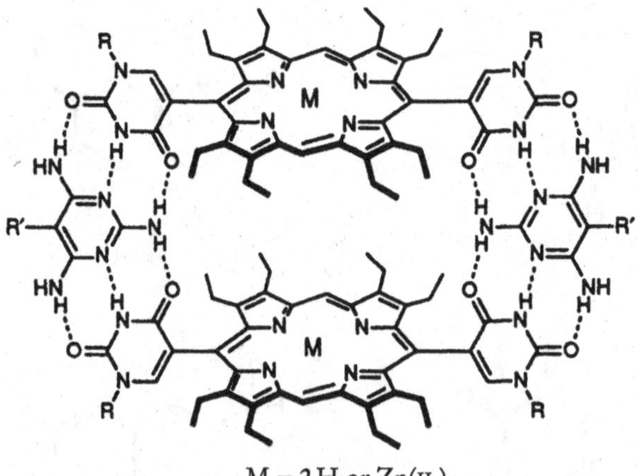

M = 2 H or Zn(II)

172

which a bipy guest molecule may bind by coordination to the Zn(II) sites in a synergic fashion [9.133]. Suitably designed cyclic peptides self-assemble into supramolecular tubular structures of specific internal diameter [8.186].

With increasing control being achieved over molecular programming of supramolecular structure generation through hydrogen bonding, the self-assembly of a variety of linear, two- or three-dimensional architectures may be realized. The fact that the processes also take place in solution and not only in the solid state is of special interest. Designed self-assembly thus opens roads towards the generation of organized entities in the liquid phase.

9.4.2 Molecular Recognition-Directed Assembly of Organized Phases

It is clear that intermolecular effects strongly influence the properties of materials. The point is that one may make use of them in a controlled fashion to induce specific changes when and where desired. The self-assembly of membranes, molecular layers, films, vesicules, etc. incorporates interactions such as hydrophobic effects, hydrogen bonding, electrostatic forces and surface binding [7.1–7.13, 7.45, 7.87, 9.134–9.141], which may be used to produce specific structural and functional properties.

Supramolecular interactions play a crucial role in particular in the formation of liquid crystals and in the determination of their features [7.13, 9.142]. They may involve organic groups or also metal ions, leading then to a variety of *metallomesogens* [9.143]. Hexagonal columnar mesophases possessing channel-like architectures are generated from components containing polyether rings [8.196, 9.142] and hydrogen-bonding groups [9.144, 9.145]. Similar effects operate in biomesogens [9.109, 9.112, 9.116, 9.146], in liquid crystalline polymers [8.196, 9.142, 9.147] and in various polyassociated microstructures (micelles, tubules, sheets, liquid or solid fibres) formed from amphiphilic components [9.148].

The association between molecular units that by themselves are not mesogenic could lead to the generation of a supramolecular species presenting liquid crystalline behaviour. It then might be possible to take advantage of selective interactions so that the mesogenic supermolecule would form only from complementary components. This would amount to macroscopic expression of molecular recognition; indeed recognition processes occurring at the molecular level would be displayed at the macroscopic level of the material by the induction of a mesomorphic phase that may be termed supramolecular and "informed", since it is conditioned by the molecular information present in its components. Such a process involves a phase change which, being a highly cooperative process, also corresponds to an *amplification* of molecular recognition and information from the microscopic to the macroscopic level. This has been accomplished in the generation of supramolecular mesophases and liquid crystalline polymers from complementary molecular units [9.149].

Mesophases Induced by Association of Complementary Molecular Components. A common type of molecular species that form thermotropic liquid crystals possesses an axial rigid core fitted with flexible chains at each end. One may then imagine splitting the central core into two complementary halves ∈ and ∋, whose association would generate the mesogenic supermolecule, as schematically represented in Figure 38.

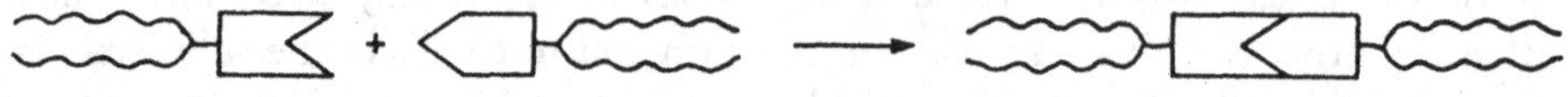

Fig. 38.   Formation of a mesogenic supermolecule from two complementary components.

This was realized with derivatives of the complementary heterocyclic groups 2,6-diaminopyridine P and uracil U bearing long aliphatic chains [9.150]. Whereas the pure compounds did not show liquid crystalline behaviour, 1:1 mixtures gave a metastable mesophase of columnar hexagonal type. The existence of the latter was attributed to the formation of a mesomorphic supermolecule **173** via association of the complementary components.

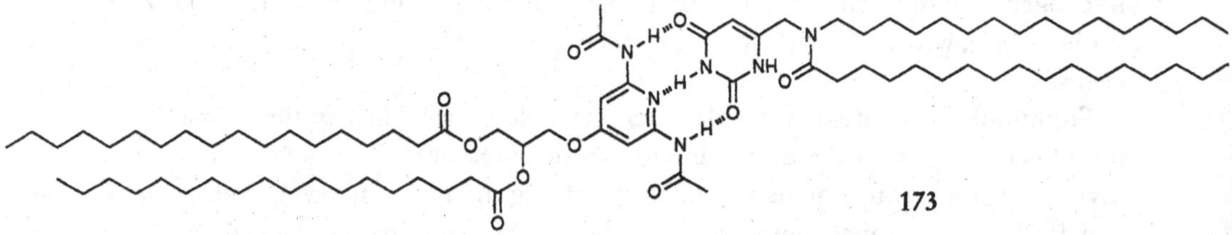

173

Numerous extensions may be envisaged, such as the introduction of various central cores, in particular those already known to yield molecular liquid crystals, the incorporation of light-sensitive or electro-sensitive units, the potential use for detection devices, and the extension to various recognition components of biological nature (for biomesogens see 9.109, 9.112, 9.115, 9.146).

Self-Organization of Supramolecular Liquid Crystalline Polymers from Complementary Components. If two (or more) complementary units ∈ or ∋ are grafted on a template T, mixing $T\in_m$ with the complementary $T\ni_m$ may lead to the hetero-self-assembly of a linear or cross-linked, main-chain supramolecular "copolymer" species $(T\in_m, T\ni_m)$, whose existence is conditioned by the molecular recognition-directed association between the ∈ and ∋ groups.

Figure 39 represents schematically such a process in the case of two-site (ditopic) complementary components $T\in_2$ and $T\ni_2$. The resulting supramolecular polymeric material $((T\in_2, T\ni_2)_n$ may present liquid-crystalline properties if suitable chains are grafted onto the components. One may note that the mixed site species ∈ T∋ repre-

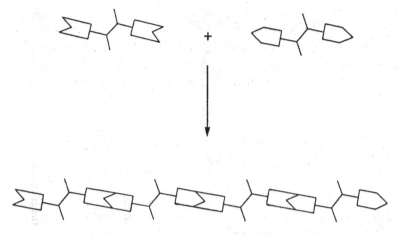

Fig. 39. Formation of a polymeric supramolecular species by association of two complementary ditopic components.

sents a self-complementary component capable of undergoing homo-self-assembly into a polymeric $(\in T \ni)_n$ entity, in a way reminiscent of the self-assembly of the protein coat of the tobacco mosaic virus from its polypeptide subunits [9.5].

Condensation of the complementary groups P and U with long chain derivatives of L-, D- or *meso*(M)–tartaric acid yields substances LP_2, LU_2, MP_2, MU_2 etc., that each contain two identical units capable of undergoing association via triple hydrogen bonding [9.151]. It was found that whereas the individual species LP_2, LU_2, DP_2, DU_2, MP_2 and MU_2 were solids, the mixtures $(LP_2 + LU_2)$, $(DP_2 + LU_2)$ and $(MP_2 + MU_2)$ give thermotropic mesophases presenting an exceptionally wide domain of liquid crystallinity (from $< 25\,°C$ to $220–250\,°C$) and a hexagonal columnar structure, with a total column diameter of about 37–38 Å.

It is reasonable to assume that the complementary units form the expected triply hydrogen bonded pairs, so that the entirely different behaviour of the pure compounds and of the 1:1 mixtures may be attributed to the spontaneous association of the complementary components into a polymolecular entity based on hydrogen bonding. The overall process may then be described as the *self-assembly of a supramolecular liquid-crystalline polymer* based on molecular recognition (Figure 40). The resulting species $(TP_2, TU_2)_n$ is represented schematically by structure **174**.

The X-ray diffraction patterns show that the materials obtained from the various configurational isomers of tartaric acid have different architectures. Both are hexagonal columnar mesophases, but, whereas the data for $(LP_2, LU_2)_n$ are consistent with columns formed by three polymeric strands having a triple helix superstructure (Figure 41), those for the $(MP_2, MU_2)_n$ mixture fit a model built on three strands in a zig-zag conformation. The LD mixture has another arrangement again.

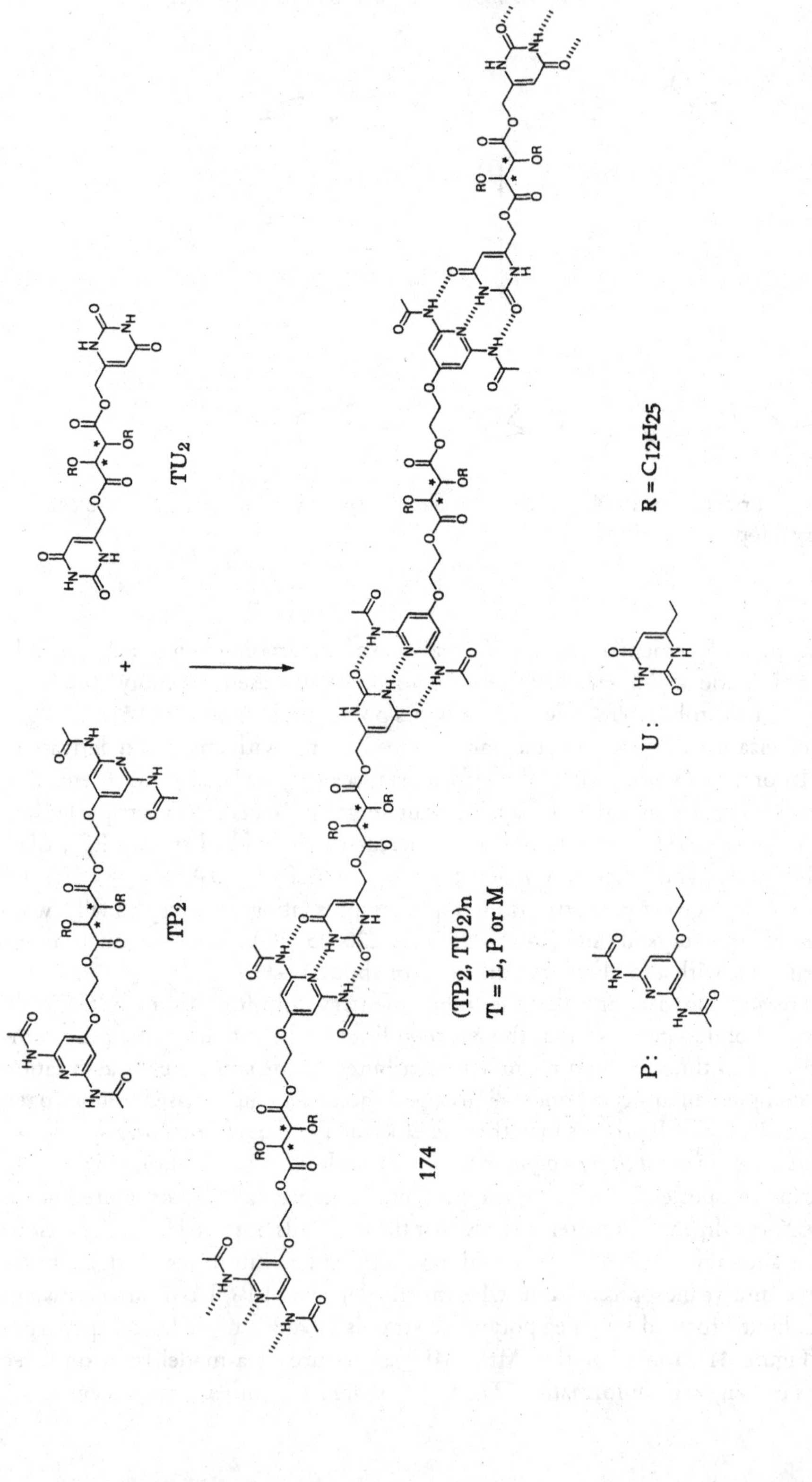

Fig. 40. Self-assembly of the polymolecular supramolecular species (TP$_2$, TU$_2$)$_n$ **174** from the complementary chiral components TP$_2$ and TU$_2$ via hydrogen bonding; T represents L-, D- or *meso*(M)-tartaric acid; R = C$_{12}$H$_{25}$.

$$R = n\text{-}C_{12}H_{25}$$

174

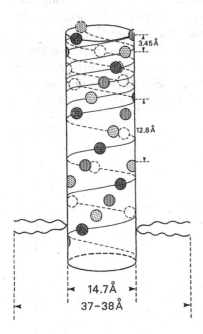

3.45 Å

12.8 Å

14.7 Å

37–38 Å

Fig. 41. Schematic representation of the columnar triple-helical superstructure derived from the X-ray data for (LP$_2$, LU$_2$)$_n$; each spot represents a PU or UP base pair; spots of the same type belong to the same supramolecular strand; the dimensions are compatible with an arrangement of the PTP and UTU components along the strands indicated (see also text); the aliphatic chains stick out of the cylinder, more or less perpendicularly to its axis; a single helical strand and the full triple helix are respectively represented at the bottom and at the top of the column [9.152].

These three materials make clear the profound influence of chirality on the super-structure formed.

A central question is that of the size and the polydispersity of the polymeric supramolecular species produced. Of course, their size is expected to increase with concentration. The molecular weight polydispersity depends on the stability constants for successive associations.

Electron microscopy studies provided further information about the nature of these materials [9.152]. Increasing the concentration of equimolar solutions of LP_2 and LU_2 results in the progressive assembly of supramolecular–polymolecular entities up to very large ones. The process involves nucleation to give small nuclei, then growth to filaments and finally lateral association to tree-like structures, strings and fibres. The species formed are helical; their helicity is right-handed, being induced by the chirality of the components and transferred to the larger entities. The primary filament is the triple helical species consisting of three helically-wound supramolecular strands. $(DP_2 + DU_2)$ mixtures yielded left-handed helical superstructures. No helicity was found for the *meso* compounds (Figure 42). Helicity has been observed in a number of polyassociated supramolecular species (see references 15–20 in [9.151]).

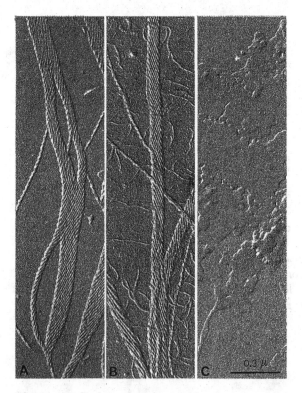

Fig. 42. Helical textures observed by electron microscopy for the materials formed by the mixture (A) $LP_2 + LU_2$ and (B) $DP_2 + DU_2$; (C) represents $MP_2 + MU_2$.

The racemic mixture of all four components LP$_2$, LU$_2$, DP$_2$ and DU$_2$ yielded long superhelices of opposite handedness that coexisted in the same sample. This points to the occurrence of spontaneous resolution through *chiral selection* in molecular recognition directed self-assembly of supramolecular liquid crystalline polymers. Racemate resolution induced by chain length has been reported in the formation of right-handed and left-handed helices by D- and L-gluconamides bearing *N*-octyl and *N*-dodecyl chains, respectively [9.148c].

Such chiral selection features of self-organized entities are of general significance in connection with the questions of spontaneous resolution and of chirality amplification. They amount to a *self-resolution* process.

These results illustrate how extended supramolecular–polymolecular entities build up through molecular recognition directed polyassociation of complementary components. They also show that molecular chirality is transduced into supramolecular helicity, which is expressed at the level of the material on nanometric and micrometric scales, amounting to a sort of *size amplification* of chirality.

Columnar mesophases of rectangular section have been obtained by combining the monotopic components of **173** with the TP$_2$ and TU$_2$ units; their occurrence may be attributed to the formation of mixed 2/1 supermolecules **175** [9.153].

175

The mixed sites species TPU would represent a self-complementary component capable of homo-self-assembly into a (TPU)$_n$ supramolecular entity.

The introduction of rigid molecular units into macromolecular species has been extensively pursued in view of the novel physico–chemical properties that the resulting rigid rods may present. Self-assembling rigid components may be designed by attaching recognition groups to a rigid core. The combination of two such complementary components, **176** and **177**, leads to the formation of a lyotropic mesophase based on the self-assembly of a rigid rod supramolecular system [9.154].

Hairy rigid rod polymers, in which flexible side chains are attached to a rigid core, present attractive properties [9.155]. A supramolecular version of such materials may be the triple helical supramolecular species described above (Figure 41), which presents the features of a hairy cylinder. Similarly, the components **176** and **177** bearing long R chains yield self-assembled, supramolecular hairy rigid rods.

176

177

Cross-Linking of Polymolecular Self-Assembled Systems. Supramolecular Automorphogenesis. Extending further the procedures of polymer chemistry to supramolecular species one may envisage devising components containing multiple recognition groups as cross-linking agents for self-assembled structures.

Tripode type species bearing three equivalent recognition subunits may be expected to establish two-dimensional networks when mixed with the linear polyassociated species described above [9.156]. An equimolar mixture of complementary ternary components such as those shown in Figure 43 (Z = U and Z = P) could lead to the spontaneous generation of tree-like species that represent a self-assembled supramolecular version of the dendrimers and arborols. Such spontaneous structure generation by recognition-directed self-assembly represents processes of supramolecular morphogenesis (be it of organic or inorganic nature; see Section 7.3). Termination control is a particularly important design feature for the formation of a well-defined discrete architecture.

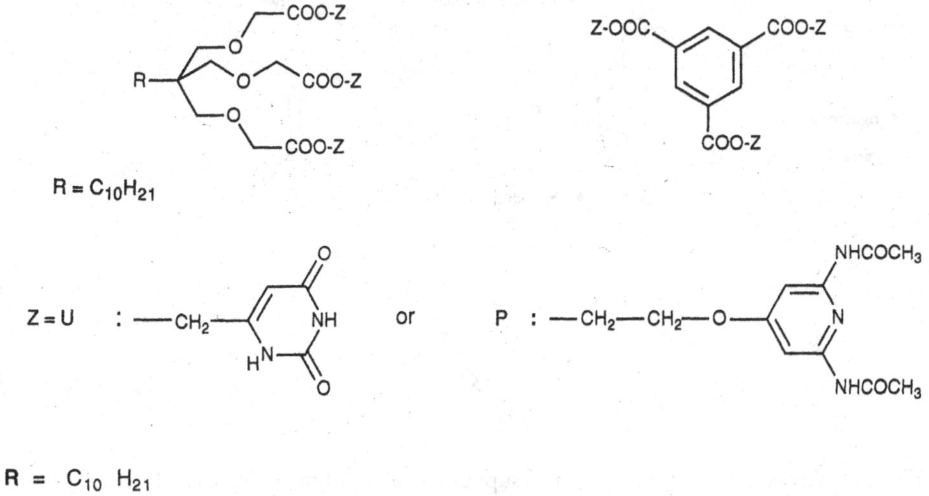

R = C$_{10}$H$_{21}$

R = C$_{10}$ H$_{21}$

Fig. 43. Ternary recognition components for the cross-linking of supramolecular polymeric species [9.149].

9.4.3 Supramolecular Polymer Chemistry

A rich domain emerges from the combination of polymer chemistry with supramolecular chemistry, defining a supramolecular polymer chemistry [9.149, 9.157]. It involves the designed manipulation of molecular interactions (hydrogen bonding, donor–acceptor effects, etc.) and recognition processes to generate main-chain (or side-chain) supramolecular polymers by the self-assembly of complementary monomeric components (or by association via lateral groups). In view of the lability of these associations, such entities present features of "living" polymers capable of growing or shortening, of rearranging their interaction patterns, of exchanging components, of undergoing annealing, healing and adaptation processes. Figure 44 displays some of the various types of polymeric superstructures that represent supramolecular versions of various species and procedures of molecular polymer chemistry. Recognition effects are expected to play a major role in the assembly and self-organization processes that determine the properties resulting from the combination of polymer structure and of liquid crystallinity [9.11, 9.157–9.159].

Supermolecules built from small molecules mainly involve intermolecular interactions. On the other hand, with macromolecules the supramolecular association may be either intermolecular, occurring between the large molecules, or intramolecular involving recognition sites located either in the main chain or in side-chain appendages, thus leading to chain folding and structuration of the macromolecular

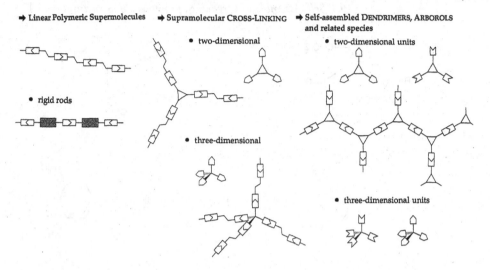

Fig. 44. An aspect of the panorama of supramolecular polymer chemistry [9.149].

entity. The latter supramolecular–intramolecular aspect is a major problem for bio-logical macromolecules and the processes controlling protein folding represent an active field of investigation [9.160].

The directed manipulation of intermolecular interactions (hydrogen bonding, van der Waals forces, metal coordination) gives access to a *supramolecular engineering* of molecular assemblies and of polymers (see, for instance, [7.10–7.13, 7.44, 9.142, 9.157, 9.161–9.163]) through the design of instructed monomeric and poly-meric species. It leads to the development of a supramolecular materials chemistry (see Section 9.8).

9.4.4 Molecular Recognition-Directed Self-Assembly of Ordered Solid-State Structures

The control of the arrangement of molecules in the solid state depends on the inter-molecular interactions and on packing factors [9.164]. It is a problem of supra-molecular nature the solution to which lies in the understanding of the factors that determine solid-state organization (see, for instance, [7.24, 7.39, 2.124, 9.101–9.105, 9.165]). The solid is thus a supermolecule, a very large supermolecule indeed, whose formation is based on molecular recognition and self-organization processes. Accordingly, the crystal represents the ultimate of the extended but periodic supra-molecular entity [9.166]. Solid state polymorphs may then be considered as supra-molecular isomers and the conversion of one polymorphic form into another one as a *supramolecular isomerization*. Similar considerations hold for changes such as the solid state to liquid-crystal transition or the interconversion of different types of mesophases.

Molecular recognition effects provide an entry into *crystal engineering* and the designed generation of organic solid state architectures. Hydrogen bonding patterns may direct structure formation in the solid as well as in solution, generating specific supramolecular crystal architectures held together by networks of H-bonds as, for instance, in associations through alcohol and amine [9.167] or carboxylic acid and amide [9.168] groups. Thus, the formation of given structures may be programmed more or less strictly depending on the arrangements permitted by the recognition groups employed ([9.3, 9.149, 9.167]; Section 9.4.1).

An approach to the generation of molecular order is based on the recognition-directed spontaneous assembly of a supramolecular strand from complementary molecular components, each of which presents two identical recognition sites (Figure 45). The interaction of two such units may produce organized polymolecular strands in solution, in a mesomorphic phase, or in the solid state by cocrystallization. All residues of the same type are expected to be located on the same side of the strand (Figure 45), thus providing spontaneous like/unlike *sorting* (into two substrands) and *orientation* of the molecular components in the supramolecular arrangement.

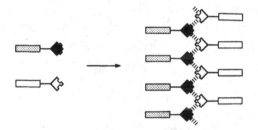

Fig. 45.   Schematic representation of the formation of an ordered supramolecular strand by the molecular-recognition-induced association of two different molecular units; each unit contains a group that possesses two identical binding sites complementary to those of the other unit [9.119].

2,4,6-Triaminopyrimidine and barbituric acid derivatives are complementary components presenting the required features, since they should be able to form two arrays of three hydrogen bonds with each other. However, as pointed out above, based solely on hydrogen bonding recognition, their association could yield either a linear or a macrocyclic supramolecular structure (Section 9.4.2; Figure 37).

Mixtures of the two species form associations in solution and 1:1 crystals by cocrystallization. In the crystal the components are arranged in mixed supramolecular strands **178** (Figure 46) in which each unit forms six hydrogen bonds with its two complementary neighbors. As a consequence of molecular recognition, expressed by the hydrogen-bonding pattern between the two components, all like residues A or B are indeed located on the same side of the strand [9.119].

178   179

Thus, the self-assembly spontaneously induces sorting of like/unlike species and generates long-range order, orientation, and left/right differentiation at the supramolecular level. Since the crystal packing consists of sheets of polar strands related to each other by translation, the entire three-dimensional structure becomes polar.

Homo-assembly of derivatives of the self-complementary recognition group **168** can only yield a linear ribbon or tape structure **179**. This is indeed the case, as shown

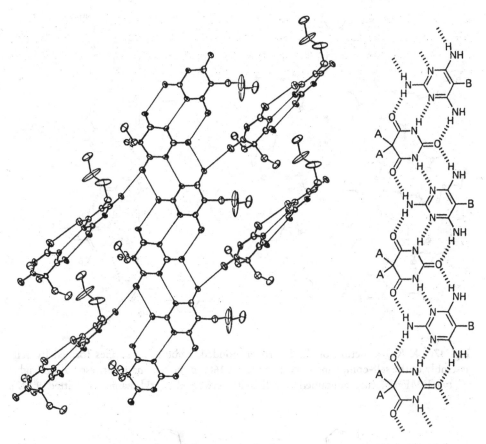

Fig. 46. X-Ray structure (left) of the cocrystal of the 1:1 species (right) formed by two complementary components derived from barbituric acid BA (A = ethyl) and 2,4,6-triamino-pyrimidine TAP (B = butyl) [9.119].

by the crystal structure of the octyl derivative (Figure 47) [9.122]. In this case the left and right sides of the structure are however, identical, contrary to the case of hetero-assembly (Figure 46). The process may be considered as the operation of a strictly programmed molecular system that contains the information required for the self-organization of precisely this supramolecular architecture. Both species **178** and **179** display unlimited base-pairing; they also represent a kind of self-assembled "ladder" polymer of supramolecular nature, as well as "hairy" ribbon like structures.

An interesting question is whether or not component selection occurs in the course of the self-assembly if a mixture is used, in particular in the case of chiral compounds. Thus, the assembly of a bis-L-tryptophan TAP derivative by di-*n*-butyl-BA was studied both as the optically pure LL compound and as a racemic mixture with the DD enantiomer [9.169]. The crystal structure of the latter showed

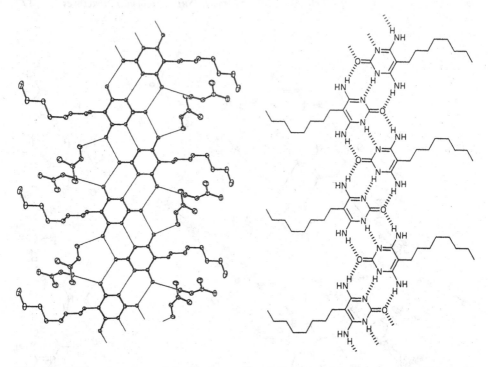

Fig. 47. X-Ray structure of the hydrogen-bonded, ribbon-like species formed by self-assembly of the self-complementary component **168**. In the schematic representation (right) the two DMF molecules contained in each cavity between the octyl chains are omitted [9.122].

180   R = CO₂CH₃

homochiral strands of "crinkled" tape or undulating ribbon-type (see **180**), indicating that chiral *self-resolution* occured on self-assembly of the supramolecular structure, a process reminiscent of that found in the helical strands considered above (see Section 9.4.2, [9.152]).

Of special interest is the generation of *functional arrays* on the basis of the self-assembly of photo-, electro- or iono-active components. For instance, the arrangement produced in **178** and **179** could confer specific properties (optical, electronic, ionic, magnetic, etc.) on the material at the macroscopic level resulting from the oriented series of groups along a strand. Such a process may allow the designed engineering of (non-centrosymmetric) cocrystals of organic molecules possessing non-linear optical [9.170], electron transfer or energy transfer properties.

In particular, the formation of ordered arrays of photoactive units like porphyrins or of ion binding sites (e.g. macrocyclic polyethers; see also [8.202]) may induce directional electron and energy transfer or ion channel features [9.171a, 9.172]. Such effects are of interest for optoelectronic information storage and for ionic devices.

Other recognition groups could generate many different patterns (ribbons, rings, cylinders, strips, belts, friezes, etc.), leading to a recognition-based "molecular patterning". The self-assembly of such supramolecular species also occurs in solution, thus giving access to functional organized structures in the liquid state, or in a mesophase.

9.4.5 Physico–Chemical Methods of Investigation

The development of physico–chemical methods for the characterization of the species formed by spontaneous association is of crucial importance for the development of research on self-assembly. Whereas solid state structures may be determined by the powerful X-ray crystallography method, there is a great need for procedures allowing the investigation of organized species formed *in solution*, in particular whether they are the same as in the solid or not. Vapor phase osmometry, membrane osmometry, and gel permeation chromatography provide useful information, but they generally yield molecular weights with degrees of error that can be as large as a single assembly unit; in addition, they only provide an average of all species in solution.

Overall observation and imaging of the self-assembled entities on a solid support can be performed by electron microscopy (Figure 42) and by scanning tunnelling or atomic force microscopy (see Section 9.9).

Global information on solutions may be obtained from light- or neutron-scattering experiments.

Spectral methods, spectroscopic titration and NMR data (chemical shifts on binding, intermolecular Nuclear Overhauser Effects, relaxation and correlation times) give access to important parameters of the equilibria present and of the structural and dynamic features (see also Section 4.5).

Mass spectrometry techniques may allow the direct detection of the various species formed in solution. Thus, electrospray mass spectrometry has been used to investigate the progressive build up of metallosupramolecular structures such as the capped **154** [9.94a], the cylindrical **156** [9.94b] as well as helicate complexes [9.94b, 9.95b].

The complexation of metal cations by an ion binding crown ether site borne by a triaminotriazine unit allowed the characterization by ESMS of the labile species formed through hydrogen bonding with a complementary barbituric acid derivative. In particular, the self-assembly of a six-component supramolecular macrocycle (of the type shown in Figure 37) is revealed when a solution of the two components is treated with an alkali salt. This points to a general and powerful ion labelling electrospray mass spectrometry IL-ESMS method, based on the use of components fitted with an ion binding site, for the investigation of self-assembly processes of discrete neutral superstructures in solution [9.171b].

In order to fully characterize self-assembly processes, it will be necessary to further develop analytical methods providing information on the composition and structure of the species formed in solution as well as on the thermodynamics of the equilibria in which they take part.

9.5   Self-Recognition. Instructed System Paradigm

In addition to high efficiency, selectivity and cooperativity, another basic feature characterizing programmed supramolecular processes is *"self-recognition"* — the recognition of like from unlike, of self from non-self — embodied in the spontaneous selection and preferential assembly of like components in a mixture.

With respect to inorganic self-assembly this would involve preferential binding of like metal ions by like ligands in a mixture of ligands and ions. Indeed, selective formation of helicates was obtained in mixtures of oligo-bipyridine strands of type **128–131** and **148** with suitable metal ions [9.173].

On treatment with copper(I) ions, mixtures of the strands **132–135** spontaneously yield the corresponding double helicates without significant crossover (Figure 48).

Similarly, when a mixture of the two tris-bipyridine ligands **129** and **148** is allowed to react simultaneously with copper(I) and nickel(II) ions, only the double helicate **132** and the triple helicate **149** are formed (Figure 49). Thus, parallel operation of two programmed molecular systems leads to the clean self-assembly of two well-defined helical complexes from a mixture of their four components in a process involving the assembly of altogether 11 particles of four different types into two supramolecular species.

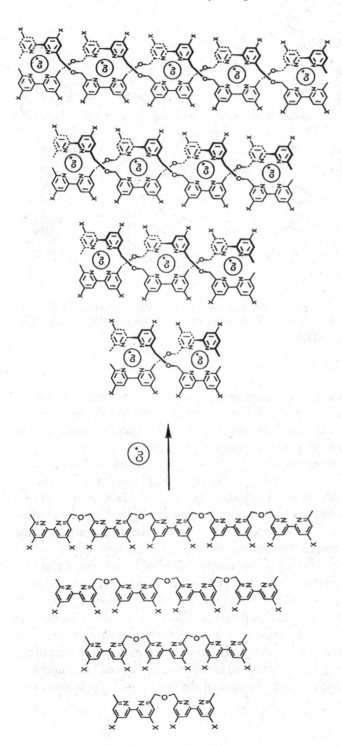

Fig. 48. Self-recognition in the self-assembly of the double helicates **132–135** (X = H) from a mixture of the oligobipyridine strands **128–131** (X = CONEt₂) and Cu(I) ions; BF₄⁻ or PF₆⁻ anions omitted [9.173].

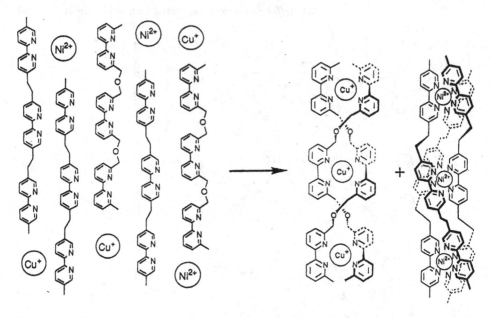

Fig. 49. Self-recognition in the self-asssembly of the double helicate **133** and the triple helicate **149** from a mixture of the oligobipyridine strands **129** and **148** and of Cu(I) and Ni(II) ions (ClO$_4^-$ anions omitted) [9.173].

In both experiments, the desired helicates are generated from a mixture of starting compounds by self-assembly with self-recognition; it involves the spontaneous selection and preferential binding of like metal ions by like ligand strands in a mixture to selectively assemble into the corresponding helicates.

Such self-recognition processes as well as the multicomponent self-assembly of the cylindrical molecular cage **156** [9.93] (see Section 9.3.2) involve the interplay of three structural factors and two thermodynamic factors. The first comprise: (1) the structural features of the ligands (nature, number and arrangement of the binding subunits; nature and position of the spacers); (2) the coordination geometries of the metal ions; (3) the steric and conformational effects within the different assembled species resulting from the various possible combinations of ligands and metal ions in a given mixture. The two thermodynamic factors are: (1) the energy-related principle of "maximal site occupancy", which implies that the system evolves towards the species or the mixture of species that presents highest occupancy of the binding sites available on both the ligand and the ions; it corresponds to the formation of the highest number of coordination bonds and therefore to the more stable state of the system; full site occupancy (site saturation) is achieved in "closed" architectures; (2) the entropy factor, which favours the state of the system with the largest number of product species.

These considerations also apply to systems where binding involves interactions other than metal coordination, such as hydrogen bonding or donor–acceptor forces. Such is the case for the chiral selection occurring in the course of the self-assembly of homochiral helical strands (Section 9.4.2) and ribbons (Section 9.4.4) through hydrogen bonding.

The self-recognition processes considered here belong to the realm of programmed supramolecular systems. They may be considered to result from the information processing that takes place in programmed systems based on instructed components operating through specific algorithms following given codes. The side-by-side formation of double- and triple-helical species is the consequence of the simultaneous operation of two different programs with no interference or crossover.

In a broader perspective, these results point to the emergence of a new outlook involving a change in paradigm, from "pure compounds" to "instructed mixtures", from "unicity" (pure substance) to "multiplicity + information" (mixture of instructed components and programme). Rather than pursuing mere chemical purity of a compound or a material, one would seek the design of instructed components that, as mixtures, would lead through self-processes to the spontaneous and selective formation of the desired (functional) superstructures. This may recall the build-up of complex species (displaying highly integrated functions) that takes place side-by-side in the self-assembly of the machinery of the living cell.

Along this line of thought, selective molecular and supramolecular reagents and catalysts may be considered to function as "instructed" reactive species possessing the ability to perform a transformation on a given substrate in a mixture of compounds. Similar considerations hold for transport processes.

One may venture to predict that this *instructed mixture paradigm* will define a major line of development of chemistry in the years to come: the spontaneous but controlled build-up of structurally organized and functionally integrated supramolecular systems from a preexisting "soup" of instructed components following well-defined programmes and interactional algorithms.

Encoded Combinatorial Libraries. Molecular Diversity Methods. Such an evolution might be brought into parallel with the recent development, via procedures of both chemical synthesis and molecular biology, of powerful *molecular diversity methods* that combine the generation of large repertoires of molecules with efficient selection procedures to obtain encoded combinatorial libraries from which one may retrieve compounds presenting specific properties [9.174–9.176], in particular recognition features (such as ATP [9.176a] or tripeptide binding [9.176b] or also base-pair sequence recognition in double-helical nucleic acids) for drug and effector design. The techniques of amplification by replication used in these methods would bear relation to the spontaneous generation of the target superstructures by the operation of self-processes.

In this respect, the very active development of the chemistry of molecular recognition makes the chemist able to design artificial receptor molecules for the selective binding and retrieval of a given substrate in a mixture of many different molecules. Any selective receptor designed for a given substrate operates in this way, as, to give just two examples, in the case of the selective extraction of an alkali cation from a mixture of several cations by a cryptand or of the selective binding of a given diammonium ion among a collection of them by a macrotricyclic ligand (see Sections 2.3 and 4.2).

Rather than striving to design as accurately as possible an effector for selectively operating on a given receptor, the "diversity" methods rely on evaluating the recognition features and activities of libraries of substrates. The goal is thus not targeted design but exhaustive coverage of effector structures by the screening of vast collections of compounds. The challenges lie in the development (1) of methods for diversity generation exploring as completely as possible size, shape and interaction space, (2) of efficient screening strategies, (3) of highly sensitive analytical techniques for separation and characterization, (4) of means for encoding via attachment of a specific label or tag (for instance of oligonucleotide type); (5) and of amplification procedures for multiplying the retrieved substance(s). Diversity techniques represent a powerful means for optimizing the efficiency of screening for drug discovery. They may also reveal unexpected activities when compounds bind to (allosteric regulation) sites in receptors that are not used by the natural effectors.

These property-oriented "discovery" technologies are specifically supramolecular since they are based on interactions between species. They present common features with procedures such as affinity chromatography, vaccine discovery and isolation by immunoprecipitation or induction of microbiological resistance. In their full expression, they involve three main steps: diversity generation, effector selection and identification through encoding.

One may imagine extending this type of methodology to reactivity, catalysis and transport by generating suitable libraries for the discovery of novel synthetic reagents, reactions, catalysts [9.176c] and carriers as well as for the exploration of product preparation through supramolecular assistance to synthesis (see Section 9.6).

9.6 Supramolecular Synthesis, Assistance and Replication

The contribution of supramolecular chemistry to chemical synthesis has two main aspects: the production of the non-covalent supramolecular species themselves, that is directly expressed in self-assembly processes, and the use of supramolecular features to assist in the synthesis of covalent molecular structures.

9.6.1 Supramolecular Synthesis

Supramolecular, non-covalent, synthesis consists in the generation of supramolecular architectures through the designed assembly of molecular components directed by the physico–chemical features of intermolecular forces; like molecular, covalent, synthesis, it requires strategy, planning and control.

Supramolecular synthesis thus comprises two stages: (1) the synthesis of the instructed molecular components by the formation of strong, kinetically non-labile covalent bonds; (2) the generation of the supramolecular entities through the spontaneous association of these components in a pre-determined arrangement by means of comparatively weak and kinetically labile non-covalent interactions following an intermolecular plan [1.1, 1.29], which may involve a given sequence and hierarchy of steps (see Section 9.2). It implies the conception of components containing already at the start the supramolecular project through a built-in programme. It strives to gain control over intermolecular bonds and events within a supramolecular strategy, in much the same way as molecular synthesis strives to control covalent bond formation.

In the realm of synthetic chemistry, supramolecular synthesis thus pursues similar endeavours in planning and control on the intramolecular level as molecular synthesis does on the intramolecular one; since it also requires the correct storage of an intermolecular project into a covalent framework, it necessitates expertise at both levels.

As pointed out in Chapter 1, chemistry has just entered the confines of the supramolecular world and great feats in synthetic power for the elaboration of ever more complex non-covalent architectures lie ahead, just as there has been a rich and exciting but long way of tribulations from the synthesis of urea in 1828 to those of vitamin B_{12} [1.3, 1.4], palytoxin [9.177a], or calicheamicin [9.177b], to cite just a few major achievements in recent times [9.177c].

9.6.2 Supramolecular Assistance to Synthesis

Supramolecular association may be put into action as an aid to the synthesis of complex covalent species, for positioning the components by templating [9.13] and self-assembly, so that subsequent reactions, deliberately performed on the preassembled species or occurring spontaneously within them, will lead to the generation of the desired architecture; dismantling of the intermolecular connection(s) may follow, thus liberating the covalently linked structure. This amounts to *supramolecular assistance to synthesis*, which may in particular become *self-replication* if spontaneous reproduction of one of the initial species takes place by binding, positioning and condensation of its parts by itself.

Intermolecular interactions and metal coordination have in particular been used very effectively for the synthesis of novel organic and inorganic entities such as *ro-*

taxanes, *catenanes* and *knots*, that would be of difficult access otherwise, leading in particular to the development of a *topological chemistry* [8.280–8.282, 9.76, 9.178].

Inorganic templating and *self-assembly* provide coordination compounds whose geometries make possible the synthesis of complex structures, namely of cyclic multiporphyrin arrays [9.13a, 9.179], of inorganic rotaxanes [9.97a, 9.180], of multi-catenates and catenands (see **181**) [8.281, 8.282] and even of molecular knots (see **182**) [8.282, 9.77, 9.181] (in **181** and **182**: a) with, b) without Cu(I) template).

Organic templating and *self-assembly* may similarly be used to position partners for subsequent reaction, amounting to a template-assisted organic synthesis that may involve *self-templating* effects [8.283, 9.182]. In an early example, the dimerization through hydrogen bonding of a 2-pyridone containing remote double bonds was shown to increase the yields of photocyclization [9.183a]. The encapsulated guest

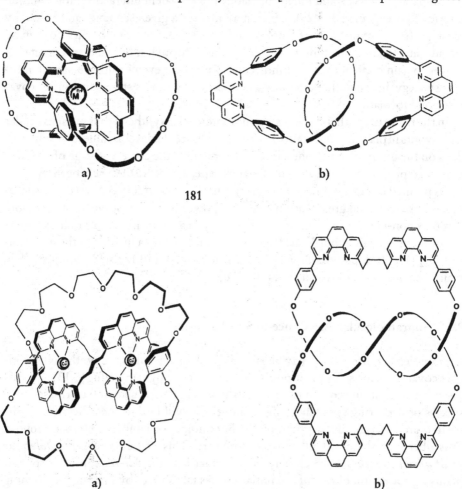

a)   b)

181

a)   b)

182

strongly affects the efficiency of carceplex formation [9.183b]. Rotaxanes and multi-catenanes (see **183**) [8.284] have been synthesized based on the spontaneous threading of donor–acceptor partners [8.283–8.287]. Similarly, making use of the substrate inclusion features of cyclodextrins for inserting linear components through their cavity, catenated cyclodextrins were obtained [9.184] as well as rotaxanes and poly-rotaxanes of cyclodextrins [9.185–9.190]. Rotaxanes [9.191] and catenanes [9.192] based on other macrocyclic systems have also been prepared using in particular H-bonding assistance. The threading of several cyclodextrin units on a polymer chain afforded "molecular necklace" polyrotaxanes that could be connected into nano-tubes by covalently condensing the cyclodextrin units [9.189, 9.193]. The synthesis of such mechanically interlocked superstructures is entirely dependent on the pre-formation by self-assembly of a threaded supramolecular species.

183

Inorganic [9.194a] or organic templating has been used for the generation by cross-linking polymerization, of polymers displaying molecular recognition through imprinting (see also Section 7.1). Polymolecular assemblies provide another effective means for the confinement and organization of reactive species as in the case of polymerization in liquid crystals and vesicles (for instance [7.9–7.13]). Similar processes may occur in self-assembled strands such as **178** or **179** bearing reactive groups in the side chains (see Section 9.6.3) [9.194b]. Finally, the control of the *topo-chemistry* of reactions in the solid state [9.195] may make use of appropriate posi-tioning of the reactive partners through recognition effects.

One may expect that by taking advantage of suitably designed supramolecular features it will be possible to generate a variety of highly complex architectures that would not be accessible otherwise (or only with low efficiency). Such supramolec-ular assistance adds a new direction with powerful means to organic synthesis.

9.6.3 Replication. Self-Replication

The template-directed self-assembly of species containing reactive groups provides pathways towards systems displaying copying, information transfer and replication. Condensation of the components within the self-assembled entity will generate a predetermined covalent structure. A special case is the reproduction of the template itself by self-replication. Replicating molecules are of interest both chemically and biologically in view of their relation to the origin of life [9.196-9.198]. Reactions occurring in organized media (molecular layers, mesophases, vesicles) [7.9–7.13, 7.35] offer an entry into the field and molecular imprinting processes (see Section 7.1) represent a way of copying the information required for recognition of the template.

Template-directed polymerization present features of replication processes [9.196]. If one of the components in self-assembled strands such as **178** or **179** were to contain reactive groups capable of undergoing condensation with each other spontaneously or by use of an external reagent, the covalently linked linear molecules formed would represent a copy of the other strand. This may be envisaged, for instance, with recognition groups (such as the TAP groups in Figure 46) fitted with side chains bearing functional groups that could undergo cross-linking polymerization or condensation; thus peptide bond formation would yield a peptide analogue bearing recognition groups for ordered assembly [9.194b]. Structures such as **178** and **179**, and the processes for which they may serve as backbone, are reminiscent of aspects of information transfer and replication (through a "negative", complementary imprint) found in biological systems.

Self-replication takes place when a molecule catalyses its own formation by acting as template for the constituents, which react to generate a copy of the template. Such systems display autocatalysis and may be termed informational or non-informational depending on whether or not replication involves the conservation of a sequence of information [9.196]. A problem is the occurrence of product inhibition when the dimer of the template, formed after the first condensation round, is too stable to be easily dissociated by the incoming components for a new cycle.

Several minimal self-replicating systems have been developed in which the template is generated from two components. The first one consisted in the replication of a self-complementary or palindromic hexanucleotide CCGCGG from two trinucleotides CCG and CGG in the presence of a condensing agent [9.197, 9.199]. Nucleotide analogues were also used [9.200, 9.201]. The more recent ones involve (1) the formation of an amide bond between two building blocks undergoing selective hydrogen bonding with the template (see **184**, [9.202]), (2) an amine + aldehyde to imine condensation between components interacting with the template via ion-pairing between an amidinium cation and a carboxylate anion (see **185**, [9.203]. Self-replication of oligonucleotides in reverse micelles has been reported [9.204].

184   185

In self-replicating systems employing three starting constituents competition between constituents can occur [9.205]. Such processes are on the way to systems displaying information transfer, whereas the two-components ones are non-informational. A shift from parabolic kinetics to exponential growth of the template concentration is required for a selection process to take place [9.197]. The evidence for self-replication on the basis of template-directed autocatalysis as in **184** requires detailed mechanistic investigation on the origin of the catalytic effects observed [9.206].

Replicating micelles have been realized by producing within the micelle, through a chemical (or enzymatic) reaction, the same surfactant as that constituting the micelle, which therefore grows and redistributes the surfactant by dividing into new micelles [9.207].

The investigation of self-replicating systems will be subject to increasing activity. For instance, it would be interesting to achieve self-replication of the double-helical metal complexes, the helicates (see Section 9.3.1 [9.208]). Also, spontaneous generation of the building blocks to allow growth, exponential expansion, information transfer, evolution and confinement in vesicles, represent important goals, for instance, towards artificial cells (see also [7.51]) or virus-like species. The information processing that operates in molecular recognition and self-assembly will play a major role, coupled to chemical reactions for connecting the building blocks. There is after all the fascination of realizing artificial, even if still very primitive, systems mimicking life-like behaviour.

9.7 Supramolecular Chirality and Self-Assembly

Self-assembly, self-recognition and replication may involve chiral components, as discussed on several instances above. This leads to some more general considerations about the role of molecular chirality in supramolecular species.

Chirality is expressed on both the molecular and the supramolecular levels. Like a molecule, a supermolecule may exist in enantiomeric or diastereomeric forms. Supramolecular chirality results both from the properties of the components and from the way in which they associate.

Thus, a supermolecule may be chiral either (1) in the simplest case, because at least one component is asymmetric, or (2) more interestingly, when the interaction between achiral components is disymmetrizing, yielding a chiral association, as may occur in crystal growth [7.39, 9.209]. The former occurs in the binding of enantiomeric receptors and substrates and in the chiral discrimination between diastereomeric supermolecules (see Section 2.5). The latter takes place when the symmetry element present in each component is destroyed in the association, e.g. when the planes of symmetry of the components are perpendicular in the supermolecule. This is, for instance, the case in the chiral two-component species 186 formed by association of an achiral imide 186a with an achiral aminoadenine derivative 186b. It

186a

186b

186

raises the intriguing possibility of a spontaneous resolution in the solid state by crystallization of homochiral supermolecules **186**. Such processes of generation of chirality from achiral components may lead to extended chiral assemblies and have intriguing implications with respect to the emergence of optical activity on earth and in biological systems (see also [7.39b].

On the other hand, the association of two enantiomeric components yields an achiral supermolecule of *meso* type; this may occur through a symmetrical bridging molecule in order to take care of the identical interaction sites at symmetry-related positions, as shown in the three-component supermolecule **187a**; the corresponding diastereomeric DL pair may also be obtained (see **187b**). Thus, achiral components can associate into a chiral supermolecule and chiral components can give an achiral supermolecule.

187a   187b

Molecular chirality also affects the way in which self-assembly from chiral components occurs and the nature of the resulting supermolecular architecture. Three cases may be distinguished:

(1)   *asymmetric induction* in self-assembly of a chiral structure, as realized in the high induction of helicity in the helicate **147** from ligand strands containing asymmetric centres [9.68];

(2)   *enantioselective self-assembly*, i.e., *self-resolution*, when two homochiral supermolecules are formed from a mixture of enantiomeric components by spontaneous selection of components of the same chirality; this is the case in the formation of the triple helical strands **174** of opposite helicity [9.152] (see Section 9.4.2) and of the homochiral supramolecular ribbons **180** [9.169a] from a racemic mixture of the components; in the solid state, a number of enantiomorphous crystals are derived from racemic compounds as well as from achiral molecules held in a chiral conformation [9.209]; the formation of such species involves discriminating interactions between enantiomeric species [9.210a];

(3)   *chirality directed self-assembly*, in which the architecture of the supramolecular species depends on the chiral features of the components, differ-

188

189

190

ent superstructures being generated from enantiomerically pure components and from racemic mixtures; this is a particularly interesting process,
since it amounts to chirality control over supramolecular entities; thus,
self-assembly of the chiral component **188** through formation of four
hydrogen bonds can in principle yield either a homochiral ring architecture **189** or a heterochiral strand **190** [9.210b].

Finally, one may note that catenated and knotted structures display *topological
chirality* [8.280, 8.282c, 9.76, 9.178].

9.8 Supramolecular Materials. Nanochemistry

The interactions between their constituents have of course always been implicitly present and operating in the determination of the structure and properties of materials. Based on the studies of oligo-associates, supermolecules, and of polymolecular organized assemblies, there emerges the ability to *explicitly* take into account and manipulate the non-covalent forces that hold the components of a material together. These interactions and the recognition processes that they underlie, allow the design of materials and the programming of their build-up, *"Aufbau"*, from suitable units by self-assembly of the final architecture. The collective physico–chemical features (electronic, optical, dynamical, mechanical, etc.) are determined by, and may be steered through, both the properties of the individual components and the interactions between them.

Thus, recognition-directed association, self-assembly and self-organization open up new perspectives in materials chemistry towards an area of supramolecular materials whose features depend on molecular information. *Supramolecular engineering* gives access to the controlled generation of well-defined polymolecular architectures and patterns in molecular layers, films, membranes, micelles, gels, mesophases and solids as well as in large inorganic entities, polymetallic coordination architectures and coordination polymers (see also Chapter 7).

Inorganic supramolecular materials and *composites* possessing defined architetures [7.18, 7.28, 7.33, 9.211, 9.212] become accessible through controlled synthesis via strategies employing recognition features and mild reactions, thus opening the way to a "soft" inorganic materials chemistry [9.213]. The engineering of solid state materials may generate either structurally molecular (with boundaries between the units) or non-molecular (extended, without boundaries) compounds. With nanosize structures, novel properties may be obtained, as is the case, for instance, for quantum confinement in nanoparticulate metals and semiconductors [9.214]. Self-assembly of inorganic architectures based on organometallic building blocks [9.215] yields various types of frameworks such as Sb [9.216] and Te [9.217] chains, polymeric chains of metal complexes [9.218–9.220], honeycomb [9.221] and diamond [9.222] arrays, infinite scaffoldings based on tetrahedral and octahedral [9.223], square-planar [9.224] or chiral [9.225] coordination centres, on porphyrin groups [9.226] or on metallo-cyclic units [9.227], frameworks of metal chalcogenides [9.214, 9.228] possessing a double-helical structure [9.229]; attractive three-dimensional networks of fully interlocked supramolecular rings formed from components of either inorganic (Mn(II)Cu(II) based metallocycles [9.230]) or organic (hydrogen-bonded trimesic acid [9.116b] and adamantane tetracarbocylic [9.231] units) nature. Bioinorganic materials and biominerals [9.211, 9.232] may be obtained using supramolecular assemblies as support, as in the case of the synthesis of nanosize inorganic particles in protein cages [9.232].

Solid-state inorganic materials present tunnels, cages and micropores (see also [7.34]), whose size and shape may be tailored by the choice of the components and of the connecting centres and that serve as (selective) hosts for various species. Tetraphenylporphyrins and their metal complexes act as "sponges", forming programmable lattice clathrates that include selected guest molecules [9.233]. A special case is the fabrication of nanowires and related entities on the basis of carbon nanotubes [9.9b, 9.234]. Liquid-crystalline complexes, metallomesogens [9.143], represent another important category of supramolecular inorganic materials capable of spontaneous long range organization.

Organic supramolecular materials may be devised on the basis of molecular components of various structures bearing recognition units [9.149, 9.235]. As shown above, liquid crystals and liquid crystalline polymers of supramolecular nature presenting various supramolecular textures are generated by the self-assembly of complementary subunits.

The role of interactions and recognition processes in the supramolecular engineering and properties of materials based on polymolecular assemblies has already been mentioned (Chapter 7; Section 9.4). Organic nanotubes are formed by self-assembly of a suitably designed cyclic peptide unit [8.186]. Molecular recognition is also displayed by imprinted polymers [7.34, 7.35, 9.236] as well as polymer membranes [9.237] and vesicles [7.51–7.53, 9.238] bearing recognition groups. Polymerizable components organized in assemblies allow the synthesis of various polymerized structures [7.9–7.13], for instance of two-dimensional polymers [9.239]. Gelation [7.8c,d, 9.240] results from the formation of large three-dimensional networks involving interactions such as hydrogen bonding [9.241, 9.242a] and may be caused by molecular recognition induced polyassociation of small molecules (see the TAP–BA pair; Figure 46) [9.242b, 9.243]. Networks formed by self-assembled actin filaments [9.244a] and lipid–protein interactions [9.244b] play a basic role in the physico–chemical behaviour of cells and vesicles.

Crystallization amounts by nature to the self-assembly of very large, boundaryless supramolecular species. Its control is a goal of major importance in order to be able to generate solid-state materials of specific structural and physical properties (see also Sections 7.1, 7.2, 9.4.4; [7.39–7.42, 9.105, 9.245]). Supramolecular effects play a crucial role. Directional growth of materials may be induced by a template and involve molecular recognition [9.246], occurring by epitaxy [9.247] or on oriented thin films [9.248].

Molecular recognition-directed processes represent a powerful entry into *supramolecular solid state chemistry* and *crystal engineering* [9.101–9.103, 9.249, 9.250]. The ability to control the way in which molecules associate allows the designed generation of supramolecular architectures in the solid state. Modification of surfaces with recognition units leads to extended exoreceptors (Section 7.2) displaying selective surface binding on the microscopic level, and to *recognition-controlled adhesion* on the macroscopic scale, pointing thus to the potential of putting supramolecular effects to use in adhesion science [9.251]. Components derived from biological structures may yield a variety of *biomaterials* [9.211, 9.252a] of basic and applied interest such as *biomesogens*, i.e. liquid crystals derived from biological molecules [9.109, 9.112, 9.115, 9.146], *biominerals* [9.211, 9.232, 9.246], nanosize architectures based on nucleic acid frameworks [9.252b] or protein arrays [9.252c].

Molecular recognition events provide a means of performing programmed *materials engineering and processing* (of biomimetic [9.211, 9.252] or abiotic type); they lead to *self-assembling nanostructures*, organized and functional species of nanometric dimensions that define a *supramolecular nanochemistry* and bridge the gap between microscopic molecular events and macroscopic features [1.9, 7.18, 9.3, 9.254]. Developments towards high-resolution imaging through spontaneous assembly may be envisaged [9.253]. By increasing the size of its entities, nanochemistry works its way upward towards microlithography and microphysical engineering, which, by further and further miniaturization, strive to produce ever smaller elements.

"Intelligent", functional supramolecular materials, network engineering and polymolecular patterning are the subject of increasing activity in chemical research. The development of advanced materials may take full advantage of the control provided by information-dependent supramolecular processes for the production of large scale architectures in a sort of molecular and supramolecular *tectonics* [9.211] leading to a *nanotechnology* and *nanomaterials* of organic or inorganic nature [1.9, 7.18, 8.278, 9.3, 9.254]. It is important to note that technologies resorting to self-organization processes should in principle be able to bypass microfabrication procedures by making use of the spontaneous formation of the desired superstructures and devices from suitably instructed and functional building blocks. There is indeed a rich palette of structures and properties to be generated by blending supramolecular chemistry with materials science!

9.9   Chemionics

Sets of instructed components endowed with photoactive, electroactive, ionoactive or switching features constitute programmed systems capable of generating functional (photonic, electronic, ionic) supramolecular devices by recognition-directed self-assembly into well defined architectures, patterns and networks possessing novel optical, electrical, ionic, etc. properties (Figure 50). Such components and devices define areas of molecular and supramolecular photonics, electronics and ionics belonging to an intriguing and rather futuristic field of chemistry that may be termed "*chemionics*" [1.1, 1.7, 1.9], the design and operation of programmed systems for information and signal handling at the molecular and supramolecular levels.

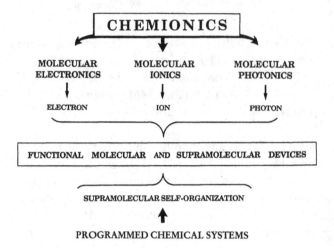

Fig. 50. Chemionics as the chemistry of recognition-directed and self-organised photonic, electronic and ionic molecular and supramolecular devices generated by means of functional programmed chemical systems.

Components and molecular devices such as molecular wires, channels, resistors, rectifiers, diodes, and photosensitive elements might be assembled into nanocircuits and combined with organised polymolecular assemblies to yield systems capable ultimately of performing functions of storage, detection, processing, amplification, and transfer of signals and information by means of various mediators (photons, electrons, protons, metal cations, anions, molecules) with coupling and regulation. Such perspectives lie, of course, in a distant future, but along the way they may be expected to yield numerous spin-offs and they do represent ultimate goals towards which work may already be planned and realized.

The reading of molecular information and the operation of molecular devices require ways and means of *addressing* molecular and supramolecular species. Despite the difficult problems that one may apprehend, encouraging and exciting developments may be anticipated. Indeed, scanning tunneling microscopy (STM) and atomic force microscopy (AFM) are providing extraordinary manipulative power at the atomic and molecular scale [9.255–9.259]; they offer procedures for handling both structural and interactional features and, beyond, they open ways towards addressing molecular functionality. Thus, a variety of processes have been realized in recent years, such as the following: single atom [9.260, 9.261] or molecule [9.262] positioning; tunnel diode effects on the atomic scale [9.263, 9.264]; optical absorption [9.265] or emission [9.266] at molecular resolution; operation of an atomic switch [8.295]; combination of the spatial resolution of STM with the time resolution of ultrafast optics for the investigation of dynamic phenomena on the atomic scale [9.267]; nanometer-scale reversible recording [9.268]; current rectification [9.269]; surface investigations of hydrophobic forces [9.270], of adhesion [9.271, 9.272] of acid-base properties [9.273] and of electrochemical effects in particular with metal complexes [9.274–9.276]; molecular recognition interactions [9.277], in particular between complementary nucleosides [9.277b] and oligonucleotides [9.277c], with potential application to DNA base sequencing [9.277d]; a light source smaller than the optical wavelength has been investigated [9.278]. These powerful methods and their offspring will play a major role in the development of the various aspects of chemionics.

As already pointed out (end of Chapter 8), the construction of multidevice arrays will require the development of appropriate experimental procedures and theoretical methods for the addressing, detection and response analysis of multicomponents–multifunction systems.

Atomic and molecular manipulation by means of STM or AFM represents the ultimate in size reduction for the fabrication and operation of nanodevices. Thus, in terms of size, *"there's plenty of room at the bottom"* as the celebrated aphorism of Richard Feynman goes [9.279]. However, the key word of supramolecular chemistry is not size but information and its route is towards complexity. Indeed, considering the ability of supramolecular species to spontaneously form from their components, bypassing microfabrication, and to accomplish intricate tasks on the basis of the encoded information and instructions, thus reaching higher levels of organization and behaviour, it is clear that through supramolecular chemistry *"there's even more room at the top"*!

10 Perspectives

10.1 From Structure to Information. The Challenge of Instructed Chemistry

In chemistry, like in other areas, the language of information is extending that of constitution and structure as the field develops towards more and more complex architecture and behaviour. And supramolecular chemistry is paving the way towards comprehending chemistry as an *information science* (see Sects. 2.1, 8.1, 9.2). In the one hundred years since 1894, molecular recognition has evolved from Emil Fischer's "Lock and Key" image [1.11] of the age of mechanics towards the *information paradigm* of the age of electronics and communication. This change in paradigm will profoundly influence our perception of chemistry, how we think about it, how we perform it. Instructed chemistry extends from selectivity in the synthesis and reactivity of molecular structures to the organization and function of complex supramolecular entities. The latter rely on sets of instructed components capable of performing on mixtures specific operations that will lead to the desired substances and properties by the action of built-in self-processes.

Supramolecular chemistry has started and developed as defined by its basic object, the chemistry of species generated by non-covalent interactions. Through recognition and self-processes it has led to the concepts of (passive and active) information and of programmed systems, becoming progressively the chemistry of molecular information, its storage at the molecular level, its retrieval, transfer and processing at the supramolecular level.

The outlook of supramolecular chemistry is toward a general *science of informed matter*, bringing forward in chemistry the third component of the basic trilogy matter–energy–information (see also [10.1]).

Chemical systems may store information either in an *analog* fashion, in the structural features (size, shape, nature and disposition of interaction sites, etc. [1.27]) of a molecule or a supermolecule, or in a *digital* fashion, in the various states or connectivities of a chemical entity. Information theory has been applied to the description of the features of molecular machines [10.2]. The evaluation of the information content of a recognition process based on structural sensing in receptor–substrate pairs requires an assessment of the relevant molecular characteristics. Recognition is not an absolute but a relative notion. It results from the structural (and eventually dynamical) information stored in the partners and is defined by the fidelity of its reading, which rests on the difference in free energy of interaction between states, represented by different receptor–substrate combinations (see also Sect. 2.1). It is thus not a yes/no process but is relative to a threshold level separating states and making them distinct. It depends on free energy and consequently on temperature. The parameter kT could be a possible reference quantity against which to evaluate threshold values, differences between states and reading accuracy. Both analogical and digital processing of chemical information depend on such factors.

The decrease in entropy implied in information storage in a covalent structure is (over)compensated by the entropy increase occurring in the course of the stepwise synthesis of the "informed" molecule in question.

Digital storage and retrieval of chemical information is found in the nucleic acids, where the basic digital operation is a two state 2/3 process (2 versus 3 hydrogen bonds in A:T and G:C base pairs, respectively) corresponding to the usual 0/1 commutation of electronic computers [10.3]. It may also be envisaged for multisite receptors or multiredox systems possessing distinct states of site occupation or of oxidation, as, for instance in calcium binding to the four sites of calmodulin [9.19] or in redox modification of the four porphyrin groups of cytochrome c_3 [8.137] (see Sect. 9.2). The possibility of information bit storage has been envisaged for donor (D)–acceptor (A) systems, for instance, in DAD and ADA architectures [10.4].

A system of intriguing potential is represented by the inorganic two-dimensional grids, such as **166, 167** (see Section 9.3.3) and related entities. One may note their resemblance to grids based on quantum dots [10.5] that are of much interest in microelectronics. They may be considered to consist of *ion dots* of still smaller size than quantum dots and that do not necessitate microfabrication but form spontaneously by self-assembly. Such architectures may foreshadow multistate digital supramolecular chips for information storage in and retrieval from inscribed patterns that might be addressable by light or electrically. Different states could, in principle, be characterized either by different local features at a given x, y coordinate, in ion dot fashion, or by specific overall optical or oxidation levels. Should adjoining inscribed geometrical or electronic patterns (for instance, n and $n \pm 1$ oxidation levels) correspond to similar information values, characteristics of interest for fuzzy logic [8.301] approaches would result. The intersection $A \cap B$ between two sets A and B corresponding to values $n_A \pm x$ and $n_B \pm x$ of a given parameter n (such as the oxidation state) could represent a *fuzzy set* where the response of the system would not be fully determined.

Addressing such grid-like structures may rely on techniques of STM or AFM type (see Section 9.9). Inducing ± 1 redox changes at specific locations in a single unit would then correspond to a sort of *single electronics* [8.117] at ion dots. In this context one may also note that an elongated electron-accepting conjugated molecule could represent a type of quantum well [10.6, 10.7] into which electrons may be introduced one by one.

The grid type arrangements also pose the intriguing question of performing matrix algebra operations on these inorganic superstructures. Extension into three dimensions through stacks of grids would lead to layered arrays in foliated spaces.

Molecular and supramolecular devices incorporated into ultra-micro circuits represent potential hardware components of eventual systems that might qualify as molecular computers, whose highly integrated architecture and operation would involve parallel rather than sequential processing [8.100, 8.102, 10.2, 10.8, 10.9]. Such species may form by the self-assembly of suitably instructed subunits so that computing via self-assembly may be envisaged [10.8b, 10.9]. On the biological side, the fabrication of components for sensory and motor protheses could be considered.

As already noted (Sect. 9.1), entities resulting from self-assembly and self-organization of a number of components may undergo self-correction and adaptation. Such features might also explain why large multisite protein architectures are formed by the association of several smaller protein subunits rather than from a single, long polypeptide [10.10].

Beyond programmed systems (Section 9.2) the next step in complexity consists in the design of chemical *"learning" systems*, systems that are not just instructed but can be trained, that possess *self-modification* ability and *adaptability* in response to external stimuli. This opens perspectives towards systems that would undergo *evolution*, i.e., progressive change of internal structure under the pressure of environmental factors. It implies also the passage from closed systems to *open systems* that are connected spatially and temporally to their surroundings.

10.2 Steps Towards Complexity

The progression from elementary particles to the nucleus, the atom, the molecule, the supermolecule and the supramolecular assembly represents steps up the ladder of complexity. Particles interact to form atoms, atoms to form molecules, molecules to form supermolecules and supramolecular assemblies, etc. At each level novel features appear that did not exist at a lower one. Thus a major line of development of chemistry is towards complex systems and the emergence of complexity.

Very active research has been devoted to the development of complexity measures that would allow the quantitative characterization of a complex system [10.11]. In the present context, complexity is not just described by the number of states, the multiplicity of a system, as defined in information science, or by the characteristics of the graphs representing a molecule or an assembly of molecules [10.12], or by structural complexity [10.13]. Complexity implies and results from multiple components *and* interactions between them with integration i.e., long range correlation, coupling and feedback. It is interaction between components that makes the whole more than the sum of the parts and leads to collective properties. Thus, the complexity of an organised system involves three basic features:

$$\text{Complexity} = (\text{Multiplicity})(\text{Interaction})(\text{Integration}) = MI_2$$

The species and properties defining a given level of complexity result from, and may be explained on the basis of, the species belonging to the level below and of their multibody interaction [10.14], e.g., supramolecular entities in terms of molecules, cells in terms of supramolecular entities, tissues in terms of cells, organisms in terms of tissues and so on up to the complexity of behaviour of societies and ecosystems [10.11b, 10.15]. For example, in the self-assembly of a virus shell local information in the subunits is sufficient to "tell" the proteins where to bind in order to generate the final polyproteinic association [10.16], thus going up a step in complexity from the molecular unit to the supramolecular architecture. Ultimately, one will have to go ever deeper and wider so as to link the structures and functions from the atom to the organism, along a hierarchy of levels defining the architecture of complexity [10.17].

The novel features that appear at each level of complexity and characterize it, do not and even *cannot* conceptually *exist* at the level below but may be explained in terms of MI_2, from the simplest particle to the highly complex multibody, multi-interactive societies of living organisms. Such an attitude is not reductionist, it is not a reduction of a level to the lower level(s) but an *integration*, connecting one level to the others by integrating species and interactions to describe and explain increasing complexity of behaviour (see also [10.17, 10.18]).

A simple but telling illustration is for instance the boiling point of a liquid. A single molecule of water has no boiling point, the concept of boiling itself does not, cannot even, exist for it. Only for a population of interacting water molecules is there such a thing as a boiling point, or a freezing point, or any other *collective property*.

A corollary is the question of how many individuals it takes to form a collectivity and to display collective properties: how many molecules of water to have a boiling point, how many atoms to form a metal, how many components to display a phase transition? Or, how do boiling point, metallic behaviour, phase transition, etc., depend on and vary with the number of components and the nature of their interaction(s)? In principle, any finite number of components leads to a collective behaviour that is only an approximation, however close it may well be, an asymptotic approach to the "true" value of a given property for an infinite number of units.

The path from the simple to the complex in behavioural space corresponds to that from the single to the collective and from the individual to the society in population space. A sum of individuals becomes a collectivity when there are interactions and at each level novel interactions appear leading to higher complexity. With respect to molecular chemistry, we come back to considering supramolecular chemistry as a sort of molecular sociology (see Chapter 1).

One may note that there are instances in which a given object directly influences, if not determines, the behaviour of a large organism of much higher complexity, as in the case of psychotropes, where molecules directly affect the psychical state. These may be considered as *"complexity-shunt"* processes due to interactions and connections that (apparently) bypass intermediate levels of complexity. Another aspect resides in the ways and means to control the behaviour of complex systems [10.19].

The global features of condensed matter may be described by phenomenological physical laws. An understanding of these macroscopic events will ultimately require their explanation in terms of the underlying molecular and supramolecular features, i.e., in terms of the chemical nature of the microscopic components and of their interactions. For instance, how is viscosity or a phase change related to the constituting molecules? How do individual partners synergetically [9.8d] cooperate to produce macroscopic spatial, temporal or functional features, inducing a transition from chaos [9.7, 9.8, 10.20] to order through self-organization? How does turbulent flow vary with the type of multibody interactions between the particles? Or how is structuration in an energy flow determined by molecular features of the components and their supramolecular interactions?

There is here a very exciting and fundamental field of investigation for supramolecular chemistry concerning the emergence of order and complexity, the passage from the microscopic to the macroscopic, from the isolated to the collective, with the aim of providing an etiological explanation of the phenomenological description. It requires bridging the gap between and integration of the points of view of the physicist and chemical physicist on the one hand and of the structural and synthetic chemist on the other.

Thus, the horizon of supramolecular chemistry lies on the road towards complexity, from the single molecule towards collective properties of adaptive multibody systems of interacting components.

Thomas Mann has himself set the goal (!) when he wrote: *"Irgendwann musste die Teilung zu* "Einheiten" *führen, die zwar zusammengesetzt, aber noch nicht organisiert, zwischen Leben und Nichtleben vermittelten,* Molekülgruppen, *den Übergang bildend zwischen Lebensordnung und bloßer Chemie"* [10.21]. *("At some point the partition must have led to units, which, although assembled but not yet organized, mediated between life and non-life, groups of molecules, forming the transition between life order and mere chemistry" .)*

A challenge is set, that of showing that there is no such a thing as "just" chemistry, but that through control over supramolecular structures, functions and organization, a bridge is built and continuity is provided between the animate and the inanimate, between life and non-life.

10.3 Chemistry and Biology, Creativity and Art

The highest level of complexity is that expressed in that highest form of matter, living matter, life [10.22], which itself culminates in the brain, the plasticity of the neural system, epigenesis, consciousness and thought.

As has been apparent at various instances in the present text, chemistry and notably supramolecular chemistry entertain a double relationship with biology. Numerous studies are concerned with substances and processes of a biological or *biomimetic* nature. There has been a profound evolution by which the chemist appropriates and diverts the power of the natural chemical processes of biology to the goals of chemistry, for instance, in the use of enzymes as reagents, the generation of catalytic antibodies, the control of gene expression, the development of molecular diversity techniques, etc. Conversely, the scrutinization of biological processes by chemists has provided understanding on a precise molecular basis and ways for acting on them by means of suitably designed substances. Thus, the cultures of chemistry and biology [10.23] are intimately linked and coming closer and closer together.

On the other hand, the challenge for chemistry lies in the development of *abiotic*, non-natural systems, figments of the imagination of the chemist, displaying desired structural features and carrying out functions *other* than those present in biology with (at least) comparable efficiency and selectivity. Subject only to the constraints one chooses to impose and not to those of the living organism, abiotic chemistry is free to invent new substances and processes. The field of chemistry is indeed broader than that of the systems actually realized in nature.

In the preceding chapters, reference has often been made to biological molecules and processes, for both biomimetic and abiotic purposes. The emphasis on molecular information in supramolecular chemistry finds its response in the operation of biological systems. There is a "molecular logic of living organisms" [10.24].

The future path of chemistry will be shaped by both inside and outside forces (see, for instance, [9.177c, 10.25]). Its evolution towards increasing diversity and towards increasing complexity also takes biological phenomena as points of reference. The specificity of chemistry may be stressed by comparing biology and chemistry with respect to these two basic parameters, complexity and breadth or diversity. As presented in Figure 51, biology is of extreme complexity, however, the substances on which it is based belong to defined classes, and although tremendously rich are nevertheless limited in variety of types. Chemistry, on the other hand, is still of very low complexity compared with biology but its breadth, the diversity of its substances, is infinite, being limited only by the imagination of the chemist in endlessly combining and recomposing the basic bricks of chemical architectures, thus filling in the unlimited white area in the complexity–diversity diagram.

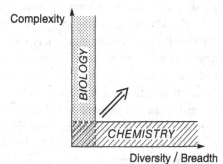

Fig. 51. A comparison between chemistry and biology with respect to the two parameters complexity and breadth or diversity.

The chemist finds illustration, inspiration and stimulation in natural processes, as well as confidence and reassurance since they are proof that such highly complex systems can indeed be achieved on the basis of molecular components. One might say that science, notably chemistry, relies on the biological world through an existence axiom: the mere fact that biological systems and, in particular, human beings exist, demonstrates the fantastic complexity of structure and function that the molecular world can present; it shows that such a complexity can indeed exist despite our *present* inability to understand how it operates and how it has come about. So to say, if we did not exist we would be unable to imagine ourselves! And the molecular world of biology is only one of all the possible worlds of the universe of chemistry, that await to be created at the hands of the chemist.

With respect to the *frontiers of life* itself three basic questions may be asked: How? Where? Why?

The first concerns the origin of life on earth as we know it, of our biological world. The second considers the possibility of extraterrestrial life, within or beyond the solar system. The third wonders why life has taken the form we know; it has as corollary the question whether other forms of life can (and do) exist: is there "artificial life"?; it also implies that one might try to set the stage and implement the steps that would allow, in a distant future, the creation of artificial forms of life.

Such an enterprise, which one cannot (and should not) at the present stage outline in detail except for the initial steps, rests on the presupposition that there may be more than one, several expressions of the processes characterizing life. It thus invites the exploration of the *"frontiers of other lifes"* and of the *chemical evolution* of living worlds.

Questions have been addressed about which one may speculate, let one's imagination wander, perhaps even set paths for future investigations. However, where the answers lie is not clear at present and future chemical research towards ever more complex systems will uncover new modes of thinking and new ways of acting that we at present do not know about and may even be unable to imagine.

The perspectives are definitely very (too?) wide and it will be necessary to distinguish the daring and visionary from the utopic and illusory! On the other hand, we may feel like progressing in a countryside of high mountains: the peaks, the goals are visible and identifiable or may become so as progress is made, but we do not yet know how to reach them. We may find landslides, rockfalls, deep crevices, tumultuous streams along the way, we may have to turn around and try again, but we must be confident that we will eventually get there. We will need the courage to match the risks, the persistence to fill in the abyss of our ignorances and the ambition to meet the challenges, remembering that *"Who sits at the bottom of a well to contemplate the sky, will find it small"* (*Han Yu*, 768–824).

But to chemistry, the skies are wide open, for if it is a science, it is also an art. By the beauty of its objects, of course, but also in its very essence, by its ability to invent the future and to endlessly recreate itself..

Like the artist, the chemist engraves into matter the products of creative imagination. The stone, the sounds, the words do not contain the works that the sculptor, the composer, the writer express from them [10.26]. Similarly, the chemist creates original molecules, new materials and novel properties from the elements provided by nature, indeed entire new worlds, that did not exist before they were shaped at the hands of the chemist, like matter is shaped by the hand of the artist, as so powerfully rendered by *Auguste Rodin* (see cover picture) [10.27].

Indeed chemistry possesses this creative power as stated by *Marcelin Berthelot*: *"La chimie crée son objet"*. (*"Chemistry creates its object"*) [10.28]. It does not merely fabricate objects, but creates its own object. It does not preexist, but is invented as progress is made. It is not just waiting to be discovered, but is to be created.

The essence of chemical science finds its full expression in the words of that epitome of the artist–scientist *Leonardo da Vinci*: *"... dove la natura finisce di produrre le sue spezie, l'uomo quivi comincia con le cose naturali, con l'aiutorio di essa natura, a creare infinite spezie ..."*. (*"Where nature finishes producing its own species, man begins, using natural things and with the help of this nature, to create an infinity of species ..."*.) [10.29].

The essence of chemistry is not only to discover but to invent and, above all, *to create*. The book of chemistry is not only to be read but to be written! The score of chemistry is not only to be played but to be composed!

Bibliography and Notes

Chapter 1

[1.1]   J.-M. Lehn, *Leçon Inaugurale*, Collège de France, Paris, 1980; *Le Débat*, 1982, *18* 46; *Interdisciplinary Science Rev.* 1985, *10*, 72.
[1.2]   F. Wöhler, *Poggendorfs Ann. Physik* 1828, *12*, 253.
[1.3]   R. B. Woodward, *Pure Appl. Chem.* 1968, *17*, 519.
[1.4]   A. Eschenmoser, *Quart. Rev.* 1970, 24, 366; *Chem. Soc. Rev.* 1976, *5*, 377; *Nova Acta Leopoldina* 1982, *55*, 5.
[1.5]   See the list of monographs and special series in Appendix A.
[1.6]   An attempt at a comprehensive description of the present status of the field, at least from the chemist's viewpoint, is being made in the multivolume *Comprehensive Supramolecular Chemistry*, (Eds.: J. L. Atwood, J. E. D. Davies, D. D. MacNicol and F. Vögtle), Pergamon, in preparation.
[1.7]   J.-M. Lehn, *Angew. Chem.* 1988, *100*, 91; *Angew. Chem. Int. Ed. Engl.* 1988, *27*, 89.
[1.8]   J.-M. Lehn, *Science* 1985, *227*, 849.
[1.9]   J.-M. Lehn, *Angew. Chem.* 1990, *102*, 1347; *Angew. Chem. Int. Ed. Engl.* 1990, 29, 1304.
[1.10] P. Ehrlich, *Studies on Immunity*, Wiley, New York, 1906.
[1.11] E. Fischer, *Ber. Deutsch. Chem. Ges.* 1894, *27*, 2985.
[1.12] A. Werner, *Zeitschr. Anorg. Chem.* 1893, *3*, 267.
[1.13] J.-M. Lehn, in *Perspectives in Coordination Chemistry*, (Eds.: A. F. Williams, C. Floriani and A. E. Merbach), VHCA, Basel, and VCH, Weinheim, 1992, p. 447.
[1.14] P. Pfeiffer, *Organische Molekülverbindungen*, Enke, Stuttgart, 1927.
[1.15] a) K. L. Wolf, H. Frahm and H. Harms, Z. *Phys. Chem.* 1937, *Abt. B 36*, 237; b) K. L. Wolf, H. Dunken and K. Merkel, Z. *Phys. Chem.* 1940, *Abt. B 46*, 287; c) K. L. Wolf and R. Wolff, *Angew. Chem.* 1949, *61*, 191.
[1.16] A. L. Lehninger, *Naturwiss.* 1966, *53*, 57.
[1.17] R. Barthes, *Leçon Inaugurale*, Collège de France, 1977, p. 9.
[1.18] For some reflections on chemistry and language see: J.-M. Lehn, *Traduire* 1983, *116*, 63; in *Quelles langues pour la science ?*, (Ed.: B. Cassen), La Découverte, Paris, 1990, p. 31.
[1.19] For reflections on representation in chemistry, see: R. Hoffmann and P. Laszlo, *Angew. Chem.* 1991, *103*, 1; *Angew. Chem. Int. Ed. Engl.* 1991, *30*, 1.

[1.20]   J.-M. Lehn, in Proceedings of the Centenary of the Geneva Conference *Organic Chemistry: Its Language and Its State of the Art*, (Ed.: M.V. Kisakürek), VHCA, Basel, and VCH, Weinheim, **1993**, p. 77.

[1.21]   a) Yu. A. Ovchinnikov, V. T. Ivanov and A. M. Skrob, *Membrane Active Complexones*, Elsevier, New York, **1974**; b) B. C. Pressman, *Annu. Rev. Biochem.* **1976**, *45*, 501.

[1.22]   a) H. Brockmann and H. Geeren, *Justus Liebigs Ann. Chem.* **1957**, *603*, 217; b) M. M. Shemyakin, N.A. Aldanova, E. I. Vinogradova and M. Yu. Feigina, *Tetrahedron Lett.* **1963**, 1921; c) C. Moore and B. C. Pressman, *Biochem. Biophys. Res. Commun.* **1964**, *15*, 562; d) B.C. Pressman, *Proc. Natl. Acad. Sci. USA* **1965**, *53*, 1077; e) P. Mueller and D. O. Rudin, *Biochem. Biophys. Res. Commun.* **1967**, *26*, 398; f) T. E. Andreoli, M. Tieffenberg and D. C. Tosteson, *J. Gen. Biol.* **1967**, *50*, 2527; g) M. M. Shemyakin, Yu. A. Ovchinnikov, V. T. Ivanov, V. K. Antonov, A. M. Skrob, I. I. Mikhaleva, A. V. Evstratov and G. G. Malenkov, *ibid.* **1967**, *29*, 834; h) B. C. Pressman, E. J. Harris, W. S. Jagger and J. H. Johnson, *Proc. Natl. Acad. Sci. USA* **1967**, *58*, 1949.

[1.23]   a) J. Beck, H. Gerlach, V. Prelog and W. Voser, *Helv. Chim. Acta* **1962**, *45*, 620; b) Z. Stefanác and W. Simon, *Chimia* **1966**, *20*, 436; *Microchem. J.* **1967**, *12*, 125; c) B. T. Kilbourn, J. D. Dunitz, L. A. R. Pioda and W. Simon,, *J. Mol. Biol.* **1967**, *30*, 559.

[1.24]   C. J. Pedersen, *J. Am. Chem. Soc.* **1967**, *89*, 7017.

[1.25]   C. J. Pedersen, *Angew. Chem.* , **1988**, *100*, 1053; *Angew. Chem. Int. Ed. Engl.*, **1988**, *27*, 1053.

[1.26]   a) B. Dietrich, J.-M. Lehn and J.-P. Sauvage, *Tetrahedron Lett.* **1969**, 2885, 2889; b) B. Dietrich, J.-M. Lehn, J.-P. Sauvage and J. Blanzat, *Tetrahedron* **1973**, *29*, 1629; B. Dietrich, J.-M. Lehn and J.-P. Sauvage, *ibid.* **1973**, *29*, 1647.

[1.27]   J.-M. Lehn, *Struct. Bonding* **1973**, *16*, 1.

[1.28]   D. J. Cram and J. M. Cram, *Science* **1974**, *183*, 803.

[1.29]   J.-M. Lehn, *Pure Appl. Chem.* **1978**, *50*, 871.

[1.30]   The term "super-molecule" was coined in the present context in 1973; see ref. [1.27] p. 2.

[1.31]   J.-M. Lehn, J. Simon and J. Wagner, *Angew. Chem.* **1973**, *85*, 621, 622; *Angew. Chem. Int. Ed. Engl.* **1973**, *12*, 578, 579.

[1.32]   F. Cramer, *Einschlussverbindungen*, Springer, Berlin, **1954**.

[1.33]   J. E. D. Davies, W. Kemula, H. M. Powell and N. O. Smith, *J. Inclusion Phenom.* **1983**, *1*, 3.

[1.34]   A. Pullman, in *Environmental Effects on Molecular Structure and Properties*, (Ed.: B. Pullman), D. Reidel, Dordrecht, **1976**, p. 1.

[1.35]   M. Badertscher, M. Welti, P. Portmann and E. Pretsch, *Topics Curr. Chem.* **1986**, *136*, 17.

[1.36]   T. E. Sloan, in *Comprehensive Coordination Chemistry*, Vol. *1*, (Eds.: G. Wilkinson, R.D. Gillard and J.A. McCleverty), Pergamon, Oxford, **1987**, Ch. 3.

[1.37]   E. Kauffmann, J.L. Dye, J.-M. Lehn and A.I. Popov, *J. Am. Chem. Soc.* **1980**, *102*, 2274.

[1.38]   *Computer Simulation of Chemical and Biomolecular Systems*, (Eds.: D. L. Beveridge and W. C. Jorgensen), *Ann. N. Y. Acad. Sci. Vol. 482*, **1986**.

[1.39]   *Modelling of Molecular Structures and Properties*, (Ed.: J.-L. Rivail), Studies in Physical and Theoretical Chemistry, **1990**, *Vol. 71*.

[1.40]   *Modern Techniques in Computational Chemistry*, (Ed.: E. Clementi), MOTECC-89, **1989**; MOTECC-90, **1990**; MOTECC-91, **1991**; ESCOM, Leiden.

[1.41]   *Molecular Modelling für Anwender*, (Ed.: R. W. Kunz), Teubner, Stuttgart, 1991.
[1.42]   J. L. Toner, in [A.21], p. 77.
[1.43]   P. A. Kollman, *Acc. Chem. Res.* 1985, *18*, 105.
[1.44]   A. T. Brünger and M. Karplus, *Acc. Chem. Res.* 1991, *24*, 54.
[1.45]   a) G. Wipff, *J. Coord. Chem.* 1992, *27*, 7; b) G. Wipff, P. Kollman and J.-M. Lehn, *J. Mol. Struct.* 1983, *93*, 153.
[1.46]   P. Hobza and R. Zahradnik, *Intermolecular Complexes*, Elsevier, Amsterdam, 1988.
[1.47]   W. L. Jorgensen, *Acc. Chem. Res.* 1989, *22*, 184.
[1.48]   W. F. van Gunsteren and H. J. C. Berendsen, *Angew. Chem.* 1990, *102*, 1020; *Angew. Chem. Int. Ed. Engl.* 1990, *29*, 992.
[1.49]   For molecular modelling of coordination compounds, see: G. R. Brubaker and D. W. Johnson, *Coord. Chem. Rev.* 1984, *53*, 1; B. P. Hay, *ibid.* 1993, *126*, 177.
[1.50]   *Molecular Mechanics and Modelling*, *Chem. Rev.* 1993, *93* (7).

Chapter 2

[2.1]   M. Eigen and L. De Maeyer, *Naturwiss.* 1966, *53*, 50.
[2.2]   a) *Concepts and Applications of Molecular Similarity*, (Eds.: M.A. Johnson and G.M. Maggiora), Wiley, New-York, 1990; b) D. Farin and D. Avnir in *Characterization of Porous Solids*, (Eds.: K. K. Unger et al.), Elsevier, Amsterdam, 1988, 421.
[2.3]   a) See for instance: Z. Simon, *Angew. Chem.*, 1974, *86*, 802; *Angew. Chem. Int. Ed. Engl.*, 1974, *13*, 719; b) for a discussion of electrostatics see: S.-C. Tam and R.J.P. Williams, *Struct. Bonding*, 1985, *63*, 104; c) G. Eisenman, in *Ion Selective Electrodes* (Ed.: R. A. Durst), Natl. Bur. Stand., Special Publ. 314, 1969, Ch. 1 .
[2.4]   a) H.-J. Schneider, *Angew. Chem.* 1991, *103*, 1419; *Angew. Chem. Int. Ed. Engl.* 1991, *30*, 1417; b) H.-J. Schneider, T. Blatter, U. Cuber, R. Juneja, T. Schiestel, U. Schneider, I. Theis and P. Zimmermann, in [A.24], p. 29; c) H.-J. Schneider, in [A.36], p. 412; d) H.-J. Schneider, V. Rüdiger and O. A. Raevsky, *J. Org. Chem.* 1993, *58*, 3648; e) G. Náray-Szabó, *J. Molec. Recogn.* 1993, *6*, 205.
[2.5]   a) J. J. Hopfield, *Proc. Natl. Acad. Sci USA* 1974, *71*, 4135; 1980, *77*, 5248; b) J. Ninio, *Biochimie* 1975, *57*, 587; c) F. Cramer and W. Freist, *Acc. Chem. Res.* 1987, *20*, 79.
[2.6]   To name such chemical entities, a term rooted in Greek and Latin, which would also be equally suggestive in French, English, German and possibly (!) other languages, was sought; "cryptates" appeared particularly suitable for designating a complex in which the cation was contained inside the molecular cavity, the crypt, of the ligand termed "cryptand".
[2.7]   D. E. Koshland, *Adv. Enzym.* 1960, *22*, 45; *Ann. Rev. Biochem.* 1968, *37*, 672.
[2.8]   An idea about the respective role of collection and orientation may be gained from examining the energies calculated for a series of $[Li(NH_3)_n]^+$ complexes of different geometries and of the corresponding $(NH_3)_n$ units in the identical arrangements. The latter are a measure of the intersite repulsive energy for bringing together two, three, or four amine binding sites into a polydentate ligand of given coordination geometry. It is found that the total collection energies are appreciably larger than the organization energies represented by the changes from one geometry to another. Ab initio computations; J.-M. Lehn, R. Ventavoli, unpublished results; see also R. Ventavoli, 3e Cycle Thesis, Université Louis Pasteur, Strasbourg 1972; and ref. 23 in [1.7].

[2.9]   D. J. Cram, *Angew. Chem.* **1986**, *98*, 1041; *Angew. Chem. Int. Ed. Engl.* **1986**, *25*, 1039.

[2.10]   D. J. Cram, *Angew. Chem.* **1988**, *100*, 1041; *Angew. Chem. Int. Ed. Engl.* **1988**, *27*, 1009.

[2.11]   J.-M. Lehn in *Biomimetic Chemistry*, (Eds.: Z. I. Yoshida and N. Ise), Kodansha, Tokyo/Elsevier, Amsterdam **1983**, p. 163.

[2.12]   See, for instance: E. Weber and F. Vögtle, *Inorg. Chim. Acta* **1980**, *45*, L65; ref. A.16, p. 6.

[2.13]   M. R. Truter, *Struct. Bonding* **1973**, *16*, 71.

[2.14]   N. S. Poonia and A. V. Bajaj, *Chem. Rev.* **1979**, *79*, 389; I. Goldberg, in [A.21], p. 359.

[2.15]   Earlier observations had suggested that polyethers interact with alkali cations. See, for instance, H. C. Brown, E. J. Mead and P. A. Tierney, *J. Am. Chem. Soc.* **1957**, *79*, 5400; J. L. Down, J. Lewis, B. Moore and G. Wilkinson, *J. Chem. Soc.* **1959**, 3767; suggestions had also been made for the design of organic ligands, see R. J. P. Williams, *The Analyst (London)* **1953**, *78*, 586; *Quart. Rev. Chem. Soc.* **1970**, *24*, 331.

[2.16]   C. J. Pedersen, H. K. Frensdorff, *Angew. Chem.* **1972**, *84*, 16; *Angew. Chem. Int. Ed. Engl.* **1972**, *11*, 16.

[2.17]   J.-M. Lehn, *Acc. Chem. Res.* **1978**, *11*, 49.

[2.18]   a) J. J. Christensen, D. J. Eatough and R. M. Izatt, *Chem. Rev.* **1974**, *74*, 351; b) R. M. Izatt, J. S. Bradshaw, S. A. Nielsen, J. D. Lamb and J. J. Christensen, *ibid.* **1985**, *85*, 271; c) R. M. Izatt, K. Pawlak, J. S. Bradshaw and R. L. Bruening, *Chem. Rev.* **1991**, *91*, 1721; d) G. W. Gokel and J. E. Trafton, in [A.13], p. 253; G. W. Gokel, *Chem. Soc. Rev.* **1992**, *21*, 39; e) T. M. Fyles, in [A.13], p. 203.

[2.19]   a) S. Bradshaw and P. E. Stott, *Tetrahedron* **1980**, *36*, 461; b) J. Jurczak and M. Pietraszkiewicz, *Topics Curr. Chem.* **1985**, *130*, 183; J. Jurczak and R. Ostaszewski, *J. Coord. Chem. Sec. B* **1992**, *27*, 201; c) for macropolycyclic polyethers and related compounds, see: H. An, J. S. Bradshaw and R. M. Izatt, *Chem. Rev.* **1992**, *92*, 543.

[2.20]   a) R. Winkler, *Structure Bonding* **1972**, *10*, 1; b) E. M. Eyring and S. Petrucci, in [A.22] p. 179.

[2.21]   D. K. Cabbiness and D. W. Margerum, *J. Am. Chem. Soc.* **1969**, *91*, 6540; F. P. Hinz and D. W. Margerum, *Inorg. Chem.* **1974**, *13*, 2941.

[2.22]   R. D. Hancock and A. E. Martell, *Comments Inorg. Chem.* **1988**, *6*, 237.

[2.23]   D. H. Busch and N. A. Stephenson, *Coord. Chem. Rev.* **1990**, *100*, 119.

[2.24]   B. Metz, D. Moras and R. Weiss, *J. Chem. Soc. Chem. Commun.* **1970**, 217; F. Mathieu, B. Metz, D. Moras and R. Weiss, *J. Am. Chem. Soc.* **1978**, *100*, 4412 and references cited therein.

[2.25]   a) D. Parker, *Adv. Inorg. Chem. Radiochem.* **1983**, *27*, 1; b) B. Dietrich, in [A.18], **1984**, *Vol. 3*, p. 337.

[2.26]   a) P. G. Potvin and J.-M. Lehn in [A.17], **1987**, *Vol. 3*, p. 167; b) K. B. Mertes and J.-M. Lehn, in *Comprehensive Coordination Chemistry*, (Eds.: G. Wilkinson, R. D. Gillard and J. A. McCleverty), Pergamon, Oxford, **1987**, *2*, 915.

[2.27]   J.-M. Lehn and J.-P. Sauvage, *J. Am. Chem. Soc.* **1975**, *97*, 6700.

[2.28]   a) E. Kauffmann, J.-M. Lehn and J. P. Sauvage, *Helv. Chim. Acta* **1976**, *59*, 1099; b) for an exhaustive review of thermodynamic data see: Y. Inoue, Y. Lin and T. Hakushi, in [A.22], p. 1.

[2.29]   a) J. Cheney and J.-M. Lehn, *J. Chem. Soc. Chem. Commun.* **1972**, 487; b) P. B. Smith, J. L. Dye, J. Cheney and J.-M. Lehn, *J. Am. Chem. Soc.* **1981**, *103*, 6044; c) H.-J. Brügge, D. Carboo, K. von Deuten, A. Knöchel, J. Kopf and W. Dreissig, *J. Am. Chem. Soc.* **1986**, *108*, 107.

[2.30]   a) R. Pizer, *J. Am. Chem. Soc.* **1978**, *100*, 4239; b) R. W. Alder, *Tetrahedron* **1990**, *46*, 683.

[2.31]   B. Dietrich, J.-M. Lehn and J.-P. Sauvage, *J. Chem. Soc. Chem. Commun.* **1973**, 15.

[2.32]   a) J.-C. Bünzli and D. Wessner, *Coord. Chem. Rev.* **1984**, *60*, 191; b) J.-C. Bünzli, in *Handbook on the Physics and Chemistry of Rare Earths*, (Eds.: K. A. Gschneider, Jr., and L. Eyring), Elsevier, Amsterdam, **1987**, Ch. 60, 321; c) L. M. Vallarino, as in b) p. 443.

[2.33]   G.-y. Adachi and Y. Hirashima, in [A.22], p. 701.

[2.34]   D. V. Dearden, H. Zhang, I.-H. Chu, P. Wong and Q. Chen, *Pure Appl. Chem.* **1993**, *65*, 423

[2.35]   I.-H. Chu, H. Zhang, and D. V. Dearden, *J. Am. Chem. Soc.* **1993**, *115*, 5736.

[2.36]   a) G. Wipff, in [1.39] p. 143; b) G. Wipff and J.-M. Wurtz, *New J. Chem.* **1989**, *13*, 807.

[2.37]   For a semi-empirical approach see: W. E. Morf and W. Simon, *Helv. Chim. Acta* **1971**, *54*, 2683.

[2.38]   J. L. Atwood, in [A.22], p. 581.

[2.39]   T. W. Bell, in [A.27], p. 303; T. W. Bell, P. J. Cragg, M. G. B. Drew, A. Firestone, A. D.-I. Kwok, J. Liu, R. T. Ludwig and A. T. Papoulis, *Pure Appl. Chem.* **1993**, *65*, 361.

[2.40]   a) E. G. Gillan, C. Yeretzian, K. S. Min, M. M. Alvarez, R. L. Whetten and R. B. Kaner, *J. Phys. Chem.* **1992**, *96*, 6869; b) H. R. Rose, I. G. Dance, K. J. Fisher, D. R. Smith, G. D. Willett and M. A. Wilson, *J. Chem. Soc. Chem. Commun.* **1993**, 1361.

[2.41]   a) D. H. Busch, *Helv. Chim. Acta, Fasc. extraord. A. Werner* **1967**, 174; b) N.F. Curtis, *Coord. Chem. Rev.* **1968**, *3*, 3; c) D. St. C. Black and H. J. Hartshorn, *Coord. Chem. Rev.* **1972–1973**, *9*, 219.

[2.42]   K. E. Krakowiak, J. S. Bradshaw and D. J. Zamecka-Krakoviak, *Chem. Rev.* **1989**, *89*, 929; T. A. Kaden, in [A.30], **1984**, *121*, 157.

[2.43]   E. Kimura, *Tetrahedron*, **1992**, *48*, 6175.

[2.44]   A. Bencini, A. Bianchi, P. Paoletti and P. Paoli, *Pure Appl. Chem.* **1993**, *65*, 381.

[2.45]   S. R. Cooper, *Acc. Chem. Res.* **1988**, *21*, 141.

[2.46]   S. R. Cooper and S. C. Rawle, *Struct. Bonding* **1990**, *72*, 1.

[2.47]   A. J. Blake and M. Schröder, *Adv. Inorg. Chem.* **1990**, *35*, 1.

[2.48]   G. Reid and M. Schröder, *Chem. Soc. Rev.* **1990**, *19*, 239.

[2.49]   J.-M. Lehn and F. Montavon, *Helv. Chim. Acta* **1976**, *59*, 1566; *ibid.* **1978**, *61*, 67.

[2.50]   M. Micheloni, *J. Coord. Chem.* **1988**, *18*, 3; *Comments Inorg. Chem.* **1988**, *8*, 79.

[2.51]   B. Dietrich, J.-M. Lehn and J.-P. Sauvage, *J. Chem. Soc. Chem. Commun.* **1970**, 1055.

[2.52]   B. Dietrich, J.-M. Lehn, J. Guilhem, and C. Pascard, *Tetrahedron Lett.* **1989**, *30*, 4125.

[2.53]   K. G. Ragunathan and P. K. Bharadwaj, *J. Chem. Soc. Dalton Trans.* **1992**, 1653.

[2.54]   a) A. M. Sargeson, *Pure Appl. Chem.* **1984**, *56*, 1603; b) G. A. Bottomley, I. J. Clark, I. I. Creaser, L. M. Engelhardt, R. J. Geue, K. S. Hagen, J. M. Harrowfield, G. A. Lawrance, P. A. Lay, A. M. Sargeson, A. J. See, B. W. Skelton, A. H. White and F. R. Wilner, *Aust. J. Chem.* **1994**, *47*, 143.

[2.55]   a) K. N. Raymond and T. M. Garrett, *Pure Appl. Chem.* **1988**, *60*, 1807; b) K. N. Raymond, *Coord. Chem. Rev.* **1990**, *105*, 135.

[2.56]   a) F. Ebmeyer and F. Vögtle, *Bioorg. Chem. Frontiers* **1990**, *1*, 143; b) for acyclic ligands, see also: A. Shanzer, J. Libman and S. Lifson, *Pure Appl. Chem.* **1992**, *64*, 1421.

[2.57] a) C. Seel and F. Vögtle, *Angew. Chem.* **1992**, *104*, 542; *Angew. Chem. Int. Ed. Engl.* **1992**, *31*, 528; b) in [A.27], p. 191; c) Y. Sun and A. E. Martell, *Tetrahedron* **1990**, *46*, 2725.

[2.58] S. M. Nelson, *Pure Appl. Chem.* **1989**, *52*, 2461; *Inorg. Chim. Acta* **1982**, *62*, 39.

[2.59] D. E. Fenton and P. A. Vigato, *Chem. Soc. Rev.* **1988**, *17*, 69.

[2.60] D. H. Busch and C. Cairns, in [A.17], **1987**, *Vol. 3*, p. 1.

[2.61] J. Jazwinski, J.-M. Lehn, D. Lilienbaum, R. Ziessel, J. Guilhem and C. Pascard, *J. Chem. Soc. Chem. Commun.* **1987**, 1691.

[2.62] D. MacDowell and J. Nelson, *Tetrahedron Lett.* **1988**, *29*, 385; V. McKee, M. R. J. Dorrity, J. F. Malone, D. Marrs and J. Nelson, *J. Chem. Soc. Chem. Commun.* **1992**, 383.

[2.63] a) J. Hunter, J. Nelson, C. Harding, M. McCann and V. McKee, *J. Chem. Soc. Chem. Commun.* **1990**, 1148; b) C. Harding, V. McKee and J. Nelson, *J. Am. Chem. Soc.* **1991**, *113*, 9684; c) P. D. Beer, O. Kocian, R. J. Mortimer and P. Spencer, *J. Chem. Soc. Chem. Commun.* **1992**, 602.

[2.64] N. W. Alcock, W.-K. Lin, C. Cairns, G. A. Pike and D. H. Busch, *J. Am. Chem. Soc.* **1989**, *111*, 6630.

[2.65] L. F. Lindoy, in [A.17], *Vol. 3*, p. 53.

[2.66] K. R. Adams and L. F. Lindoy, in [A.27], p. 69.

[2.67] R. D. Hancock and A. E. Martell, *Chem. Rev.* **1989**, *89*, 1875.

[2.68] a) E. Kimura, in [A.27], p. 81; b) K. S. Hagen, *Angew. Chem.* **1992**, *104*, 804; *Angew. Chem. Int. Ed. Engl.* **1992**, *31*, 764.

[2.69] R. J. P. Williams, *J. Molecular Catal.* **1985**, *30*, 1.

[2.70] G. R. Newkome, J. D. Sauer, J. M. Roper and D. C. Hager, *Chem. Rev.* **1977**, *77*, 513.

[2.71] J.-C. Rodriguez-Ubis, B. Alpha, D. Plancherel and J.-M. Lehn, *Helv. Chim. Acta* **1984**, *67*, 2264.

[2.72] U. Lüning, R. Baumstark, C. Wangnick, M. Müller, W. Schyja, M. Gerst and M. Gelbert, *Pure Appl. Chem.* **1993**, *65*, 527.

[2.73] H. Dürr, S. Bossmann, H. Kilburg, H. P. Trierweiler and R. Schwarz, in [A.24], p. 453.

[2.74] B. Alpha, J.-M. Lehn and G. Mathis, *Angew. Chem.* **1987**, *99*, 259; *Angew. Chem. Int. Ed. Engl.* **1987**, *26*, 266.

[2.75] N. Sabbatini, M. Guardigli and J.-M. Lehn, *Coord. Chem. Rev.* **1993**, *123*, 201.

[2.76] a) S-P Huang and M. G. Kanatzidis, *Angew. Chem.* **1992**, *104*, 799; *Angew. Chem. Int. Ed. Engl.* **1992**, *31*, 787; b) A. Müller, K. Hovemeier and R. Rohlfing, *Angew. Chem.* **1992**, *104*, 1214; *Angew. Chem. Int. Ed. Engl.* **1992**, *31*, 1192; c) H. Reuter, *Angew. Chem.* **1992**, *104*, 1210; *Angew. Chem. Int. Ed. Engl.* **1992**, *31*, 1185.

[2.77] M. S. Lah, B. R. Gibney, D. L. Tierney, J. E. Penner-Hahn and V. L. Pecorato, *J. Am. Chem. Soc.* **1993**, *115*, 5857.

[2.78] J.-M. Lehn, *Pure Appl. Chem.* **1980**, *52*, 2303.

[2.79] J. L. Dye, *Scientific Amer.* **1977**, *237*, 92; J. L. Dye, *Angew. Chem.* **1979**, *91*, 613; *Angew. Chem. Int. Ed. Engl.* **1979**, *18*, 587; J. L. Dye, M. G. DeBacker, *Annu. Rev. Phys. Chem.* **1987**, *38*, 271.

[2.80] a) J. L. Dye, *Chemtracts, Inorg. Chem.* **1993**, *5*, 243; b) M. J. Wagner and J. L. Dye, *Annu. Rev. Mater. Sci.* **1993**, *23*, 223.

[2.81] J. L. Dye, *Science* **1990**, *247*, 663; J. L. Dye and R.-H. Huang, *Chem. Brit.* **1990**, 239.

[2.82] J. D. Corbett, *Chem. Rev.* **1985**, *85*, 383.

[2.83] L. Echegoyen, A. DeCian, J. Fischer and J.-M. Lehn, *Angew. Chem.* **1991**, *103*, 884; *Angew. Chem. Int. Ed. Engl.* **1991**, *30*, 838.

[2.84] a) L. Echegoyen, E. Pérez-Cordero, J.-B. Regnouf de Vains, C. Roth and J.-M. Lehn, *Inorg. Chem.* **1993**, *32*, 572; b) unpublished observations; c) for instance, the analogies of cryptatium type species with reduced fullerene carbon frameworks may be noted [2.83].

[2.85] a) G. W. Gokel and H. D. Durst, *Synthesis* **1976**, 168; b) C. L. Liotta, in [A.15], p. 111; c) R. A. Bartsch and J. Závada, *Chem. Rev.* **1980**, *80*, 453.

[2.86] F. Montanari, D. Landini and F. Rolla, *Topics Curr. Chem.*, **1982**, *101*, 147.

[2.87] a) W. P. Weber and G. W. Gokel, *Phase Transfer Catalysis in Organic Synthesis*, Springer, Berlin, **1977**; b) for triphase catalysis see: S. L. Regen, *Nouv. J. Chim.* **1982**, *6*, 629.

[2.88] E. V. Dehmlow and S. S. Dehmlow, *Phase Transfer Catalysis*, VCH, Weinheim, **1993**.

[2.89] a) E. Blasius and K.-P. Janzen, *Topics Curr. Chem.* **1982**, *98*, 163; b) J. Smid and R. Sinta, *Topics Curr. Chem.* **1984**, *121*, 105.

[2.90] a) K. G. Heumann, *Topics Curr. Chem.* **1985**, *127*, 77; b) Z. Chen and L. Echegoyen, in [A.27], p. 27.

[2.91] M. Takagi and H. Nakamura, *J. Coord. Chem.* **1986**, *15*, 53.

[2.92] I. M. Kolthoff, *Anal. Chem.* **1979**, *51*, 1R.

[2.93] E. Graf and J.-M. Lehn, *J. Am. Chem. Soc.* **1975**, *97*, 5022; b) *Helv. Chim. Acta* **1981**, *64*, 1040.

[2.94] E. Graf, J.-P. Kintzinger, J.-M. Lehn and J. LeMoigne, *J. Am. Chem. Soc.* **1982**, *104*, 1672.

[2.95] B. Dietrich, J.-P. Kintzinger, J.-M. Lehn, B. Metz and A. Zahidi, *J. Phys. Chem.* **1987**, *91*, 6600.

[2.96] a) E. Graf, J.-P. Kintzinger and J.-M. Lehn, unpublished results; E. Graf, Thèse de Doctorat, Université Louis Pasteur, Strasbourg, **1979**; b) for a water cryptate see also: C. Bazzicalupi, A. Bencini, A. Bianchi, V. Fusi, P. Paoletti and B. Valtancoli, *J. Chem. Soc. Perkin Trans. 2* **1994**, 815.

[2.97] E. Graf and J.-M. Lehn, *J. Am. Chem. Soc.* **1976**, *98*, 6403.

[2.98] H. Takemura, T. Shinmyozu and T. Inazu, *J. Am. Chem. Soc.* **1991**, *113*, 1323.

[2.99] a) P. Kebarle, *Annu. Rev. Phys. Chem.* **1977**, *28*, 445; b) M. Meot-Ner (Mautner), *J. Am. Chem. Soc.* **1983**, *105*, 4912; *ibid.* **1992**, *114*, 3312; in [A.37], p. 31 and references therein.

[2.100] D. J. Cram and J. M. Cram, *Acc. Chem. Res.* **1978**, *11*, 8.

[2.101] D. J. Cram and K. N. Trueblood, *Topics Curr. Chem.* **1981**, *98*, 43.

[2.102] a) J. F. Stoddart, *Chem. Soc. Rev.* **1979**, *8*, 85; *Annu. Rep. Prog. Chem.* **1983**, *Sect. B*, 353; b) J. F. Stoddart, *Topics Stereochem.* (Eds.: E. L. Eliel and S. H. Wilen), Wiley, New York, **1987**, *17*, 207.; c) M. Pietraszkiewicz and N. Spencer, *J. Coord. Chem.* **1992**, *27*, 115.

[2.103] a) F. De Jong and D. N. Reinhoudt in *Adv. Phys. Org. Chem.* (Eds.: V. Gold and D. Bethell), **1980**, *17*, 279; b) R. M. Izatt, C. Y. Zhu, P. Huszthy and J. S. Bradshaw, in [A.27], p. 207; c) for the binding of alkylammonium ions by the rigid cucurbituril receptor see: W. L. Mock and N.-Y. Shih, *J. Org. Chem.* **1986**, *51*, 4440; *J. Am. Chem. Soc.* **1988**, *110*, 4706; P. Cintas, *J. Incl. Phenom. Molec. Reco. Chem.* **1994**, *17*, 205.

[2.104] R. C. Hayward, *Chem. Soc. Rev.* **1983**, *12*, 285.

[2.105] I. O. Sutherland, *Chem. Soc. Rev.* **1986**, *15*, 63.

[2.106] J.-M. Lehn and P. Vierling, *Tetrahedron Lett.* **1980**, *21*, 1323.

[2.107] D. J. Cram, R. C. Helgeson, L. R. Sousa, J. M. Timko, M. Newcomb, P. Moreau, F. De Jong, G. W. Gokel, D. H. Hoffman, L. A. Domeier, S. C. Peacock, K. Madan and L. Kaplan, *Pure Appl. Chem.* **1975**, *43*, 327.

[2.108]   V. Prelog, *Pure Appl. Chem.* **1978**, *50*, 893.

[2.109]   a) J.-P. Behr, J.-M. Lehn and P. Vierling, *J. Chem. Soc. Chem. Commun.* **1976**, 621; *Helv. Chim. Acta* **1982**, *65*, 1853; b) J.-P. Behr and J.-M. Lehn, *Helv. Chim. Acta* **1980**, *63*, 2112.

[2.110]   J.-P. Behr, J.-M. Lehn, D. Moras and J.-C. Thierry, *J. Am. Chem. Soc.* **1981**, *103*, 701.

[2.111]   J. S. Brodbelt and C.-C. Liou, *Pure Appl. Chem.* **1993**, *65*, 409.

[2.112]   a) H. M. Colquhoun, J. F. Stoddart and D. J. Williams, *Angew. Chem.* **1986**, *98*, 483; *Angew. Chem. Int. Ed. Engl.* **1986**, *25*, 487; b) J. F. Stoddart and R. Zarzycki, ref. [A.22], p. 631.

[2.113]   J. F. Stoddart and R. Zarzycki, *Rec. Trav. Chim. Pays-Bas* **1988**, *107*, 515.

[2.114]   a) R. Ballardini, M. T. Gandolfi, L. Prodi, M. Ciano, V. Balzani, F. H. Kohnke, H. Shahriari-Zavareh, N. Spencer and J. F. Stoddart, *J. Am. Chem. Soc.* **1989**, *111*, 7072; b) L. Prodi, R. Ballardini, M. T. Gandolfi, V. Balzani, J.-P. Desvergne and H. Bouas-Laurent, *J. Phys. Chem.* **1991**, *95*, 2080.

[2.115]   J. C. Metcalfe, J. F. Stoddart and G. Jones, *J. Am. Chem. Soc.* **1977**, *99*, 8317; J. Krane and O. Aune, *Acta Chem. Scand.* **1980**, *B 34*, 397.

[2.116]   J.-M. Lehn, P. Vierling and R. C. Hayward, *J. Chem. Soc. Chem. Commun.* **1979**, 296.

[2.117]   K. Madan and D. J. Cram, *J. Chem. Soc. Chem. Commun.* **1975**, 427; J. W. H. M. Uiterwijk, S. Harkema, J. Geevers and D. N. Reinhoudt, *ibid.* **1982**, 200.

[2.118]   M. T. Reetz, C. M. Niemeyer, M. Hermes and R. Goddard, *Angew. Chem.* **1992**, *104*, 1053; *Angew. Chem. Int. Ed. Engl.* **1992**, *31*, 1017.

[2.119]   a) R. Foster, *Organic Charge-Transfer Complexes*, Academic Press, London, 1969; b) *Molecular Association*, (Ed.: R. Foster), Academic Press, London, **1975**, Vol. 1 and 2; c) T. Dahl, *Acta Chem. Scand.* **1994**, *48*, 95.

[2.120]   F. Vögtle, W. M. Müller and W. H. Watson, *Topics Curr. Chem.* **1984**, *125*, 131.

[2.121]   G. A. Jeffrey and W. Saenger, *Hydrogen Bonding in Biological Structure*, Springer, Berlin, 1991.

[2.122]   K. Morokuma, *Acc. Chem. Res.* **1977**, *10*, 294.

[2.123]   P. A. Kollman, *Acc. Chem. Res.* **1977**, *10*, 365.

[2.124]   M. C. Etter, *Acc. Chem. Res.* **1990**, *23*, 120.

[2.125]   A. R. Fersht, *Trends Biol. Sci.* **1987**, *12*, 301.

[2.126]   J. Rebek, *Topics Curr. Chem.* **1988**, *149*, 189; *Acc. Chem. Res.* **1990**, *23*, 399.

[2.127]   J. Rebek, *Angew. Chem.* **1990**, *102*, 261; *Angew. Chem. Int. Ed. Engl.* **1990**, *29*, 245.

[2.128]   A. D. Hamilton in [A.26], p. 1; *Bioorg. Chem. Frontiers* **1991**, 2, 115.

[2.129]   S. S. Yoon and W. C. Still, *J. Am. Chem. Soc.* **1993**, *115*, 823.

[2.130]   H.-J. Schneider, *Angew. Chem.* **1993**, *105*, 890; *Angew. Chem. Int. Ed. Engl.* **1993**, *32*, 848.

[2.131]   a) P. B. Dervan, R. S. Youngquist and J. P. Sluka, in *Stereochemistry of Organic and Bioorganic Transformations*, (Eds.: W. Bartmann and K. B. Sharpless), VCH, Heidelberg, 1987, p. 221; b) C. Hélène and G. Lancelot, *Prog. Biophys. Molec. Biol.* **1982**, *39*, 1; c) A. Travers, *DNA-Protein Interactions*, Chapman Hall, London, 1993.

[2.132]   R. U. Lemieux, *Chem. Soc. Rev.* **1989**, *18*, 347.

[2.133]   M. S. Searle and D. H. Williams, *J. Am. Chem. Soc.* **1992**, *114*, 10690; M. S. Searle, D. H. Williams, and U. Gerhard, *ibid.* **1992**, *114*, 10697.

[2.134]   C. S. Wilcox, J. C. Adrian, Jr., T. H. Webb and F. J. Zawacki, *J. Am. Chem. Soc.* **1992**, *114*, 10189.

[2.135]   W. L. Jorgensen and J. Pranata, *J. Am. Chem. Soc.* **1990**, *112*, 2008.

Chapter 3

[3.1]   a) F. Vögtle, H. Sieger and W. M. Müller, *Topics Curr. Chem.* **1981**, *98*, 107;
        b) K. Saigo, *Kagaku to Kogyo (Osaka)* **1982**, *35*, 90.

[3.2]   J.-L. Pierre and P. Baret, *Bull. Soc. Chim. Fr.* **1983**, *II*, 367.

[3.3]   E. Kimura, *Topics Curr. Chem.* **1985**, *128*, 113.

[3.4]   F. P. Schmidtchen, *Topics Curr. Chem.* **1986**, *132*, 101; *Nachr. Chem. Tech. Lab.*
        1988, 36, 8.

[3.5]   B. Dietrich, *Pure Appl. Chem.* **1993**, *65*, 1457.

[3.6]   a) H. E. Katz, *J. Org. Chem.* **1985**, *50*, 5027; *Organometal.* **1987**, 6, 1134;
        b) M. Newcomb, A. M. Madonik, M. T. Blanda and J. K. Judice, *Organometallics*
        1987, *6*, 145; c) M. T. Blanda, J. H. Horner and M. Newcomb, *J. Org. Chem.* **1989**,
        54, 4626; d) M. Newcomb, J. H. Horner, M. T. Blanda and P. J. Squattrito, *J. Am.
        Chem. Soc.* **1989**, *111*, 6294.

[3.7]   a) A. L. Beauchamp, M. J. Olivier, J. D. Wuest and B. Zacharie, *J. Am. Chem. Soc.*
        1986, *108*, 73; b) X. Yang, C. B. Knobler and M. F. Hawthorne, *Angew. Chem.*
        1991, *103*, 1519, *Angew. Chem. Int. Ed. Engl.* **1991**, *30*, 1507; c) X. Yang, Z. Zheng,
        C. B. Knobler and M. F. Hawthorne, *J. Am. Chem. Soc.* **1993**, *115*, 193; d) R. N.
        Grimes, *Angew. Chem.* **1993**, *105*, 1350, *Angew. Chem. Int. Ed. Engl.* **1993**, *32*,
        1289; e) V. B. Shur, I. A. Tikhonova, A. I. Yanovsky, Yu. T. Struchkov, P. V. Petro-
        vskii, S. Yu. Panov, G. G. Furin and M. E. Vol'pin, *J. Organomet. Chem.* **1991**, *418*,
        C29.

[3.8]   W. D. Clark, T. Y. Lin, S. D. Maleknia and R. J. Lagow, *J. Org. Chem.* **1990**, *55*,
        5933; b) W. B. Franham, D. C. Roe, D. A. Dixon, J. C. Calabrese and R. L. Har-
        low, *J. Am. Chem. Soc.* **1990**, *112*, 7707; c) T.-Y. Lin, W.-H. Lin, W. D. Clark, R. J.
        Lagow, S. B. Larson, S. H. Simonsen, V. M. Lynch, J. S. Brodbelt, S. D. Maleknia
        and C.-C. Liou, *J. Am. Chem. Soc.* **1994**, *116*, 5172.

[3.9]   C. H. Park and H. E. Simmons, *J. Am. Chem. Soc.* **1968**, *90*, 2431.

[3.10] a) F. Schmidtchen and G. Müller, *J. Chem. Soc. Chem. Commun.* **1984**, 1115;
        b) F. P. Schmidtchen, *Chem. Ber.* **1981**, *114*, 597.

[3.11] a) J.-M. Lehn, E. Sonveaux and A. K. Willard, *J. Am. Chem. Soc.* **1978**, *100*, 4914;
        b) B. Dietrich, J. Guilhem, J.-M. Lehn, C. Pascard and E. Sonveaux, *Helv. Chim.
        Acta* **1984**, *67*, 91.

[3.12] For other macrobicyclic receptors see also M. W. Hosseini and J.-M. Lehn, *Helv.
        Chim. Acta* **1988**, *71*, 749; B. Dietrich, M. W. Hosseini, J.-M. Lehn and R. B. Ses-
        sions, *Helv. Chim. Acta* **1985**, *68*, 289.

[3.13] a) B. Dietrich, M. W. Hosseini, J.-M. Lehn and R. B. Sessions, *J. Am. Chem. Soc.*
        1981, *103*, 1282; *Helv. Chim. Acta* **1983**, *66*, 1262; b) S. Boudon, A. DeCian, J.
        Fischer, M. W. Hosseini, J.-M. Lehn and G. Wipff, *J. Coord. Chem.* **1991**, *23*, 113;
        c) P. V. Bernhardt, G. A. Lawrance, B. W. Skelton and A. H. White, *Aust. J. Chem.*
        1989, *42*, 1035; d) A. Bencini, A. Bianchi, P. Dapporto, E. Garcia-Espanā, M.
        Micheloni, P. Paoletti and P. Paoli, *J. Chem. Soc., Chem. Commun.* **1990**, 753; A.
        Bencini, A. Bianchi, P. Dapporto, E. Garcia-Espanā, M. Micheloni, J. A. Ramirez,
        P. Paoletti and P. Paoli, *Inor. Chem.* **1992**, *31*, 1902.

[3.14] a) J. Cullinane, R. I. Gelb, T. N. Margulis and L. J. Zompa, *J. Am. Chem. Soc.* **1982**,
        104, 3048; b) E. Kimura, M. Kodama and T. Yatsunami, *J. Am. Chem. Soc.* **1982**,
        104, 3182; c) J. F. Marecek and C. J. Burrows, *Tetrahedron Lett.* **1986**, *27*, 5943.

[3.15] M. W. Hosseini, J.-M. Lehn, M. P. Mertes, *Helv. Chim. Acta* **1983**, *66*, 2454.

[3.16] M. W. Hosseini and J.-M. Lehn, *Helv. Chim. Acta* **1987**, *70*, 1312; see also H. R.
        Wilson and R. J. P. Williams, *J. Chem. Soc. Faraday Trans. I*, **1987**, *83*, 1885.

[3.17]   C. Oriol-Audit, M. W. Hosseini and J.-M. Lehn, *Eur. J. Biochem.* **1985**, *151*, 557.

[3.18]   a) J. L. Sessler, M. Cyr, H. Furuta, V. Král, T. Mody, T. Morishima, M. Shionoya and S. Weghorn, *Pure Appl. Chem.* **1993**, *65*, 393; b) J. L. Sessler, H. Furuta and V. Král, *Supramol. Chem.* **1993**, *1*, 209.

[3.19]   a) B. Dietrich, D. L. Fyles, T. M. Fyles and J.-M. Lehn, *Helv. Chim. Acta* **1979**, *62*, 2763; b) B. Dietrich, T. M. Fyles, J.-M. Lehn, L. G. Pease and D. L. Fyles, *J. Chem. Soc. Chem. Commun.* **1978**, 934.

[3.20]   a) A. Echavarren, A. Galán, J.-M. Lehn and J. de Mendoza, *J. Am. Chem. Soc.* **1989**, *111*, 4994; b) A. Gleich, F. P. Schmidtchen, P. Mikulcik and G. Müller, *J. Chem. Soc. Chem. Commun.* **1990**, 55; c) A. Galán, E. Pueyo, A. Salmerón and J. de Mendoza, *Tetrahedron Lett.* **1991**, *32*, 1827; d) H. Furuta, D. Magda and J. L. Sessler, *J. Am. Chem. Soc.* **1991**, *113*, 978.

[3.21]   F. Peter, M. Gross, M. W. Hosseini and J.-M. Lehn, *J. Electroanal. Chem.* **1983**, *144*, 279.

[3.22]   E. Garcia-España, M. Micheloni, P. Paoletti and A. Bianchi, *Inorg. Chim. Acta* **1985**, *102*, L9; A. Bianchi, E. Garcia-España, S. Mangani, M. Micheloni, P. Orioli and P. Paoletti, *J. Chem. Soc. Chem. Commun.* **1987**, 729.

[3.23]   M. F. Manfrin, L. Moggi, V. Castelvetro, V. Balzani, M. W. Hosseini and J.-M. Lehn, *J. Am. Chem. Soc.* **1985**, *107*, 6888; F. Pina, L. Moggi, M. F. Manfrin, V. Balzani, M. W. Hosseini and J.-M. Lehn, *Gazz. Chim. Ital.* **1989**, *119*, 65.

[3.24]   J.-M. Lehn, *Pure Appl. Chem.* **1980**, *52*, 2441.

[3.25]   R. J. Motekaitis, A. E. Martell, B. Dietrich and J.-M. Lehn, *Inorg. Chem.*, **1984**, *23*, 1588; R. J. Motekaitis, A. E. Martell and I. Murase, *ibid.* **1986**, *25*, 938; A. Evers, R. D. Hancock and I. Murase, *ibid.* **1986**, *25*, 2160; D. E. Whitmoyer, D. P. Rillema and G. Ferraudi, *J. Chem. Soc. Chem. Commun.* **1986**, 677.

[3.26]   a) A. E. Martell, *J. Incl. Pheno. Molec. Recog. Chem.* **1989**, *7*, 99; b) A. E. Martell, in [A.26], p. 145; in [A.27], p. 99; D. Chen and A. E. Martell, *Tetrahedron Lett.* **1991**, *47*, 6895; c) A. E. Martell, R. J. Motekaitis, D. Chen and I. Murase, *Pure Appl. Chem.* **1993**, *65*, 959.

[3.27]   W. Xu, J. J. Vittal and R. J. Puddephatt, *J. Am. Chem. Soc.* **1993**, *115*, 6456.

[3.28]   J.-P. Kintzinger, J.-M. Lehn, E. Kauffmann, J. L. Dye and A. I. Popov, *J. Am. Chem. Soc.* **1983**, *105*, 7549.

[3.29]   D. Heyer and J.-M. Lehn, *Tetrahedron Lett.* **1986**, *27*, 5869.

[3.30]   T. Fujita and J.-M. Lehn, *Tetrahedron Lett.* **1988**, *29*, 1709.

[3.31]   P. D. Beer, J. W. Wheeler and C. Moore, in [A.28], p. 105.

[3.32]   D. Kaufmann and A. Otten, *Angew. Chem.* **1994**, *106*, 1917; *Angew. Chem. Int. Ed. Engl.* **1994**, *33*, 1832.

[3.33]   a) T. P. Lybrand, J. A. McCammon and G. Wipff, *Proc. Natl. Acad. Sci. USA* **1986**, *83*, 833; b) B. Owenson, R. D. MacElroy and A. Pohorille, *J. Mol. Struct. (Theochem)* **1988**, *179*, 467; *J. Am. Chem. Soc.* **1988**, *110*, 6992.

Chapter 4

[4.1]   J.-M. Lehn in *IUPAC Frontiers of Chemistry* (Ed.: K. J. Laidler), Pergamon, Oxford, **1982**, p. 265.

[4.2]   a) J.-M. Lehn, S. H. Pine, E. I. Watanabe and A. K. Willard, *J. Am. Chem. Soc.* **1977**, *99*, 6766; b) P. K. Coughlin, J. C. Dewan, S. J. Lippard, E. I. Watanabe and J.-M.

Lehn, *ibid.* 1979, *101*, 265 c) C. A. Salata, M.-T. Youinou and C. J. Burrows, *Inorg. Chem.* 1991, *30*, 3454.

[4.3]   M. A. Pérez, and J. M. Bermejo, *Anal. Quimica,* 1993, 105; C. Acerete, J. M. Bueno, L. Campayo, P. Navarro, M. I. Rodriguez-Franco, and A. Samat, *Tetrahedron* 1994, *50*, 4765.

[4.4]   J. Comarmond, P. Plumeré, J.-M. Lehn, Y. Agnus, R. Louis, R. Weiss, O. Khan and I. Morgenstern-Badarau, *J. Am. Chem. Soc.* 1982, *104*, 6330.

[4.5]   J. de Mendoza, E. Mesa, J.-C. Rodriguez-Ubis, P. Vázquez, F. Vögtle, P.-M. Windscheif, K. Rissanen, J.-M. Lehn, D. Lilienbaum and R. Ziessel, *Angew. Chem.* 1991, *103*, 1365; *Angew. Chem. Int. Ed. Engl.* 1991, *30*, 1331.

[4.6]   J.-P. Gisselbrecht and M. Gross, *Adv. Chem. Series, No. 201,* ACS Washington DC, 1982, 109.

[4.7]   J.-P. Gisselbrecht, M. Gross, A. H. Alberts and J.-M. Lehn, *Inorg. Chem.* 1980, *19*, 1386.

[4.8]   A. Carroy and J.-M. Lehn, *J. Chem. Soc. Chem. Commun.* 1986, 1232.

[4.9]   J.-P. Lecomte, J.-M. Lehn, D. Parker, J. Guilhem and C. Pascard, *J. Chem. Soc. Chem. Commun.* 1983, 296.

[4.10]   J. Comarmond, B. Dietrich, J.-M. Lehn and R. Louis, *J. Chem. Soc. Chem. Commun.* 1985, 74.

[4.11]   a) S. C. Lee and R. H. Holm, *Angew. Chem.* 1990, *102*, 868; *Angew. Chem. Int. Ed. Engl.* 1990, *29*, 840; b) B. Krebs and G. Henkel, *Angew. Chem.* 1991, *103*, 785; *Angew. Chem. Int. Ed. Engl.* 1991, *30*, 769; c) J.-F. You, G. C. Papaefthymiou and R. H. Holm, *J. Am. Chem. Soc.* 1992, *114*, 2697.

[4.12]   a) H. (Y) Okuno, K. Uoto, T. Tomohiro and M.-T. Youinou, *J. Chem. Soc., Dalton Trans.* 1990, 3375; b) for the assembly of an Mn_4 cluster inside a macrocycle see: V. McKee and W. B. Sheppard, *J. Chem. Soc. Chem. Commun.* 1985, 158.

[4.13]   F. Kotzyba-Hibert, J.-M. Lehn and P. Vierling, *Tetrahedron Lett.* 1980, *21*, 941.

[4.14]   F. Kotzyba-Hibert, J.-M. Lehn and K. Saigo, *J. Am. Chem. Soc.* 1981, *103*, 4266.

[4.15]   C. Pascard, C. Riche, M. Cesario, F. Kotzyba-Hibert and J.-M. Lehn, *J. Chem. Soc. Chem Commun.* 1982, 557.

[4.16]   I. O. Sutherland, *J. Incl. Pheno. Molec. Reco. Chem.* 1989, *7*, 213.

[4.17]   I. O. Sutherland, in [A.26], p. 65.

[4.18]   M. W. Hosseini and J.-M. Lehn, *J. Am. Chem. Soc.* 1982, *104*, 3525; *Helv. Chim. Acta* 1986, *69*, 587.

[4.19]   J.-M. Lehn, R. Méric, J.-P. Vigneron, I. Bkouche-Waksman and C. Pascard, *J. Chem. Soc. Chem. Commun.* 1991, 62.

[4.20]   a) E. Fan, S. A. Van Arman, S. Kincaid and A. D. Hamilton, *J. Am. Chem. Soc.* 1993, *115*, 369; b) L. Owens, C. Thilgen, F. Diederich and C. B. Knobler, *Helv. Chim. Acta* 1993, *76*, 2757; c) S. S. Flack, J.-L. Chaumette, J. D. Kilburn, G. J. Langley and M. Webster, *J. Chem. Soc. Chem. Commun.* 1993, 399.

[4.21]   S. Mallik, R. D. Johnson and F. H. Arnold, *J. Am. Chem. Soc.* 1993, *115*, 2518.

[4.22]   a) F. P. Schmidtchen, *J. Am. Chem. Soc.* 1986, *108*, 8249; b) J. Rebek, Jr., D. Nemeth, P. Ballester and F.-T. Lin, *ibid.* 1987, *109*, 3474.

[4.23]   a) J.-M. Lehn, J. Simon and A. Moradpour, *Helv. Chim. Acta* 1978, *61*, 2407; b) J. Simon, *Thèse de Doctorat d'Etat,* Université Louis Pasteur, Strasbourg, 1976; see also compound 28 in [2.104], p. 305.

[4.24]   a) F. P. Schmidtchen, *J. Org. Chem.* 1986, *51*, 5161; b) J. Rebek, Jr., B. Askew, D. Nemeth and K. Parris, *J. Am. Chem. Soc.* 1987, *109*, 2432; c) M. T. Reetz, C. M. Niemeyer and K. Harms, *Angew. Chem.* 1991, *103*, 1515, 1517; *Angew. Chem. Int. Ed. Engl.* 1991, *30*, 1472, 1474.

[4.25]   a) T. H. Webb and C. S. Wilcox, *Chem. Soc. Rev.* **1993**, *22*, 383; b) A. Galán, D. Andreu, A. M. Echavarren, P. Prados and J. de Mendoza, *J. Am. Chem. Soc.* **1992**, *114*, 1511.

[4.26]   a) E. Ruoslahti and M. D. Pierschbacher, *Science* **1987**, *238*, 491; b) J. Travis, *Science* **1993**, *260*, 906.

[4.27]   M. W. Hosseini, A. J. Blacker and J.-M. Lehn, *J. Am. Chem. Soc.* **1990**, *112*, 3896.

[4.28]   a) M. L. Bender and M. Komiyama, *Cyclodextrin Chemistry*, Springer, Berlin, **1978**; b) W. Saenger, in [A.18], **1984**, *Vol. 2*, 231; c) R. J. Bergeron, in [A.18], **1984**, *Vol. 3*, 391.

[4.29]   J. Szejtli, *Cyclodextrin Technology*, Kluwer, Dordrecht, **1988**.

[4.30]   a) C. Tanford, *The Hydrophobic Effect*, 2nd edn., Wiley, New York, **1980**; b) W. Blokzijil and J. B. F. N. Engberts, *Angew. Chem.* **1993**, *105*, 1610, *Angew. Chem. Int. Ed. Engl.* **1993**, *32*, 1545.

[4.31]   I. Tabushi and K. Yamamura, *Topics Curr. Chem.* **1983**, *113*, 145.

[4.32]   a) J. Franke and F. Vögtle, *Topics Curr. Chem.* **1986**, *132*, 135; b) A. P. Davis, *Chem. Soc. Rev.* **1993**, *22*, 243.

[4.33]   F. Diederich, *Angew. Chem.* **1988**, *100*, 372; *Angew. Chem. Int. Ed. Engl.* **1988**, *27*, 362.

[4.34]   a) Y. Murakami, J. Kikuchi and T. Ohno, in [A.26], p. 109; b) Y. Murakami, O. Hayashida, T. Ito and Y. Hisaeda, *Pure Appl. Chem.* **1993**, *65*, 551.

[4.35]   a) F. H. Kohnke, J. P. Mathias and J. F. Stoddart, *Topics Curr. Chem.* **1993**, *165*, 1; b) A. Godt, V. Enkelmann and A.-D. Schlüter, *Angew. Chem. Int. Ed. Engl.* **1989**, *28*, 1680; c) R. M. Cory, C. L. McPhail and A. J. Dikmans, *Tetrahedron Lett.* **1993**, *34*, 7533; d) J. Benkhoff, R. Boese, F.-G. Klärner and A. E. Wigger, *Tetrahedron Lett.* **1994**, *35*, 73; e) W. Josten, D. Karbach, M. Nieger, F. Vögtle, K. Hägele, M. Svoboda and M. Przybylski, *Chem. Ber.* **1994**, *127*, 767; f) M. Pollmann and K. Müllen, *J. Am. Chem. Soc.* **1994**, *116*, 2318.

[4.36]   a) C. D. Gutsche, *Topics Curr. Chem.* **1984**, *123*, 1; *Acc. Chem. Res.* **1983**, *16*, 161.; b) R. Ungaro and A. Pochini, in [A.24], p. 57; c) E. van Dienst, W. I. I. Bakker, J. F. J. Engbersen, W. Verboom and D. N. Reinhoudt, *Pure Appl. Chem.* **1993**, *65*, 387; d) V. Böhmer and W. Vogt, *ibid.* **1993**, *65*, 403; e) Z. Asfari, J. Weiss, S. Pappalardo and J. Vicens, *ibid.* **1933**, *65*, 585; f) S. Shinkai, *Tetrahedron* **1993**, *49*, 8933; g) L. C. Groenen and D. N. Reinhoudt, in [A.28], p. 51.

[4.37]   D. J. Cram, *Science* **1983**, *219*, 1177.

[4.38]   a) H. L. Anderson, R. P. Bonar-Law, L. G. Mackay, S. Nicholson and J. K. M. Sanders, in [A.28] p. 359; b) L. G. Mackay, R. P. Bonar-Law and J. K. M. Sanders, *J. Chem. Soc. Perkin Trans. 1* **1993**, 1377; c) H. S. Ham, D. Liu and C. J. Burrows, *J. Incl. Phenom.* **1987**, *5*, 117; d) J. Zhang, D. J. Pesak, J. J. Ludwick, J. S. Moore, *J. Am. Chem. Soc.* **1994**, *116*, 4227; Z. Wu, S. Lee, J. S. Moore, *J. Am. Chem. Soc.* **1992**, *114*, 8730.

[4.39]   a) A. Collet, *Tetrahedron* **1987**, *43*, 5725; b) A. Collet, J.-P. Dutasta, B. Lozach and J. Canceill, *Topics Curr. Chem.* **1993**, *165*, 103.

[4.40]   D. J. Cram, *Nature* **1992**, *356*, 29; M. E. Tanner, C. B. Knobler and D. J. Cram, *J. Org. Chem.* **1992**, *57*, 40.

[4.41]   a) D. J. Cram, M. E. Tanner and R. Thomas, *Angew. Chem.* **1991**, *103*, 1048; *Angw. Chem. Int. Ed. Engl.* **1991**, *30*, 1024; b) T. A. Robbins and D. J. Cram, *J. Am. Chem. Soc.* **1993**, *115*, 12199.

[4.42]   J. Canceill, A. Collet, J. Gabard, F. Kotzyba-Hibert and J.-M. Lehn, *Helv. Chim. Acta* **1982**, *65*, 1894.

[4.43]   a) M. Dhaenens, L. Lacombe, J.-M. Lehn and J.-P. Vigneron, *J. Chem. Soc., Chem. Commun.* **1984**, 1097; b) M. Dhaenens, J.-M. Lehn and J.-P. Vigneron, unpublished results.

[4.44] R. Méric, J.-P. Vigneron and J.-M. Lehn, *J. Chem. Soc. Chem. Commun.* **1993**, 129.

[4.45] a) M. A. Petti, T. J. Shepodd, R. E. Barrans, Jr. and D. A. Dougherty, *J. Am. Chem. Soc.* **1988**, *110*, 6825; b) D. A. Dougherty and D. A. Stauffer, *Science* **1990**, *250*, 1558; c) P. C. Kearney, L. S. Mizoue, R. A. Kumpf, J. E. Forman, A. McCurdy ad D. A. Dougherty, *J. Am. Chem. Soc.* **1993**, *115*, 9907.

[4.46] H.-J. Schneider, D. Güttes and U. Schneider, *J. Am. Chem. Soc.* **1988**, *110*, 6449.

[4.47] F. Vögtle, T. Merz and H. Wirtz, *Angew. Chem.* **1985**, *97*, 226, *Angew. Chem., Int. Ed. Engl.* **1985**, *24*, 221; T. Merz, H. Wirtz and F Vögtle, *Angew. Chem.* **1986**, *98*, 549, *Angew. Chem. Int. Ed. Engl.* **1986**, *25*, 567.

[4.48] E. T. Jarvi and H. W. Whitlock, *J. Am. Chem. Soc.* **1982**, *104*, 7196.

[4.49] M. Miyake, M. Kirisawa and K. Koga, *Tetrahedron Lett.* **1991**, *32*, 7295.

[4.50] B.-L. Poh and C. S. Lim, *Tetrahedron* **1990**, *46*, 3651.

[4.51] a) S. Shinkai, K. Araki and O. Manabe, *J. Am. Chem. Soc.* **1988**, *110*, 7214; b) S. Shinkai, K. Araki, T. Matsuda, N. Nishiyama, H. Ikeda, I. Takasu and M. Iwamoto, *J. Am. Chem. Soc.* **1990**, *112*, 9053; c) S. Shinkai, *Bioorg. Chem. Frontiers* **1990**, *1*, 161.

[4.52] C. S. Wilcox, T. H. Webb, F. J. Zawacki, N. Glagovich and H. Suh, *Supramol. Chem.* **1993**, *1*, 129.

[4.53] a) Y. Kikuchi, K. Kobayashi and Y. Aoyama *J. Am. Chem. Soc.* **1992**, *114*, 1351; K. Kobayashi, Y. Asakawa, Y. Kato and Y. Aoyama, *ibid.* **1992**, *114*, 10307; Y. Aoyama, in [A.28], p. 17; b) J. M. Coterón, C. Vicent, C. Bosso and S Penadés, *J. Am. Chem. Soc.* **1993**, *115*, 10066; c) N. Sharon and H. Lis, *Scientific Amer.* **1993**, *268*, January, p. 74.

[4.54] For dissymmetric cylindrical macrotricyclic coreceptors that bind ammonium ions see: A. D. Hamilton and P. Kazanjian, *Tetrahedron Lett.* **1985**, *26*, 5735; K. Saigo, R.-J. Lin, M. Kubo, A. Youda and M. Hasegawa, *Chem. Lett.* **1986**, 519.

[4.55] a) A. R. Bernardo, J. F. Stoddart and A. E. Kaifer, *J. Am. Chem. Soc.* **1992**, *114*, 10624; b) M. Bühner, W. Geuder, W.-K. Gries, S. Hünig, M. Koch and T. Poll, *Angew. Chem.* **1988**, *100*, 1611; *Angew. Chem. Int. Ed. Engl.* **1988**, *27*, 1553.

[4.56] a) A. J. Blacker, J. Jazwinski and J.-M. Lehn, *Helv. Chim. Acta* **1987**, *70*, 1; b) A. J. Blacker, J. Jazwinski, J.-M. Lehn and F. X. Wilhelm, *J. Chem. Soc. Chem. Commun.* **1986**, 1035. See also: A. Slama-Schwok, J. Jazwinski, A. Béré, T. Montenay-Garrestier, M. Rougée, C. Hélène and J.-M. Lehn, *Biochemistry* **1989**, *28*, 3227.

[4.57] S. C. Zimmerman, *Bioorg. Chem. Frontiers* **1991**, *2*, 33; *Topics Curr. Chem.* **1993**, *165*, 71.

[4.58] a) C.-W. Chen and H. W. Whitlock, Jr., *J. Am. Chem. Soc.* **1978**, *100*, 4921; b) R. E. Sheridan and H. W. Whitlock, Jr., *J. Am. Chem. Soc.* **1986**, *108*, 7120; **1988**, *110*, 4071.

[4.59] M. M. Conn, G. Deslongchamps, J. de Mendoza and J. Rebek, Jr., *J. Am. Chem. Soc.* **1993**, *115*, 3548.

[4.60] a) J. Jazwinski, A. J. Blacker, J.-M. Lehn, M. Cesario, J. Guilhem and C. Pascard, *Tetrahedron Lett.* **1987**, *28*, 6057; b) S. Claude, J.-M. Lehn, F. Schmidt and J.-P. Vigneron, *J. Chem. Soc., Chem. Commun.* **1991**, 1182; c) M. Zinic, P. Cudic, V. Skaric, J.-P. Vigneron and J.-M. Lehn, *Tetrahedron Lett.* **1992**, *33*, 7417; d) M.-P. Teulade-Fichou, J.-P. Vigneron and J.-M. Lehn, *Supramolecular Chem.* **1995**, in press; e) M. Dhaenens, J.-M. Lehn and J.-P. Vigneron, *J. Chem. Soc., Perkin Trans. 2* **1993**, 1379; f) A. Slama-Schwok, M.-P. Teulade-Fichou, J.-M. Lehn and J.-P. Vigneron, to be published.

[4.61]   a) T. Andersson, K. Nilsson, M. Sundahl, G. Westman and O. Wennerström, *J. Chem. Soc. Chem. Commun.* **1992**, 604; b) R. M. Williams and J. W. Verhoeven, *Recl. Trav. Chim. Pays Bas* **1992**, *111*, 531; J. L. Atwood, G. A. Koutsantonis and C. L. Raston, *Nature* **1994**, *368*, 229; c) A. L. Balch, V. J. Catalano, J. W. Lee and M. M. Olmstead, *J. Am. Chem. Soc.* **1992**, *114*, 5456.

[4.62]   a) D. N. Reinhoudt, *J. Coord. Chem.* **1988**, *18*, 21; b) D. N. Reinhoudt and H. J. den Hertog, Jr., *Bull. Soc. Chim. Belg.* **1988**, *97*, 645; c) F. C. J. M. van Veggel and D. N. Reinhoudt, in [A.24], p. 83; d) D. N. Reinhoudt, A. R. van Doorn and W. Verboom, *J. Coord. Chem.* **1992**, *27*, 91; e) A. M. Reichwein, W. Verboom and D. N. Reinhoudt, in [A.36], p 358.

[4.63]   a) J. E. Kickham, S. J. Loeb and S. L. Murphy, *J. Am. Chem. Soc.* **1993**, *115*, 7031; b) J. E. Kickham and S. J. Loeb, *J. Chem. Soc., Chem. Commun.* **1993**, 1848; c) F. Vögtle, I. Lüer, V. Balzani and N. Armaroli, *Angew. Chem.* **1991**, *103*, 1367; *Angew. Chem. Int. Ed. Engl.* **1991**, *30*, 1333.

[4.64]   a) H. L. Anderson and J. K. M. Sanders, *Angew. Chem.* **1990**, *102*, 1478; *Angew. Chem. Int. Ed. Engl.* **1990**, *29*, 1400; b) S. Anderson, H. L. Anderson and J. K. M. Sanders, *Angew. Chem.* **1992**, *104*, 921; *Angew. Chem. Int. Ed. Engl.* **1992**, *31*, 907.

[4.65]   a) J. P. Collman, P. S. Wagenknecht and J. E. Hutchison, *Angew. Chem.* **1994**, *106*, 1620; *Angew. Chem. Int. Ed.* **1994**, *33*, 1537; b) D. Dolphin, J. Hiom and J. B. Paine III, *Heterocycles* **1981**, *16*, 417.

[4.66]   a) A. D. Hamilton, J.-M. Lehn and J. L. Sessler, *J. Chem. Soc. Chem. Commun.* **1984**, 311; *J. Am. Chem. Soc.* **1986**, *108*, 5158; b) A. Slama-Schwok and J.-M. Lehn, *Biochemistry* **1990**, *29*, 7895.

[4.67]   For other metalloreceptor-type species see for instance references in [2.60, 2.105]; N. M. Richardson, I. O. Sutherland, P. Camilleri and J. A. Page, *Tetrahedron Lett.* **1985**, *26*, 3739; M. C. Gonzalez and A. C. Weedon, *Can. J. Chem.* **1985**, *63*, 602.

[4.68]   For porphyrins bearing crown-ether rings, see: a) V. Thanabal and V. Krishnan, *J. Am. Chem. Soc.* **1982**, *104*, 3643; b) G. B. Maiya and V. Krishnan, *Inorg. Chem.* **1985**, *24*, 3253; c) V. Khrisnan, *Proc. Indian Natn. Sci. Acad.* **1986**, *52*, A, 909.

[4.69]   R. M. Izatt, J. S. Bradshaw, K. Pawlak, R. L. Bruening and B. J. Tarbet, *Chem. Rev.* **1992**, *92*, 1261.

[4.70]   A. F. Danil de Namor, *Pure Appl. Chem.* **1993**, *65*, 193.

[4.71]   M. Meot-Ner (Mautner), *Acc. Chem. Res.* **1984**, *17*, 186.

[4.72]   F. Diederich, D. B. Smithrud, E. M. Sanford, T. B. Wyman, S. B. Ferguson, D. R. Carcanague, I. Chao and K. N. Houk, *Acta Chem. Scand.* **1992**, *46*, 205.

[4.73]   M. Cesario, J. Guilhem, J.-M. Lehn, R. Méric, C. Pascard and J.-P. Vigneron, *J. Chem. Soc., Chem. Commun.* **1993**, 540.

[4.74]   V. Luzzati, in *Biological Membranes*, (Ed.: D. Chapman), Academic, London, **1968**.

[4.75]   A. Skoulios, *Ann. Phys.* **1978**, *3*, 421; J. Charvolin and J.-F. Sadoc, *La Recherche* **1992**, *23*, 307.

[4.76]   C. Brevard, J.-P. Kintzinger and J.-M. Lehn, *Tetrahedron* **1972**, *28*, 2429.

[4.77]   J. R. Lyerla, Jr. and G. C. Levy, *Top. Carbon-13 Nucl. Magn. Reson. Spectrosc.* **1974**, *1*, 79; D. A. Wright, D. E. Axelson and G. C. Levy, *ibid.* **1979**, *3*, 103 and references therein; G. C. Levy R. A. Komoroski and J. A. Halstead, *J. Am. Chem. Soc.* **1974**, *96*, 5456; G. C. Levy and T. Terpstra, *Org. Magn. Reson.* **1976**, *8*, 658; W. F. Reynolds, P. Dais, A. Mar and M. A. Winnick, *J. Chem. Soc. Chem. Commun.* **1976**, 757; G. C. Levy, M. P. Cordes, J. S. Lewis, and D. E. Axelson, *J. Am. Chem. Soc.* **1977**, *99*, 5492.

[4.78]   C. Brevard and J.-M. Lehn, *J. Am. Chem. Soc.* **1970**, *92*, 4987.

[4.79] a) J. P. Behr and J.-M. Lehn, *J. Am. Chem. Soc.* **1976**, *98*, 1743; b) R. J. Bergeron and M. A. Channing, *J. Am. Chem. Soc.* **1979**, *101*, 2511.

[4.80] L. Pang and M. A. Whitehead, *Supramol. Chem.* **1992**, *1*, 81.

[4.81] J.-P. Kintzinger, F. Kotzyba-Hibert, J.-M. Lehn, A. Pagelot and K. Saigo, *J. Chem. Soc. Chem. Commun.* **1981**, 833.

Chapter 5

[5.1]   E. Fischer, in *Nobel Lectures — Chemistry 1901–1921*, Elsevier, Amsterdam, **1966**, p. 34.

[5.2]   J. B. S. Haldane, *Enzymes*, Longmann, Green and Co., London **1930**, MIT Press, Cambridge, Mass. **1965**.

[5.3]   L. Pauling, *Chem. Eng. News* **1946**, 1375.

[5.4]   R. L. Schowen, in *Transition States in Biochemical Processes*, (Eds.: R. D. Gandour and R. L. Schowen), Plenum, New York, **1988**; J. Kraut, *Science* **1988**, *242*, 533.

[5.5]   W. P. Jencks, *Catalysis in Chemistry and Enzymology*, McGraw-Hill, New York, **1969**.

[5.6]   a) F. M. Menger, *Acc. Chem. Res.* **1985**, *18*, 128; b) *ibid.* **1993**, *26*, 206 and references therein.

[5.7]   a) T. H. Fife, *Adv. Phys. Org. Chem.* **1975**, *11*, 1; b) A. Fersht, *Enzyme Structure and Mechanism*, W. H. Freeman, New York, **1977**.

[5.8]   a) R. Breslow, *Chem. Soc. Rev.* **1972**, *1*, 553; b) *Science* **1982**, *218*, 532; c) *Adv. Enzymol. Relat. Areas Mol. Biol.* **1986**, *58*, 1; d) in [A.18], **1984**, *Vol. 3*, 473.

[5.9]   V. T. D'Souza and M. L. Bender, *Acc. Chem. Res.* **1987**, *20*, 146.

[5.10] I. Tabushi, *Acc. Chem. Res.* **1982**, *15*, 66; b) in [A.18] **1984**, *Vol. 3*, 445; c) *Pure Appl. Chem.* **1986**, *58*, 1529.

[5.11] a) I. Tabushi and Y. Kuroda, *Adv. Catal.* **1983**, *32*, 417; b) O. S. Tee, *Adv. Phys. Org. Chem.* **1994**, *29*, 1.

[5.12] J.-M. Lehn, *Pure Appl. Chem.* **1979**, *51*, 979; *Ann. N. Y. Acad. Sci.* **1986**, *471*, 41.

[5.13] C. Sirlin, *Bull. Soc. Chim. Fr. II* **1984**, 5; M. W. Hosseini, *La Recherche* **1989**, *20*, 24.

[5.14] J.-M. Lehn and T. Nishiya, *Chem. Lett.* **1987**, 215.

[5.15] Y. Chao, G. R. Weisman, G. D. Y. Sogah and D. J. Cram, *J. Am. Chem. Soc.* **1979**, *101*, 4948.

[5.16] J.-M. Lehn and C. Sirlin, *J. Chem. Soc., Chem. Commun.* **1978**, 949; *New J. Chem.* **1987**, *11*, 693.

[5.17] S. Sasaki and K. Koga, *Heterocycles* **1979**, *12*, 1305.

[5.18] D. J. Cram, H. E. Katz and I. B. Dicker, *J. Am. Chem. Soc.* **1984**, *106*, 4987.

[5.19] J.-P. Behr, J.-M. Lehn, *J. Chem. Soc. Chem. Commun.* **1978**, 143.

[5.20] R. M. Kellogg, *Topics Curr. Chem.* **1982**, *101*, 111; *Angew. Chem.* **1984**, *96*, 769; *Angew. Chem. Int. Ed. Engl.* **1984**, *23*, 782.

[5.21] M. W. Hosseini, J.-M. Lehn, L. Maggiora, K. B. Mertes and M. P. Mertes, *J. Am. Chem. Soc.* **1987**, *109*, 537; M. W. Hosseini, J.-M. Lehn, K. C. Jones, K. E. Plute, K. B. Mertes and M. P. Mertes, *J. Am. Chem. Soc.* **1989**, *111*, 6330.

[5.22] For a detailed review of the work on phosphoryl transfer catalysis by macrocyclic polyamines see: M. W. Hosseini, *Bioorg. Chem. Frontiers* **1993**, *3*, 67.

[5.23] G. M. Blackburn, G. R. J. Thatcher, M. W. Hosseini and J.-M. Lehn, *Tetrahedron Lett.* **1987**, *28*, 2779.

[5.24]   a) P. Tecilla, S.-K. Chang and A. D. Hamilton, *J. Am. Chem. Soc.* 1990, *112*, 9586;
         b) E. V. Anslyn, J. Smith, D M. Kneeland, K. Ariga and F.-Y. Chu, *Supramolec.*
         Chem. 1993, *1*, 201; c) R. Gross, G. Dürner and M. W. Göbel, *Liebigs Ann. Chem.*
         1994, 49.

[5.25]   a) M. A. Podyminogin, V. V. Vlassov and R. Giegé, *Nucleic Acids Research* 1993,
         21, 5950; b) J.-M. Lehn, A. Lorrente, J.-P. Vigneron and K. Watson, unpublished
         studies; c) A. Fkyerat, M. Demeunynck, J.-F. Constant, P. Michon and J. Lhomme,
         J. Am. Chem. Soc. 1993, *115*, 9952.

[5.26]   a) D. G. Knorre and V. V. Vlassov, *Progress Nucl. Acid. Res. and Mol. Biol.* 1985, *32*,
         291; b) D. G. Knorre, V. V. Vlassov, V. F. Zarytova, A. V. Lebedev and O. S. Federova,
         Design and Targeted Reactions of Oligonucleotide Derivatives, CRC Press, Boca
         Raton, 1994.

[5.27]   P. B. Dervan, in *Nucleic Acids and Molecular Biology*, (Eds.: F. Eckstein and D. M.
         J. Lilley), Springer, Berlin, 1988, *Vol. 2*, 49; T. J. Povsic, S. A. Strobel and P. B. Der-
         van, *J. Am. Chem. Soc.* 1992, *114*, 5934.

[5.28]   a) D. S. Sigman, *Biochemistry* 1990, *29*, 9097; b) D. S. Sigman, T. W. Bruice,
         A. Mazumder and C. L. Sutton, *Acc. Chem. Res.* 1993, *26*, 98; c) D. S. Sigman,
         A. Mazumder and D. M. Perrin, *Chem. Rev.* 1993, *93*, 2295.

[5.29]   J. K. Barton, *Comments Inorg. Chem.* 1985, *3*, 321.

[5.30]   N. T. Thuong and C. Hélène, *Angew. Chem.* 1993, *105*, 697; *Angew. Chem. Int.*
         Ed. Engl. 1993, *32*, 666.

[5.31]   P. E. Nielsen, *J. Mol. Reco.* 1990, *3*, 1.

[5.32]   H. Fenniri and J.-M. Lehn, unpublished results; H. Fenniri, Thèse de Doctorat ès
         Sciences, Université Louis Pasteur, Strasbourg, 1994.

[5.33]   F. P. Schmidtchen, *Topics Curr. Chem.* 1986, *132*, 101.

[5.34]   D. J. Cram and G. D. Y. Sogah, *J. Chem. Soc. Chem. Commun.* 1981, 625.

[5.35]   R. Breslow, R. Rajagopalan and J. Schwarz, *J. Am. Chem. Soc.* 1981, *103*, 2905.

[5.36]   R. Cacciapaglia and L. Mandolini, *Pure Appl. Chem.* 1993, *65*, 533; *Chem. Soc. Rev.*
         1993, 22, 221; R. Cacciapaglia, A. Casnati, L. Mandolini and R. Ungaro, *J. Am.*
         Chem. Soc. 1992, *114*, 10956.

[5.37]   a) Y. Murakami, *Topics Curr. Chem.* 1983, *115*, 107; b) Y. Murakami, in *Proceedings*
         Intern. Symp. Org. Reactions, Kyoto, 1991; c) *Trends in Biotech.* 1992, *10*, 170.

[5.38]   F. Diederich, in [A.24], p. 167.

[5.39]   J. H. Fendler and E. J. Fendler, *Catalysis in Micellar and Macromolecular Systems*,
         Academic Press, New York, 1975.

[5.40]   C. J. O'Connor, R. E. Ramage and A. J. Porter, *Adv. Colloid Interface Sci.* 1981,
         15, 25.

[5.41]   S. Shinkai, *Prog. Polym. Sci.* 1982, *8*, 1.

[5.42]   R. Breslow, A. W. Czarnik, M. Lauer, R. Leppkes, J. Winkler and S. Zimmerman,
         J. Am. Chem. Soc. 1986, *108*, 1969.

[5.43]   B. J. Whitlock and H. W. Whitlock, Jr., *Tetrahedron Lett.* 1988, *29*, 6047.

[5.44]   L. Jimenez and F. Diederich, *Tetrahedron Lett.* 1989, *30*, 2759.

[5.45]   D. A. Stauffer, R. E. Barrans, Jr. and D. A. Dougherty, *Angew. Chem.* 1990, *102*,
         953; *Angew. Chem. Int. Ed. Engl.* 1990, *29*, 915.

[5.46]   I. Tabushi, N. Shimizu, T. Sugimoto, M. Shiozuka and K. Yamamura, *J. Am. Chem.*
         Soc. 1977, *99*, 7100.

[5.47]   a) R. Breslow and S. Singh, *Bioorg. Chem.* 1988, *16*, 408; b) R. Breslow and
         B. Zhang, *J. Am. Chem. Soc.* 1992, *114*, 5882.

[5.48]   K. D. Karlin, *Science* 1993, *261*, 701.

[5.49]   E. U. Akkaya and A. W. Czarnik, *J. Phys. Org. Chem.* 1992, *5*, 540.

[5.50]   J. Chin, *Acc. Chem. Res.* **1991**, *24*, 145; *Bioorg. Chem. Frontiers* **1991**, 2, 175.

[5.51]   E. Kimura and T. Koike, *Comments Inorg. Chem.* **1991**, *11*, 285.

[5.52]   P. Scrimin, P. Tecilla and U. Tonellato, *J. Phys. Org. Chem.* **1992**, *5*, 619.

[5.53]   a) R. Breslow, A. B. Brown, R. D. McCullough and P. W. White, *J. Am. Chem. Soc.* **1989**, *111*, 4517; b) J. T. Croves and R. Neumann, *J. Am. Chem. Soc.* **1987**, *109*, 5045.

[5.54]   H. K. A. C. Coolen, P. W. N. M. van Leeuwen and R. J. Nolte, *Angew. Chem.* **1992**, *104*, 906; *Angew. Chem. Int. Ed. Engl.* **1992**, *31*, 905.

[5.55]   a) D. R. Benson, R. Valentekovich and F. Diederich, *Angew. Chem.* **1990**, *102*, 213; *Angew. Chem. Int. Ed., Engl.* **1990**, *29*, 191; D. R. Benson, R. Valentekovich, S.-W. Tam and F. Diederich, *Helv. Chim. Acta* **1993**, *76*, 2034; b) B. Coltrain, Y. Deng, N. Herron, P. Padolik and D. H. Busch, *Pure Appl. Chem.* **1993**, *65*, 367; c) L. Weber, I. Imiolzyk, G. Haufe, D. Rehorek and H. Hennig, *J. Chem. Soc., Chem. Commun.* **1992**, 301.

[5.56]   R. Breslow, P. J. Duggan and J. P. Light, *J. Am. Chem. Soc.* **1992**, *114*, 3982.

[5.57]   a) I. G. Muller, X. Chen, A. C. Dadiz, S. E. Rokita and C. J. Burrows, *Pure Appl. Chem.* **1993**, *65*, 545; b) X. Chen, C. J. Burrows and S. E. Rokita, *J. Am. Chem. Soc.* **1992**, *114*, 322; c) J. F. Kinneary, T. M. Roy, J. S. Albert, H. Yoon, T. R. Wagler, L. Shen and C. J. Burrows, *J. Incl. Phenom., Mol. Recogn. Chem.* **1989**, *7*, 155; d) C. J. Burrows and S. E. Rokita, *Acc. Chem. Res.* **1994**, *27*, 295.

[5.58]   a) N. N. Murthy, M. Mahroof-Tahir and K. D. Karlin, *J. Am. Chem. Soc.* **1993**, *115*, 10404; b) N. M. Kostic, *Comments Inorg. Chem.* **1988**, *8*, 137.

[5.59]   a) *Catalytic Asymmetric Synthesis*, (Ed.: I. Ojima), VCH, New York, **1993**; b) *Asymmetric Catalysis*, (Ed.: B. Bosnich), Martinus Nijhoff Publ., Dordrecht, **1986**; c) J. Seyden-Penne, *Synthèse et Catalyse Asymétriques*, InterEditions/CNRS, Paris, **1994**; d) R. Noyori, *Asymmetric Catalysis in Organic Synthesis*, Wiley, Chichester, **1994**.

[5.60]   M. W. Hosseini, J.-M. Lehn, *J. Chem. Soc. Chem. Commun.* **1985**, 1155; *J. Am. Chem. Soc.* **1987**, *109*, 7047.

[5.61]   P. G. Yohannes, M. P. Mertes, K. B. Mertes, *J. Am. Chem. Soc.* **1985**, *107*, 8288.

[5.62]   a) M. W. Hosseini and J.-M. Lehn, *J. Chem. Soc., Chem. Commun.* **1988**, 397; b) M. W. Hosseini and J.-M. Lehn, *J. Chem. Soc. Chem. Commun.* **1991**, 451.

[5.63]   H. Fenniri and J.-M. Lehn, *J. Chem. Soc., Chem. Commun.* **1993**, 1819.

[5.64]   a) T. R. Kelly, G. J. Bridger and C. Zhao, *J. Am. Chem. Soc.* **1990**, *112*, 8024, b) C. J. Walter, H. L. Anderson and J. K. M. Sanders, *J. Chem. Soc., Chem. Commun.* **1993**, 458; for a similar acyl transfer process see: L. G. Mackay, R. S. Wylie and J. K. M. Sanders, *J. Am. Chem. Soc.* **1994**, *116*, 3141.

[5.65]   a) F. Diederich and H.-D. Lutter, *J. Am. Chem. Soc.* **1989**, *111*, 8438; b) H.-J. Schneider, R. Kramer and J. Rammo, *J. Am. Chem. Soc.* **1993**, *115*, 8980.

[5.66]   R. Breslow, *Supramolec. Chem.* **1993**, *1*, 111.

[5.67]   J. Wolfe, A. Muehldorf and J. Rebek, Jr., *J. Am. Chem. Soc.* **1991**, *113*, 1453.

[5.68]   S. Sasaki, M. Shionoya and K. Koga, *J. Am. Chem. Soc.* **1985**, *107*, 3371.

[5.69]   G. Zuber, C. Sirlin and J.-P. Behr, *J. Am. Chem. Soc.* **1993**, *115*, 4939.

[5.70]   E. T. Kaiser, D. S. Lawrence, *Science (Washington)* **1984**, *226*, 505.

[5.71]   See, for instance A. R. Fersht, J.-P. Shi, A. J. Wilkinson, D. M. Blow, P. Carter, M. M. Y. Waye and G. P. Winter, *Angew. Chem.* **1984**, *96*, 475; *Angew. Chem. Int. Ed. Engl.* **1984**, *23*, 467; J. A. Gerlt, *Chem. Rev.* **1987**, *87*, 1079; A. J. Russell and A. R. Fersht, *Nature* **1987**, *328*, 496.

[5.72]   F. Kohen, J. B. Kim, H. R. Lindner, Z. Eshhar and B. S. Green, *FEBS Lett.* **1980**, *111*, 427.

[5.73]   See ref. [5.5] p. 288.

[5.74]   A. Tramontano, K. D. Janda and R. A. Lerner, *Science (Washington)* **1986**, *234*, 1566; S. J. Pollack, J. W. Jacobs and P. G. Schultz, *Science (Washington)* **1986**, *234*, 1570.

[5.75]   a) R. A. Lerner, S. J. Benkovic and P. G. Schultz, *Science* **1991**, *252*, 659; b) P. G. Schultz and R. A. Lerner, *Acc. Chem. Res.* **1993**, *26*, 391; c) B. S. Green and D. S. Tawfik, *Trends Biotechnol.* **1989**, *7*, 304.

[5.76]   S. J. Benkovic, *Annu. Rev. Biochem.* **1992**, *61*, 29; D. Hilvert, *Pure Appl. Chem.* **1992**, *64*, 1103, *Acc. Chem. Res.* **1993**, *26*, 552; J. D. Stewart, L. J. Liotta and S. J. Benkovic, *Acc. Chem. Res.* **1993**, *26*, 396; J. D. Stewart and S. J. Benkovic, *Chem. Soc. Rev.* **1993**, *22*, 213.

[5.77]   F. M. Menger and M. Ladika, *J. Am. Chem. Soc.* **1987**, *109*, 3145.

Chapter 6

[6.1]   J.-M. Lehn in *Physical Chemistry of Transmembrane Ion Motions*, (Ed.: G. Spach), Elsevier, Amsterdam, **1983**, p. 181.

[6.2]   W. Simon, W. E. Morf and P. C. Meier, *Struct. Bonding* **1973**, *16*, 113; W. E. Morf, D. Amman, R. Bissig, E. Pretsch and W. Simon in [A.17], *Vol. 1*, p. 1.

[6.3]   J.-P. Behr and J.-M. Lehn, *J. Am. Chem. Soc.* **1973**, *95*, 6108.

[6.4.]   M. Kirch and J.-M. Lehn, *Angew. Chem.* **1975**, *87*, 542; *Angew. Chem. Int. Ed. Engl.* **1975**, *14*, 555; M. Kirch, *Thèse de Doctorat-ès-Sciences*, Université Louis Pasteur, Strasbourg, **1980**.

[6.5]   *Membranes, Vol. 2*, (Ed.: G. Eisenman), Marcel Dekker, New York, **1973**,.

[6.6]   W. E. Morf, D. Amman, R. Bissig, E. Prestch and W. Simon, in [A.17], **1979**, *Vol. 1*, 1; W. Simon and W. E. Morf, in [6.5], p. 329.

[6.7]   E. Grell, T. Funck and F. Eggers, in *Membranes*, (Ed.: G. Eisenman), Marcel Dekker, New-York, **1975**, *Vol. 3*, 1.

[6.8]   G. W. Liesegang and E. M. Eyring in [A.15], p. 245.

[6.9]   a) D. W. McBride, Jr., R. M. Izatt, J. D. Lamb and J. J. Christensen, in [A.18], *Vol. 3*, **1984**, 571; b) for the basic kinetics see: K. D. Neame and T. G. Richards, *Elementary Kinetics of Membrane Carrier Transport*, Blackwell, Oxford, **1972**.

[6.10]   a) H. Tsukube, *J. Coord. Chem.* **1987**, *16*, 101; b) W. F. van Straaten-Nijenhuis, F. de Jong and D. N. Reinhoudt, *Recl. Trav. Chim. Pays-Bas* **1993**, *112*, 317; c) H. C. Visser, D. N. Reinhoudt and F. de Jong, *Chem. Soc. Rev.* **1994**, *23*, 75.

[6.11]   a) T. M. Fyles, *Bioorg. Chem. Frontiers* **1991**, *1*, 71; b) in [A.25], p 59.

[6.12]   a) P. R. Brown and R. A. Bartsch, in [A.25], p. 1; b) R.A. Bartsch, I.-W. Yang, E.-G. Jeon, W. Walkowiak and W. A. Charewicz, *J. Coord. Chem.* **1992**, *27*, 75.

[6.13]   G. De Santis, M. Di Casa, L. Fabbrizzi, A Forlini, M. Licchelli, C. Mangano, J. Mocák, P. Pallavicini, A. Poggi and B. Seghi, *J. Coord. Chem.* **1992**, *27*, 39.

[6.14]   Y. Takeda, *Topics Curr. Chem.* **1984**, *121*, 1.

[6.15]   R. L. Bruening, R. M. Izatt and J. S. Bradshaw, in [A.22], p. 111.

[6.16]   J. D. Lamb, J. J. Christensen, J. L. Oscarson, B. L. Nielsen, B. W. Asay and R. M. Izatt, *J. Am. Chem. Soc.* **1980**, *102*, 6820.

[6.17]   J.-P. Behr, M. Kirch and J.-M. Lehn, *J. Am. Chem. Soc.* **1985**, *107*, 241.

[6.18]   T. M. Fyles, *Can. J. Chem.* **1987**, *65*, 884.

[6.19]   M. Castaing, F. Morel and J.-M. Lehn, *J. Membr. Biol.* **1986**, *89*, 251; M. Castaing and J.-M. Lehn, *ibid.* **1987**, *97*, 79; M. Castaing, J.-L. Kraus, P. Beaufils and J. Ricard, *Biophys. Chem.* **1991**, *41*, 203.

[6.20]   W. F. Nijenhuis, E. G. Buitenhuis, F. de Jong, E. J. R. Sudhölter and D. N. Rein-houdt, *J. Am. Chem. Soc.* **1991**, *113*, 7963.

[6.21]   J.-M. Lehn, in *The Neurobiology of Lithium, Neurosciences Res. Prog. Bull.*, (Eds.: W. E. Bunney, Jr. and D. L. Murphy), **1976**, *14/2*, 133; M. Güggi, M. Oehme, E. Pretsch and W. Simon, *Helv. Chim. Acta* **1976**, *58*, 2417; R. Margalit and G. Eisenman, *J. Membrane Biol.* **1981**, *61*, 209; R. Margalit and A. Shanzer, *Biochim. Biophys. Acta* **1981**, *649*, 441.

[6.22]   E. Bacon, L. Jung and J.-M. Lehn, *J. Chem. Res. (S)* **1980**, 136.

[6.23]   M. Newcomb, J. L. Toner, R. C. Hegelson and D. J. Cram, *J. Am. Chem. Soc.* **1979**, *101*, 4941.

[6.24]   T. B. Stolwijk, E. J. R. Sudhölter and D. N. Reinhoudt, *J. Am. Chem. Soc.* **1989**, *111*, 6321.

[6.25]   a) H. Tsukube, *Angew. Chem.* **1982**, *94*, 312; *Angew. Chem. Int. Ed. Engl.* **1982**, *21*, 304; *Angew. Chem. Suppl.* **1982**, 575; b) U. Wuthier, H. V. Pham, E. Pretsch, D. Ammann, A. K. Beck, D. Seebach and W. Simon, *Helv. Chim. Acta* **1985**, *68*, 1822; H-V. Pham, E. Pretsch, K. Fluri, A. Bezegh and W. Simon, *Helv. Chim. Acta* **1990**, *73*, 1894.

[6.26]   B. Dietrich, T. M. Fyles, M. W. Hosseini, J.-M. Lehn and K. C. Kaye, *J. Chem. Soc. Chem. Commun.* **1988**, 691.

[6.27]   J. L. Sessler, D. A. Ford, M. J. Cyr and H. Furuta, *J. Chem. Soc. Chem. Commun.* **1991**, 1733.

[6.28]   I. Tabushi, Y. Kobuke and J.-i Imuta, *J. Am. Chem. Soc.* **1981**, *103*, 6152.

[6.29]   H. Furuta, M. J. Cyr and J. L. Sessler, *J. Am. Chem. Soc.* **1991**, *113*, 6677.

[6.30]   a) T. Li and F. Diederich, *J. Org. Chem.* **1992**, *57*, 3449; b) T. Li, S. J. Krasne, B. Persson, H. R. Kaback and F. Diederich, *J. Org. Chem.* **1993**, *58*, 380.

[6.31]   G. Deslongchamps, A. Galán, J. de Mendoza and J. Rebek, Jr., *Angew. Chem.* **1992**, *104*, 58; *Angew. Chem. Int. Ed. Engl.* **1992**, *31*, 61.

[6.32]   V. Král, J. L. Sessler and H. Furuta, *J. Am. Chem. Soc.* **1992**, *114*, 8704.

[6.33]   A. D. Miller, *Nature* **1992**, *357*, 455.

[6.34]   R. C. Mulligan, *Science* **1993**, *260*, 926.

[6.35]   D. T. Curiel, E. Wagner, M. Cotten, M. L. Birnstiel, S. Agarwal, C.-M. Li, S. Loechel and P.-C. Hu, *Human Gene Therapy* **1992**, *3*, 147.

[6.36]   J.-P. Behr, *Acc. Chem. Res.* **1993**, *26*, 274.

[6.37]   a) J.-M. Lehn, C. Fouquey, L. Vergely and J.-P. Vigneron, work in progress; b) J. G. Smith, R. L. Walzem and J. B. German, *Biochim., Biophys. Acta* **1993**, *1154*, 327.

[6.38]   E. S. Matulevicius and N. N. Li, *Sep. Purif. Methods* **1975**, *4*, 73.

[6.39]   J.-M. Lehn, A. Moradpour and J.-P. Behr, *J. Am. Chem. Soc.* **1975**, *97*, 2532.

[6.40]   F. Diederich and K. Dick, *J. Am. Chem. Soc.* **1984**, *106*, 8024.

[6.41]   A. Harada and S. Takahashi, *J. Chem. Soc., Chem. Commun.* **1987**, 527.

[6.42]   a) W. F. van Straaten-Nijenhuis, A. R. van Doorn, A. M. Reichwein, F. de Jong and D. N. Reinhoudt, *J. Org. Chem.* **1993**, *58*, 2265; b) H. Furuta, K. Furuta and J. Sessler, *J. Am. Chem. Soc.* **1991**, *113*, 4706.

[6.43]   B. G. Malmström, *Acc. Chem. Res.* **1993**, *26*, 332.

[6.44]   a) J. K. Hurst and D. H. P. Thompson, *J. Membrane Sci.* **1986**, *28*, 3; b) J. K. Hurst, in *Kinetics and Catalysis in Microheterogeneous Systems*, (Eds,: M. Grätzel and K. Kalyanasundaram), Marcel Dekker, New York, **1991**, p. 183.

[6.45]   T. Matsuo, *Pure Appl. Chem.* **1982**, *54*, 1693.

[6.46]   J. N. Robinson and D. J. Cole-Hamilton, *Chem. Soc. Rev.* **1991**, *20*, 49.

[6.47]   J. J. Grimaldi and J.-M. Lehn, *J. Am. Chem. Soc.* **1979**, *101*, 1333.

[6.48]   S. S. Anderson, I. G. Lyle and R. Paterson, *Nature* **1976**, *259*, 147.

[6.49]   J. J. Grimaldi, S. Boileau and J.-M. Lehn, *Nature* **1977**, *265*, 229.

[6.50]   a) A. E. Kaifer and L. Echegoyen, in [A.22] p. 363; Z. Chen, G. W. Gokel and L. Echegoyen, *J. Org. Chem.* **1991**, *56*, 3369; b) T. Saji and I. Kinoshita, *J. Chem. Soc. Chem. Commun.* **1986**, 716.

[6.51]   a) G. de Santis, L. Fabbrizzi, M. Liccheli, A. Monichino and P. Pallavicini, *J. Chem. Soc. Dalton Trans.* **1992**, 2219 and references therein; b) I. Tabushi and S.-i. Kugimiya, *Tetrahedron Lett.* **1984**, *25*, 3723.

[6.52]   M. Ohakara and Y. Nakatsuji, *Topics Curr. Chem.* **1985**, *128*, 37.

[6.53]   A. Hriciga and J.-M. Lehn, *Proc. Natl. Acad. Sci. USA* **1983**, *80*, 6426.

[6.54]   Y. Nakatsuji, T. Inoue, M. Wada and M. Okahara, *J. Incl. Phenom. Mol. Reco. Chem.* **1991**, *10*, 379.

[6.55]   a) A. Spisni, L. Franzoni, R. Corradini, R. Marchelli and A. Dossena, *J. Molec. Reco.* **1989**, *2*, 94; b) A. Biancardi, R. Marchelli and A. Dossena, *ibid.* **1992**, *5*, 139.

[6.56]   L. M. Dulyea, T. M. Fyles and G. D. Robertson, *J. Membrane Sci.* **1987**, *34*, 87.

[6.57]   M. Okahara, Y. Nakatsuji, M. Sakamoto and M. Watanabe, *J. Incl. Phenom. Mol. Reco. Chem.* **1992**, *12*, 199.

[6.58]   M. Calvin, *Acc. Chem. Res.* **1978**, *11*, 369; M. Calvin, I. Willner, C. Laane and J. W. Otvos, *J. Photochem.* **1981**, *17*, 195.

[6.59]   D. G. Whitten, *Acc. Chem. Res.* **1980**, *13*, 83.

[6.60]   *Photochemical Conversion and Storage of Solar Energy*, (Ed.: J. S. Connolly), Academic Press, New York, **1981**.

[6.61]   J.-M. Lehn, in [6.60], p. 161; J.-M. Lehn, in *Proc. 8th Intern. Congress on Catalysis*, *Vol. 1*, Verlag Chemie, Weinheim, **1984**, p. 63.

[6.62]   R. Frank and H. Rau, *Z. Naturforsch.* **1982**, *A 37*, 1253.

[6.63]   a) S. Shinkai and O. Manabe, *Topics Curr. Chem.* **1984**, *121*, 67; b) S. Shinkai, in [A.22], p. 397; c) S. Shinkai, *Pure Appl. Chem.* **1987**, *59*, 425.

[6.64]   a) T. Osa and J. Ansai, in [A.25] p. 157; b) J.-I. Anzai and T. Osa, *Tetrahedron* **1994**, *50*, 4039.

[6.65]   a) J. D. Winkler, K. Deshayes and B. Shao, *J. Am. Chem. Soc.* **1989**, *111*, 769; b) H. Sasaki, A. Ueno, J.-i. Anzai and T. Osa, *Bull. Chem. Soc. Japan* **1986**, *59*, 1953; c) M. Inouye, M. Ueno, K. Tsuchiya, N. Nakayama, T. Konishi and T. Kitao, *J. Org. Chem.* **1992**, *57*, 5377; d) K. Kimura, T. Yamashita and M. Yokoyama, *J. Chem. Soc. Perkin Trans.* **1992**, 613.

[6.66]   a) D. W. Urry, *Topics Curr. Chem.* **1985**, *128*, 175; b) Gramicidin channel, structural data: B. A. Wallace and K. Ravikumar, *Science* **1988**, *241*, 182; D. A. Langs, *ibid.* p. 188.

[6.67]   R. Nagaraj and P. Balaram, *Acc. Chem. Res.* **1981**, *14*, 356.

[6.68]   R. O. Fox, Jr. and F. M. Richards, *Nature* **1982**, *300*, 325.

[6.69]   P. Läuger, *Angew. Chem.* **1985**, *97*, 939; *Angew. Chem. Int. Ed. Engl.* **1985**, *24*, 905.

[6.70]   F. M. Menger, in [A.24], p. 193.

[6.71]   J.-P. Behr, J.-M. Lehn, A.-C. Dock and D. Moras, *Nature* **1982**, *295*, 526.

[6.72]   U. F. Kragten, M. F. M. Roks and R. J. M. Nolte, *J. Chem. Soc., Chem. Commun.* **1985**, 1275.

[6.73]   I. Tabushi, Y. Kuroda and K. Yokota, *Tetrahedron Lett.* **1982**, *23*, 4601.

[6.74]   A. Nakano, Q. Xie, J. V. Mallen, L. Echegoyen and G. W. Gokel, *J. Am. Chem. Soc.* **1990**, *112*, 1287.

[6.75]   a) J.-H. Fuhrhop and U. Liman, *J. Am. Chem. Soc.* **1984**, *106*, 4643; b) J.-H. Fuhrhop, U. Liman and H. H. David, *Angew. Chem.* **1985**, *97*, 3371; *Angew. Chem. Int. Ed. Engl.* **1985**, *24*, 339.

[6.76]   a) J.-M. Lehn and J. Simon, *Helv. Chim. Acta* 1977, *60*, 141; b) see also: A. Nakano, Y. Li, P. Geoffroy, M. Kim, J. L. Atwood, S. Bott, H. Zhang, L. Echegoyen and G. W. Gokel, *Tetrahedron Lett.* 1989, *30*, 5099.

[6.77]   J.-M. Lehn and M. E. Stubbs, *J. Am. Chem. Soc.* 1974, *96*, 4011.

[6.78]   F. Ohseto and S. Shinkai, *Chem. Lett.* 1993, 2045.

[6.79]   a) W. Fischer, J. Brickmann and P. Läuger, *Biophys. Chem.* 1981, *13*, 105; W. Fischer and J. Brickmann, *ibid.* 1983, *18*, 323; b) for theoretical studies of ion transfer across an interface, see for instance: I. Benjamin, *Science* 1993, *261*, 1558; R. Brasseur, M. Notredame and J.-M. Ruysschaert, *Biochem. Biophys. Res. Comm.* 1983, *114*, 632.

[6.80]   B. Roux and M. Karplus, *J. Am. Chem. Soc.* 1993, *115*, 3250.

[6.81]   R. Langer, *Science* 1990, *249*, 1527.

Chapter 7

[7.1]   J. H. Fendler, *Membrane Mimetic Chemistry*, Wiley, New York, 1982.

[7.2]   a) H. Kuhn and D. Moebius, *Angew. Chem.* 1971, *83*, 672; *Angew. Chem. Int. Ed. Engl.* 1971, *10*, 620; b) H. Kuhn, *J. Photochem.* 1979, *10*, 111; *Pure Appl. Chem.* 1981, *53*, 2105.

[7.3]   D. Moebius, *Acc. Chem. Res.* 1981, *14*, 63; *Ber. Bunsenges. Phys. Chem.* 1978, *82*, 848; *Z. Phys. Chem. (Munich)* 1987, *154*, 121.

[7.4]   a) G. G. Roberts, *Adv. Phys.* 1985, 34, 475; b) A. Ulman, *Ultrathin Organic Films: From Langmuir–Blodgett to Self-Assembly*, Academic Press, Boston, 1991.

[7.5]   a) J. D. Swalen, D. L. Allara, J. D. Andrade, E. A. Chandross, S. Garoff, J. Israelachvili, T. J. McCarthy, R. Murray, R. F. Pease, J. F. Rabolt, K. J. Wynne and H. Yu, *Langmuir* 1987, *3*, 932; b) L. Netzer and J. Sagiv, *J. Am. Chem. Soc.* 1983, *105*, 674.

[7.6]   H. M. McConnell, L. K. Tamm and R. W. Weiss, *Proc. Natl. Acad. Sci. USA* 1984, *81*, 3249; R. M. Weiss and H. M. McConnell, *Nature* 1984, *310*, 47.

[7.7]   T. Kunitake, Y. Okahata, M. Shimomura, S.-i. Yasunami and K. Takarabe, *J. Am. Chem. Soc.* 1981, *103*, 5401; N. Nakashima, S. Asakuma and T. Kunitake, *ibid.* 1985, *107*, 509.

[7.8]   a) J.-H. Fuhrhop and M. Krull, in [A.24], p. 223; b) G. H. Escamilla and G. R. Newkome, *Angew. Chem.* 1994, *106*, 2013; *Angew. Chem. Int. Ed. Engl.* 1994, *33*, 1937; c) H. Hoffmann and G. Ebert, *Angew. Chem.* 1988, *100*, 933; *Angew. Chem. Int. Ed. Engl.* 1988, *27*, 902; d) H. Hoffmann, *Adv. Mater.* 1994, *6*, 116.

[7.9]   H.-H. Hub, B. Hupfer, H. Koch and H. Ringsdorf, *Angew. Chem.* 1980, *92*, 962; *Angew. Chem. Int. Ed. Engl.* 1980, *19*, 938; L. Gros, H. Ringsdorf and H. Schupp, *ibid.* 1981, *93*, 311 and 1981, *20*, 305.

[7.10]   H. Ringsdorf, B. Schlarb and J. Venzmer, *Angew. Chem.* 1988, *100*, 117; *Angew. Chem. Int. Ed. Engl.* 1988, *27*, 113.

[7.11]   G. Wegner, *Chimia* 1982, *36*, 63.

[7.12]   a) C. M. Paleos, *Chem. Rev.* 1985, *14*, 45; b) G. D. Rees and B. H. Robinson, *Adv. Mater.* 1993, *5*, 608.

[7.13]   a) *Polymerization in Organized Media*, (Ed.: C. M. Paleos), Gordon and Breach, Philadelphia, 1992; b) S. Kobayashi and H. Uyama, *Polish J. Chem.* 1994, *68*, 417.

[7.14]   a) J. M. Thomas, *Angew. Chem.* 1988, *100*, 1735; *Angew. Chem. Int. Ed. Engl.* 1988, *27*, 1673; b) for the catalytic aspects, see: J. M. Thomas, *Angew. Chem.* 1994, *106*, 963; *Angew. Chem. Int. Ed. Engl.* 1994, *33*, 913.

[7.15]   a) Y. Izumi, K. Urabe and M. Onaka, *Zeolite, Clay and Heteropolyacids in Orga-nic Reactions*, Kodansha, Tokyo; VCH, Weinheim, **1992**; b) J. M. Thomas and C. R. Theocaris, in [A.18], *Vol. 5*, **1991**, 104.

[7.16]   a) *Guidelines for Mastering the Properties of Molecular Sieves*, (Ed.: D. Bartho-meuf), Plenum, New York, **1990**; b) E. G. Derouane, *ibid.* p. 225.

[7.17]   G. A. Ozin, *Chem. Mater.* **1992**, *4*, 511.

[7.18]   G. A. Ozin, *Adv. Mater.* **1992**, *4*, 612.

[7.19]   a) C. J. Brinker and G. W. Scherer, *Sol-Gel Science, The Physics and Chemistry of Sol-Gel Processing*, Academic Press, New York, **1990**; b) C. Sanchez and F. Ribot, *New. J. Chem.* **1994**, *18*, 1007.

[7.20]   H. Dislich, *Angew. Chem.* **1971**, *83*, 428; *Angew. Chem. Int. Ed. Engl.* **1971**, *10*, 363.

[7.21]   For an extensive description see various chapters in [A.18].

[7.22]   E. Weber, *J. Mol. Graphics* **1989**, *7*, 12.

[7.23]   a) D. Worsch and F. Vögtle, in [A.32], *140*, 21; b) I. Goldberg, in [A.32], *149*, 1.

[7.24]   a) F. Toda, in [A.32], *140*, 43; b) in [A.26], *vol 2*, p. 141; c) in [A.32], *149*, 211.

[7.25]   R. Arad-Yellin, B. S. Green, M. Knossow and G. Tsoucaris, in [A.18], *Vol. 3*, p. 263.

[7.26]   R.M. Barrer, in [A.18], *Vol. 1*, p. 190.

[7.27]   R. Schöllhorn, in [A.18], *Vol. 1*, p. 248.

[7.28]   T. E. Mallouk and H. Lee, *J. Chem. Educ.* **1990**, *67*, 829.

[7.29]   a) E. G. Derouane, L. Maistriau, Z. Gabelica, A. Tuel, J. B. Nagy and R. Von Ball-moos, *Applied Catal.* **1989**, *51*, L 13; b) J. S. Beck, J. C. Vartuli, W. J. Roth, M. E. Leo-nowicz, C. T. Kresge, K. D. Schmitt, C. T.-W. Chu, D. H. Olson, E. W. Sheppard, S. B. McCullen, J. B. Higgins and J. C. Schlenker, *J. Am. Chem. Soc.* **1992**, *114*, 10834.

[7.30]   F. Dougnier, J. Patarin, J. L. Guth and D. Anglerot, *Zeolites* **1992**, *12*, 160; J. L. Guth, P. Caullet, A. Seive, J. Patarin and F. Delprato in [7.16a]; F. Delprato, L. Del-motte, J. L. Guth and L. Huve, *Zeolites* **1990**, *10*, 546; S. D. Kinrade and D. L. Pole, *Inorg. Chem.* **1992**, *31*, 4558.

[7.31]   T. J. Pinnavaia, *Science* **1983**, *220*, 365.

[7.32]   a) K. J. Shea, D. A. Loy and O. Webster, *J. Am. Chem. Soc.* **1992**, *114*, 6700; b) K. J. Shea, O. Webster and D. A. Loy, *Mat. Res. Soc. Symp. Proc.* **1990**, *180*, 975; c) D. A. Archibald and S. Mann, *Nature* **1993**, *364*, 430; d) R. Dagani, *Chem. Eng. News* **1993**, August 9, 19; e) H. R. Allcock, *Science* **1992**, *255*, 1106; *Adv. Mater* **1994**, *6*, 106.

[7.33]   a) G. Cao, H.-G. Hong and T. E. Mallouk, *Acc. Chem. Res.* **1992**, *25*, 420; b) G. Alberti, U. Costantino, F. Marmottini, R. Vivani and P. Zappelli, *Angew. Chem.* **1993**, *105*, 1396; *Angew. Chem. Int. Ed. Engl.* **1993**, *32*, 1357; c) R. C. Haus-halter and L. A. Mundi, *Materials* **1992**, *4*, 31.

[7.34]   a) G. Wulff, *ACS Symp. Ser.* **1986**, *308*, 186; b) G. Wulff, *TIBTECH* **1993**, *11*, 85.

[7.35]   a) K. Mosbach, *TIBS* **1994**, *19*, 9; b) G. Vlatakis, L. I. Andersson, R. Müller and K. Mos-bach, *Nature* **1993**, *361*, 645; c) A. Moradian and K. Mosbach, *J. Mol. Recogn.* **1989**, *2*, 167; d) B. Sellergren and K. G. I. Nilsson, *Methods Mol. Cell Biol.* **1989**, *1*, 59.

[7.36]   A. G. Amit, R. A. Mariuzza, S. E. V. Phillips and R. J. Poljak, *Science* **1986**, *233*, 747.

[7.37]   H. M. Geysen, J. A. Tainer, S. J. Rodda, T. J. Mason, H. Alexander, E. D. Getzoff and R. A. Lerner, *Science* **1987**, *235*, 1184; D. R. Burton, *TIBS* **1990**, *15*, 64.

[7.38]   a) M. M. Harding and J.-M. Lehn, unpublished work, see in [1.7]; b) S. Sakai and T. Sasaki, *J. Am. Chem. Soc.* **1994**, *116*, 1587.

[7.39]   a) L. Addadi, Z. Berkovitch-Yellin, I. Weissbuch, J. van Mil, L. J. W. Shimon, M. Lahav and L. Leiserowitz, *Angew. Chem.* **1985**, *97*, 476; *Angew. Chem. Int. Ed.*

Engl. 1985, *24*, 466; b) L. Addadi and M. Lahav, *Pure and Appl. Chem.* 1979, *51*, 1269.

[7.40]   a) L. Addadi, Z. Berkovitch-Yellin, I. Weissbuch, J. Van Mil, M. Lahav and L. Leiserowitz, in [A.19], p. 245; b) for biomineralisation see: *Biomineralization: Chemical and Biological Perspectives*, (Eds.: S. Mann, J. Webb and R. J. P. Williams, VCH, Weinheim, 1989; L. Addadi and S. Weiner, *Angew. Chem.* 1992, *104*, 159; *Angew. Chem. Int. Ed. Engl.* 1992, *31*, 153.

[7.41]   I. Weissbuch, L. Addadi, M. Lahav and L. Leiserowitz, *Science* 1991, *253*, 637.

[7.42]   R. J. Davey, S. N. Black, L. A. Bromley, D. Cottier, B. Dobbs and J. E. Rout, *Nature* 1991, *353*, 549.

[7.43]   a) M. Ahlers, W. Müller, A. Reichert, H. Ringsdorf and J. Venzmer, *Angew. Chem.* 1990, *102*, 1310; *Angew. Chem. Int. Ed. Engl.* 1990, *29*, 1269; b) L. Häussling, W. Knoll, H. Ringsdorf, F.-J. Schmitt and J. Yang, *Makromol. Chem. Macromol. Symp.* 1991, *46*, 145.

[7.44]   W. Knoll, L. Angermaier, G. Batz, T. Fritz, S. Fujisawa, T. Furuno, H.-J. Guder, M. Hara, M. Liley, K. Niki and J. Spinke, *Synthetic Metals* 1993, *61*, 5.

[7.45]   T. Kunitake, *Angew. Chem.* 1992, *104*, 742; *Angew. Chem. Int. Ed. Engl.* 1992, *31*, 709.

[7.46]   L. Häussling, B. Michel, H. Ringsdorf and H. Rohrer, *Angew. Chem.* 1990, *103*, 568; *Angew. Chem. Int. Ed. Engl.* 1990, *30*, 569.

[7.47]   a) D. Y. Sasaki, K. Kurihara and T. Kunitake, *J. Am. Chem. Soc.* 1991, *113*, 9685, b) *ibid.* 1992, *114*, 10994; c) Y. Ikeura, K. Kurihara and T. Kunitake, *J. Am. Chem. Soc.* 1991, *113*, 7342

[7.48]   N. Kimizuka, T. Kawasaki and T. Kunitake, *J. Am. Chem. Soc.* 1993, *115*, 4387.

[7.49]   a) E. M. Arnett, N. G. Harvey and P. L. Rose, *Acc. Chem. Res.* 1989, *22*, 131; b) J. G. Heath and E. M. Arnett, *J. Am. Chem. Soc.* 1992, *114*, 4500; c) P. L. Rose, N. G. Harvey and E. M. Arnett, *Adv. Phys. Org. Chem.* 1993, *28*, 45.

[7.50]   K. B. Eisenthal, *Acc. Chem. Res.* 1993, *26*, 636.

[7.51]   L. Gros, H. Ringsdorf and H. Schupp, *Angew. Chem.* 1981, *93*, 311; *Angew. Chem. Int. Ed. Engl.* 1981, *20*, 305.

[7.52]   a) J. N. Weinstein, *Pure Appl. Chem.* 1981, *53*, 2241; b) M. C. Annesini, A. Finazzi-Agrò and G. Mossa, *Chim. Oggi* 1992, 11.

[7.53]   D. D. Lasic, *Liposomes: From Physics to Applications*, Elsevier, Amsterdam, 1993.

[7.54]   H. Kitano, N. Kato and N. Ise, *J. Am. Chem. Soc.* 1989, *111*, 6809.

[7.55]   D. A. Tomalia, A. M. Naylor and W. A. Goddard III, *Angew. Chem.* 1990, *102*, 119; *Angew. Chem. Int. Ed. Engl.* 1990, *29*, 138.

[7.56]   D. A. Tomalia and H. D. Durst, *Topics Curr. Chem.* 1993, *165*, 193.

[7.57]   H.-B. Mekelburger, W. Jaworek and F. Vögtle, *Angew. Chem.* 1992, *104*, 1609; *Angew. Chem. Int. Ed. Engl.* 1992, *31*, 1571.

[7.58]   G. R. Newkome, Z.-q. Yao, G. R. Baker, V. K. Gupta, P. S. Russo and M. J. Saunders, *J. Am. Chem. Soc.* 1986, *108*, 849; G. R. Newkome, G. R. Baker, M. J. Saunders, P. S. Russo, V. K. Gupta, Z.-q. Yao, J. E. Miller and K. Bouillion, *J. Chem. Soc. Chem. Commun.* 1986, 752.

[7.59]   a) G. R. Newkome, C. N. Moorefield and G. R. Baker, *Aldrichimica Acta* 1992, *25*, 31; b) G. R. Newkome, C. N. Moorefield, G. R. Baker, A. L. Johnson and R. K. Behera, *Angew. Chem.* 1991, *103*, 1205; *Angew. Chem. Int. Ed. Engl.* 1991, *30*, 1176; c) G. R. Newkome, C. N. Moorefield, G. R. Baker and M. J. Saunders, *Angew. Chem.* 1991, *103*, 1207; *Angew. Chem. Int. Ed. Engl.* 1991, *30*, 1178.

[7.60]   J. M. J. Fréchet, *Science* 1994, *263*, 1710.

[7.61]   S. Serroni, G. Denti, S. Campagna, A. Juris, M. Ciano and V. Balzani, *Angew. Chem.* 1992, *104*, 1540; *Angew. Chem. Int. Ed. Engl.* 1992, *31*, 1493.

[7.62]   G. R. Newkome, F. Cardullo, E. C. Constable, C. N. Moorefield and A. M. W. C. Thompson, *J. Chem. Soc. Chem. Commun.* **1993**, 925.

[7.63]   G. M. Edelman, *Topobiology*, Basic Books, New York, **1988**.

[7.64]   J. H. Fendler and E. J. Fendler, *Catalysis in Micellar and Macromolecular Systems*, Academic Press, New York, **1975**.

[7.65]   C. J. O'Connor, R. E. Ramage and A. J. Porter, *Adv. Colloid Interf. Sci.* **1981**, *15*, 25.

[7.66]   a) J. Sunamoto, in *Solution Behavior of Surfactants — Theoretical and Applied Aspects*, (Eds.: K. L. Mittal and E. J. Fendler), Plenum, New York, **1982**, p. 767; b) M. P. Pileni, *J. Phys. Chem.* **1993**, *97*, 6961.

[7.67]   R. A. Moss, K. Y. Kim and S. Swarup, *J. Am. Chem. Soc.* **1986**, *108*, 788.

[7.68]   T. Kunitake and S. Shinkai, *Adv. Phys. Org. Chem.* **1980**, *17*, 435.

[7.69]   Y. Murakami, A. Nakano, A. Yoshimatsu and K. Fukuya, *J. Am. Chem. Soc.* **1981**, *103*, 728.

[7.70]   a) Y. Okahata, H.-J. Lim, G.-i. Nakamura, S. Hachiya, *J. Am. Chem. Soc.* **1983**, *105*, 4855; Y. Okahata, K. Ariga and T. Seki, *J. Am. Chem. Soc.* **1988**, *110*, 2495; b) R. Ahuja, P.-L. Caruso, D. Möbius, W. Paulus, H. Ringsdorf and G. Wildburg, *Angew. Chem.* **1993**, *105*, 1082; *Angew. Chem. Int. Ed. Engl.* **1993**, *32*, 1033.

[7.71]   a) P. C. Mitchell, *Chem. Ind.* **1991**, 308; b) D. Losset, G. Dupas, J. Duflos, J. Bourguignon and G. Queguiner, *Bull. Soc. Chim. France* **1991**, *128*, 721; c) D. Avnir, S. Braun, O. Lev and M. Ottolenghi, *Chem. Materials* **1994**, *6*, 1605.

[7.72]   H. B. Bürgi and J. D. Dunitz, *Acc. Chem. Res.* **1983**, *16*, 153.

[7.73]   a) K. Morihara, S. Kurihara and J. Suzuki, *Bull. Soc. Chim. Japan* **1988**, *61*, 3991; b) J. Heilmann and W. F. Maier, *Angew. Chem.* **1994**, *106*, 491; *Angew. Chem. Int. Ed. Engl.* **1994**, *33*, 471.

[7.74]   a) K. Morihara, E. Nishihata, M. Kojima and S. Miyake, *Bull. Chem. Soc. Japan* **1988**, *61*, 3999; b) T. Shimada, R. Hirose and K. Morihara, *Bull. Chem. Soc. Japan* **1994**, *67*, 227; c) K. Morihara, M. Kurokawa, Y. Kamata and T. Shimada, *J. Chem. Soc., Chem. Commun.* **1992**, 358; d) K. Morihara, S. Kawasaki, M. Kofuji and T. Shimada, *Bull. Chem. Soc. Japan* **1993**, *66*, 906.

[7.75]   G. Wulff, in *Biomimetic Polymers*, (Ed.: C. G. Gebelein), Plenum, New York, **1990**, p. 1; G. Wulff, in *Bioorganic Chemistry in Healthcare and Technology*, (Eds.: U. K. Pandit and F. C. Alderweireldt), Plenum, New York, **1991**, p. 55.

[7.76]   G. Wulff and J. Vietmeier, *Makromol. Chem.* **1989**, *190*, 1727.

[7.77]   D. K. Robinson and K. Mosbach, *J. Chem. Soc., Chem. Commun.* **1989**, 969.

[7.78]   a) A. Barraud, *Vacuum* **1990**, *41*, 1624; *Pour la Science* **1993**, *189*, 62; b) *Langmuir–Blodgett Films*, (Ed.: G. G. Roberts), Plenum, New York, **1990**.

[7.79]   H. Kuhn, *Thin Solid Films* **1989**, *178*, 1.

[7.80]   J.-H. Fuhrhop and J. Mathieu, *Angew. Chem.* **1984**, *96*, 124; *Angew. Chem. Int. Ed. Engl.* **1984**, *23*, 100.

[7.81]   J.-H. Fuhrhop and D. Fritsch, *Acc. Chem. Res.* **1986**, *19*, 130.

[7.82]   Y. Okahata, *Acc. Chem. Res.* **1986**, *19*, 57.

[7.83]   a) N. J. Turro, J. K. Barton and D. A. Tomalia, *Acc. Chem. Res.* **1991**, *24*, 332; b) for processes on DNA support see: S. M. Risser, D. N. Beratan and T. J. Meade, *J. Am. Chem. Soc.* **1993**, *115*, 2508; C. J. Murphy, M. R. Arkin, Y. Jenkins, N. D. Ghatlia, S. H. Bossmann, N. J. Turro and J. K. Barton, *Science* **1993**, *262*, 1025; R. F. Pasternack, A. Giannetto, P. Pagano and E. J. Gibbs, *J. Am. Chem Soc.* **1991**, *113*, 7799; A. M. Brun and A. Harriman, *J. Am. Chem. Soc.* **1994**, *116*, 10383.

[7.84]   S. V. Lymar, V. N. Parmon and K. I. Zamaraev, *Topics Curr. Chem.* **1991**, *159*, 1.

[7.85]   M. A. Fox, *Topics Curr. Chem.* **1991**, *159*, 67.

[7.86]   I. Willner and B. Willner, *Topics Curr. Chem.* **1991**, *159*, 153.

[7.87]   G. Decher, *Nachr. Chem. Tech. Lab.* **1993**, *41*, 793.
[7.88]   a) S. Palacin, A. Ruaudel-Teixier and A. Barraud, *J. Phys. Chem.* **1989**, *93*, 7195; b) F. Porteu, S. Palacin, A. Ruaudel-Teixier and A. Barraud, *J. Phys. Chem.* **1991**, *95*, 7438; c) F. Porteu, S. Palacin, A. Ruaudel-Teixier, A. Barraud, *Mol. Cryst. Liq. Cryst.* **1992**, *211*, 193.
[7.89]   I. Yamazaki, N. Tamai and T. Yamazaki, *J. Phys. Chem.* **1990**, *94*, 516.
[7.90]   H. Tachibana and M. Matsumoto, *Adv. Mater.* **1993**, *5*, 796.

Chapter 8

[8.1]   Semiology is the general theory of signs; see also [1.17]. Semiochemicals are signalling chemicals that an organism can detect in its environment; a) F. E. Regnier, *Biol. Reprod.* **1971**, *4*, 309; b) D. R. Kelly, *Chem. Brit.* **1990**, *26*, 24; c) H. J. Bestmann and O. Vostrowsky, *Chem. unser. Zeit* **1993**, *27*, 123. See also for instance: *Mammalian Semiochemistry. The Investigation of Chemical Signals between Mammals*, (Eds.: E. S. Albone and S. G. Shirley), Wiley, Chichester, **1984**.
[8.2]   J.-M. Lehn in [A.20], p. 29.
[8.3]   V. Balzani, L. Moggi and F. Scandola in [A.20], p. 1.
[8.4]   A. Caron, J. Guilhem, C. Riche, C. Pascard, B. Alpha and J.-M. Lehn, *Helv. Chim. Acta* **1985**, *68*, 1577.
[8.5]   B. Alpha, E. Anklam, R. Deschenaux, J.-M. Lehn and M. Pietraszkiewicz, *Helv. Chim. Acta* **1988**, *71*, 1042.
[8.6]   M. Cesario, J. Guilhem, C. Pascard, E. Anklam, J.-M. Lehn and M. Pietraszkiewicz, *Helv. Chim. Acta* **1991**, *74*, 1157.
[8.7]   I. Bkouche-Waksman, J. Guilhem, C. Pascard, B. Alpha, R. Deschenaux and J.-M. Lehn, *Helv. Chim. Acta* **1991**, *74*, 1163.
[8.8]   J.-M. Lehn and J.-B. Regnouf de Vains, *Helv. Chim. Acta* **1992**, *75*, 1221.
[8.9]   J.-M. Lehn, M. Pietraszkiewicz and J. Karpiuk, *Helv. Chim. Acta* **1990**, *73*, 106.
[8.10]   J.-M. Lehn and C. O. Roth, *Helv. Chim. Acta* **1991**, *74*, 572.
[8.11]   L. Prodi, M. Maestri, V. Balzani, J.-M. Lehn and C. O. Roth, *Chem. Phys. Letters* **1991**, *180*, 45.
[8.12]   B. Alpha, V. Balzani, J.-M. Lehn, S. Perathoner and N. Sabbatini, *Angew. Chem.* **1987**, *99*, 1310; *Angew. Chem. Int. Ed. Engl.* **1987**, *26*, 1266.
[8.13]   N. Sabbatini, S. Perathoner, V. Balzani, B. Alpha and J.-M. Lehn in [A.20], p. 187.
[8.14]   V. Balzani, *Gazz. Chim. Ital.* **1989**, *119*, 311.
[8.15]   B. Alpha, R. Ballardini, V. Balzani, J.-M. Lehn, S. Perathoner and N. Sabbatini, *Photochem. Photobiol.* **1990**, *52*, 299.
[8.16]   G. Blasse, G. J. Dirksen, D. Van der Voort, N. Sabbatini, S Perathoner and J.-M. Lehn, *Chem. Phys. Lett.* **1988**, *146*, 347.
[8.17]   G. Blasse, G. J. Dirksen, N. Sabbatini, S. Perathoner, J.-M. Lehn and B. Alpha, *J. Phys. Chem.* **1988**, *92*, 2419.
[8.18]   V. Balzani, J.-M. Lehn, J. van de Loosdrecht, A. Mecati, N. Sabbatini and R. Ziessel, *Angew. Chem.* **1991**, *103*, 186; *Angew. Chem. Int. Ed. Engl.* **1991**, *30*, 190.
[8.19]   a) K. Watson and J.-M. Lehn, unpublished work; K. Watson, Thèse de Doctorat, 1992, Université Louis Pasteur, Strasbourg; b) C. O. Roth and J.-M. Lehn, unpublished work; C. O. Roth, Thèse de Doctorat, 1992, Université Louis Pasteur, Strasbourg.
[8.20]   a) O. Prat, E. Lopez and G. Mathis, *Anal. Biochem.* **1991**, *195*, 283; b) E. Lopez, C. Chypre, B. Alpha and G. Mathis, *Clin. Chem.* **1993**, *39*, 196.

[8.21]   a) G. Mathis, *Clin. Chem.* **1993**, *39*, 1953; b) for related processes see also: A. Oser and G. Valet, *Angew. Chem.* **1990**, *102*, 1197; *Angew. Chem. Int. Ed. Engl.* **1990**, *29*, 117; P. R. Selvin, T. M. Rana and J. E. Hearst, *J. Am. Chem. Soc.* **1994**, *116*, 6029; P. R. Selvin and J. E. Hearst, *Proc. Natl. Acad. Sci. USA* **1994**, *91*, 10024; c) E. P. Diamandis and T. K. Christopoulos, *Anal. Chem.* **1990**, *62*, 1149A; E. P. Diamandis, *Clin. Chem.* **1991**, *37*, 1486; A. Mayer and S. Neuenhofer, *Angew. Chem.* **1994**, *106*, 1097; *Angew. Chem. Int. Ed. Engl.* **1994**, *33*, 1044.

[8.22]   P. Tundo and J. H. Fendler, *J. Am. Chem. Soc.* **1980**, *102*, 1760.

[8.23]   J.-C. G. Bünzli, P. Froidevaux and J. Harrowfield, *Inorg. Chem.* **1993**, *32*, 3306.

[8.24]   a) Z. Pikramenou and D. G. Nocera, *Inorg. Chem.*, **1992**, *31*, 532; b) R. Deschenaux, M. M. Harding and T. Ruch, *J. Chem. Soc. Perkin Trans. 2* **1993**, 1251.

[8.25]   G. Denti, S. Serroni, S. Campagna, V. Ricevuto and V. Balzani, *Coord. Chem. Rev.* **1991**, *111*, 227.

[8.26]   S. Campagna G. Denti, S. Serroni, M. Ciano, A. Juris and V. Balzani, *Inorg. Chem.* **1992**, *31*, 2982.

[8.27]   a) G. Denti, S. Campagna, S. Serroni, M. Ciano and V. Balzani, *J. Am. Chem. Soc.* **1992**, *114*, 2944; b) S. Campagna, G. Denti, S. Serroni, M. Ciano, A. Juris and V. Balzani, *Inorg. Chem.* **1992**, *31*, 2982; c) V. Balzani, A. Credi and F. Scandola, in [A.40] p. 1.

[8.28]   B. Valeur, in *Fluorescent Biomolecules. Methodologies and Applications*, (Eds.: D. M. Jameson and G. D. Reinhart), Plenum, New York, **1989**, p. 269.

[8.29]   a) J. Davila, A. Harriman and L. R. Milgrom, *Chem. Phys. Lett.* **1987**, *136*, 427; b) F. Effenberger, H. Schlosser, P. Bäuerle, S. Maier, H. Port and H. C. Wolf, *Angew. Chem.* **1988**, *100*, 274; *Angew. Chem. Int. Ed. Engl.* **1988**, *27*, 281; c) T. Nagata, A. Osuka and K. Maruyama, *J. Am. Chem. Soc.* **1990**, *112*, 3054; d) J. L. Sessler, V. L. Capuano and A. Harriman, *J. Am. Chem. Soc.* **1993**, *115*, 4618; e) S. Prathapan, T. E. Johnson and J. S. Lindsey, *J. Am. Chem. Soc.* **1993**, *115*, 7519; f) R. W. Wagner and J. S. Lindsey, *J. Am. Chem. Soc.* **1994**, *116*, 9759; g) for a dendritic type molecular antenna, see: Z. Wu and J. S. Moore, *Acta Polymer.* **1994**, *45*, 83.

[8.30]   a) P. Tecilla, R. P. Dixon, G. Slobodkin, D. S. Alavi, D. H. Waldeck and A. D. Hamilton, *J. Am. Chem. Soc.* **1990**, *112*, 9408; b) C. Turró, C. K. Chang, G. E. Leroi, R. I. Cukier and D. G. Nocera, *J. Am. Chem. Soc.* **1992**, *114*, 4013.

[8.31]   a) A. Harriman, D. J. Magda and J. L. Sessler, *J. Phys. Chem.* **1991**, *95*, 1530; b) A. Harriman, Y. Kubo and J. L. Sessler, *J. Am. Chem. Soc.* **1992**, *114*, 388; c) J. L. Sessler, B. Wang and A. Harriman, *J. Am. Chem. Soc.* **1993**, *115*, 10418.

[8.32]   a) M. N. Berberan-Santos, J. Canceill, J.-C. Brochon, L. Jullien, J.-M. Lehn, J. Pouget, P. Tauc and B. Valeur, *J. Am. Chem. Soc.* **1992** *114*, 6427; b) M. N. Berberan-Santos, J. Pouget, B. Valeur, J. Canceill, L. Jullien and J.-M. Lehn, *J. Phys. Chem.* **1993**, *97*, 11376; c) L. Jullien, J. Canceill, B. Valeur, E. Bardez and J.-M. Lehn, *Angew. Chem.* **1994**, *106*, 2582; *Angew. Chem. Int. Ed. Engl.* **1994**, *33*, 2438; d) R. W. Wagner and J. S. Lindsey, *J. Am. Chem. Soc.* **1994**, *116*, 9759.

[8.33]   B. Blanzat, C. Bathou, N. Tercier, J.-J. André and J. Simon, *J. Am. Chem. Soc.* **1987**, *109*, 6193; D. Markovitsi, I. Lécuyer and J. Simon, *J. Phys. Chem.* **1991**, *95*, 3620.

[8.34]   M. Takagi and K. Ueno, *Topics Curr. Chem.* **1984**, *121*, 39.

[8.35]   H.-G. Löhr and F. Vögtle, *Acc. Chem. Res.* **1985**, *18*, 65.

[8.36]   a) C. Reichardt, *Chem. Soc. Rev.* **1992**, 147; b) Z. Pikramenou, J. Yu, R. B. Lessard, A. Ponce, P. A. Wong and D. G. Nocera, *Coord. Chem. Reviews* **1994**, *132*, 181.

[8.37]   a) J.-P. Desvergne, F. Fages, H. Bouas-Laurent and P. Marsau, *Pure Appl. Chem.* **1992**, *64*, 1231; b) H. Bouas-Laurent, J.-P. Desvergne, F. Fages and P. Marsau, in

[A.24], p. 265; c) H. Bouas-Laurent, J.-P. Desvergne, F. Fages and P. Marsau, in *Fluorescent Chemosensors for Ion and Molecule Recognition*, (Ed.: A. W. Czarnik), ACS Symp. Ser. No. 538, **1993**.

[8.38]   S. Misumi, *Topics Curr. Chem.* **1993**, *165*, 163.

[8.39]   R. A. Bissell, A. P. de Silva, H. Q. N. Gunaratne, P. L. M. Lynch, G. E. M. Maguire and K. R. A. S. Sandanayake, *Chem. Soc. Rev.* **1992**, *21*, 187.

[8.40]   R. A. Bissell, A. P. de Silva, H. Q. N. Gunaratne, P. L. M. Lynch, G. E. M. Maguire, C. P. McCoy and K. R. A. S. Sandanayake *Topics Curr. Chem.* **1993**, *168*, 223.

[8.41]   R. Y. Tsien, *Ann. Rev. Neurosci.* **1989**, *12*, 227.

[8.42]   a) *Fluorescent Chemosensors for Ion and Molecular Recognition*, (Ed.; A. W. Czarnik), ACS Symp. Ser. No. 538, **1993**; b) A. W. Czarnik, *Acc. Chem. Res.* **1994**, *27*, 302.

[8.43]   a) W. Zazulak, E. Chapoteau, B. P. Czech and A. Kumar, *J. Org. Chem.* **1992**, *57*, 6720; b) E. Chapoteau, B. P. Czech, W. Zazulak and A. Kumar, *Clin. Chem.* **1992**, *38*, 1654; c) R. Klink, D. Bodart, J.-M. Lehn, B. Helfert and R. Bitsch, Merck Patent GmbH, *Eur. Pat. Appl.* **1983**, No. 83100281.1.

[8.44]   A. F. Sholl and I. O. Sutherland, *J. Chem. Soc. Chem. Commun.* **1992**, 1716.

[8.45]   a) K. R. A. L. Sandanayake and I. O. Sutherland, *Tetrahedron Lett.* **1993**, *34*, 3165; b) see also: M. Dolman and I. O. Sutherland, *J. Chem. Soc., Chem. Commun.* **1993**, 1793.

[8.46]   J. P. Konopelski, F. Kotzyba-Hibert, J.-M. Lehn, J.-P. Desvergne, F. Fagès, A. Castellan and H. Bouas-Laurent, *J. Chem. Soc. Chem. Commun.* **1985**, 433; F. Fages, J.-P. Desvergne, H. Bouas-Laurent, P. Marsau, J.-M. Lehn, F. Kotzyba-Hibert, A.-M. Albrecht-Gary and M. Al-Joubbeh, *J. Am. Chem. Soc.* **1989**, *111*, 8672.

[8.47]   F. Fages, J.-P. Desvergne, H. Bouas-Laurent, J.-M. Lehn, J. P. Konopelski, P. Marsau and Y. Barrans, *J. Chem. Soc. Chem. Commun.* **1990**, 655.

[8.48]   a) K. Golchini, M. Mackovic-Basic, S. A. Gharib, D. Masilamani, M. E. Lucas and I. Kurtz, *Am. J. Physiol.* **1990**, *258*, F 438; b) R. Warmuth, B. Gersch, F. Kastenholz, J.-M. Lehn, E. Bamberg and E. Grell, in *The Sodium Pump*, (Eds.: E. Bamberg and W. Schoner, Steinkopf, Darmstadt, **1994**, p. 621.

[8.49]   J. Bourson, J. Pouget and B. Valeur, *J. Phys. Chem.* **1993**, *97*, 4552.

[8.50]   G. A. Smith, T. R. Hesketh and J. C. Metcalfe, *Biochem. J.* **1988**, *250*, 227.

[8.51]   R. A. Bartsch, D. A. Babb, B. P. Czech, D. H. Desai, E. Chapoteau, C. R. Gebauer, W. Zazulak and A. Kumar, *J. Incl. Phenom. Molec. Reco. Chem.* **1990**, *9*, 113.

[8.52]   D. J. Cram, R. A. Carmack and R. C. Hegelson, *J. Am. Chem. Soc.* **1988**, *110*, 571.

[8.53]   M. McCarrick, B. Wu, S. J. Harris, D. Diamond, G. Barrett and M. A. McKervey, *J. Chem. Soc., Chem. Commun.* **1992**, 1287.

[8.54]   a) F. Fages, J.-P. Desvergne, K. Kampke, H. Bouas-Laurent, J.-M. Lehn, M. Meyer and A-M. Albrecht-Gary, *J. Am. Chem. Soc.* **1993**, *115*, 3658; b) R. Ballardini, V. Balzani, A. Credi, M. T. Gandolfi, F. Kotzyba-Hibert, J.-M. Lehn and L. Prodi, *J. Am. Chem. Soc.* **1994**, *116*, 5741.

[8.55]   a) I. Aoki, T. Harada, T. Sakaki, Y. Kawahara and S. Shinkai, *J. Chem. Soc., Chem. Commun.* **1992**, 1341; b) M. Inouye, K. Kim and T. Kitao, *J. Am. Chem. Soc.* **1992**, *114*, 778.

[8.56]   a) K. Hamasaki, H. Ikeda, A. Nakamura, A. Ueno, F. Toda, I. Suzuki and T. Osa, *J. Am. Chem. Soc.* **1993**, *115*, 5035; b) S. Minato, T. Osa and A. Ueno, *J. Chem. Soc. Chem. Commun.* **1991**, 107.

[8.57]   M. E. Huston, E. U. Akkaya and A. W. Czarnik, *J. Am. Chem. Soc.* **1989**, *111*, 8735.

[8.58]   S. A. Van Arman and A. W. Czarnik, *Supramol. Chem.* **1993**, *1*, 99.

[8.59]   S. O. Wolfbeis, in *Molecular Luminescence Spectroscopy: Methods and Applications*, (Ed.: S. G. Shulman), Wiley, New York, 1988, p. 129.

[8.60]   G. Boisdé and J. J. Perez, *La Vie des Sciences, Compt. Rend. Acad. Sci., Sér. Gén.* 1988, *5*, 303.

[8.61]   T. Goto and T. Kondo, *Angew. Chem.* 1991, *103*, 17; *Angew. Chem. Int. Ed. Engl.* 1991, *30*, 17; R. Brouillard, *Phytochemistry* 1983, *22*, 1311.

[8.62]   a) D. Gust and T. A. Moore, *Topics Curr. Chem.* 1991, *159*, 103; b) D. Gust, T. A. Moore, A. L. Moore, A. N. Macpherson, A. Lopez, J. M. DeGraziano, I. Gouni, E. Bittersmann, G. R. Seely, F. Gao, R. A. Nieman, X. C. Ma, L. J. Demanche, S.-C. Hung, D. K. Luttrull, S.-J. Lee and P. K. Kerrigan, *J. Am. Chem. Soc.* 1993, *115*, 11141; c) D. Gust, T. A. Moore and A. L. Moore, *Acc. Chem. Res.* 1993, *26*, 198; d) D. Gust, T. A. Moore and A. L. Moore, *Ing. Med. Biol.* 1994, 58.

[8.63]   a) S. V. Lymar, V. N. Parmon and K. I. Zamaraev, *Topics Curr. Chem.* 1991, *159*, 1; b) M. A. Fox, *ibid* 1991, *159*, 67.

[8.64]   *Photoinduced Electron Transfer*, (Eds.: M. A. Fox and M. Chanon), Elsevier, New York, 1988.

[8.65]   a) V. Balzani, F. Bolletta, M. T. Gandolfi and M. Maestri, *Topics Curr. Chem.* 1978, *75*, 1; b) N. Sutin, *J. Photochem.* 1979, *10*, 19; c) J. D. Petersen, *Coord. Chem. Rev.* 1985, *64*, 261; d) T. J. Meyer, in *Photochemical Processes in Organized Molecular Systems*, (Ed.: K. Honda), Elsevier, Amsterdam, 1991, p. 133.

[8.66]   I. Willner and B. Willner, *Topics Curr. Chem.* 1991, *159*, 153.

[8.67]   *Photoinduced Electron Transfer, Topics Curr. Chem.* 1990, *156*; 1990, *158*; 1991, *159*; 1992, *163*; 1993, *168*.

[8.68]   M. Calvin, *Acc. Chem. Res.* 1978, *11*, 369; J. R. Norris, Jr., and D. Meisel, *Photochemical Energy Conversion*, Elsevier, New York, 1989.

[8.69]   a) M. R. Wasielewski, M. P. O'Neil, D. Gosztola, M. P. Niemczyk and W. A. Svec, *Pure Appl. Chem.* 1992, *64*, 1319; b) M. R. Wasielewski, *Chem. Rev.* 1992, *92*, 435; c) J.-C. Chambron, A. Harriman, V. Heitz and J.-P. Sauvage, *J. Am. Chem. Soc.* 1993, *115*, 6109; J.-C. Chambron, S. Chardon-Noblat, A. Harriman, V. Heitz and J.-P. Sauvage, *Pure and Appl. Chem.* 1993, *65*, 2343.

[8.70]   a) M. Gubelmann, A. Harriman, J.-M. Lehn and J. L. Sessler, *J. Chem. Soc. Chem. Commun.* 1988, 77; *J. Phys. Chem.* 1990, *94*, 308; b) for a related process see: H. L. Anderson, C. A. Hunter and J. K. M. Sanders, *J. Chem. Soc., Chem. Commun.* 1989, 226.

[8.71]   a) M. N. Paddon-Row, *Acc. Chem. Res.* 1994, *27*, 18; b) A. Helms, D. Heiler and G. McLendon, *J. Am. Chem. Soc.* 1992, *114*, 6227; c) W. E. Jones Jr., S. M. Baxter, S. L. Mecklenburg, B. W. Erickson, B. M. Peek and T. J. Meyer, in [A.28], p. 249.

[8.72]   For an approach to a molecular shift register memory, see J. J. Hopfield, J. N. Onuchic and D. N. Beratan, *J. Phys. Chem.* 1989, *93*, 6350.

[8.73]   M. F. Manfrin, L. Moggi, V. Castelvetro, V. Balzani, M. W. Hosseini and J.-M. Lehn, *J. Am. Chem. Soc.* 1985, *107*, 6888; F. Pina, L. Moggi, M. F. Manfrin, V. Balzani, M. W. Hosseini and J.-M. Lehn, *Gazz. Chim. Ital.* 1989, *119*, 65.

[8.74]   V. Balzani, R. Ballardini, M. T. Gandolfi and L. Prodi, in [A.24] p. 371.

[8.75]   V. Balzani, N. Sabbatini and F. Scandola, *Chem. Rev.* 1986, *86*, 319.

[8.76]   A. J. Parola and F. Pina, *J. Photochem. Photobiol. A: Chem.* 1992, *66*, 337.

[8.77]   M. F. Manfrin, L. Setti and L. Moggi, *Inorg. Chem.* 1992, *31*, 2768.

[8.78]   I. Cabrera, M. Engel, L. Häussling, C. Mertesdorf and H. Ringsdorf, in [A.24] p. 311.

[8.79]   *Non-linear Optical Properties of Organic Molecules and Crystals*, (Eds.: D. S. Chemla and J. Zyss), Academic Press, New York, Vol. *1*, 1986 and Vol. *2*, 1987.

[8.80]   *Nonlinear Optical Properties of Organic and Polymeric Material*, (Ed.: D. J. Williams), A.C.S. Symp. Ser. No. 233, Washington, **1983**.

[8.81]   a) D. J. Williams, *Angew. Chem.* **1984**, *96*, 637; *Angew. Chem. Int. Ed. Engl.* **1984**, *23*, 690; b) A. F. Garito, C. C. Teng, K. Y. Wong and O. Zammani'Khamiri, *Mol. Cryst. Liq. Cryst.* **1984**, *106*, 219.

[8.82]   a) A. M. Glass, *Science* **1984**, *226*, 657; b) D. B. Neal, N. Kalita, C. Pearson, M. C. Petty, J. P. Lloyd, G. G. Roberts, M. M. Ahmad and W. J. Feast, *Synthetic Metals* **1989**, *28*, D711; c) H. Sixl, W. Groh and D. Lupo, *Adv. Mater.* **1989**, *11*, 366; d) J. Simon, P. Bassoul and S. Norvez, *New J. Chem.* **1989**, *13*, 13; e) P. N. Prasad and B. A. Reinhardt, *Chem. Mater.* **1990**, *2*, 660.

[8.83]   *Materials for Nonlinear Optics. Chemical Perpectives*, (Eds.: S. R. Marder, J. E. Sohn and G. D. Stucky), *Adv. Chem. Ser.* **1991**, 455.

[8.84]   a) W. Nie, *Adv. Mater.* **1993**, *5*, 520; b) H. S. Nalwa, *Adv. Mater.* **1993**, *5*, 341; c) S. R. Marder and J. W. Perry, *Adv. Mater.* **1993**, *5*, 804; d) I. Weissbuch, M. Lahav, L. Leiserowitz, G. R. Meredith and H. Vanherzeele, *Chem. Mater.* **1989**, *1*, 114.

[8.85]   a) J.-M. Lehn in [8.79], *Vol. 2*, p. 215; b) J.-M. Lehn in [8.83], p. 436.

[8.86]   D. F. Eaton, A. G. Anderson, W. Tam, W. Mahler and Y. Wang, *Mol. Cryst. Liq. Cryst.* **1992**, *211*, 125.

[8.87]   A. Slama-Schwok, M. Blanchard-Desce and J.-M. Lehn, *J. Phys. Chem.* **1990**, *94*, 3894.

[8.88]   M. Blanchard-Desce, I. Ledoux, J.-M. Lehn, J. Malthête and J. Zyss, *J. Chem. Soc. Chem. Commun.* **1988**, 736.

[8.89]   M. Barzoukas, M. Blanchard-Desce, D. Josse, J.-M. Lehn and J. Zyss, *Chem. Phys.* **1989**, *133*, 323.

[8.90]   a) J. Messier, F. Kajzar, C. Sentein, M. Barzoukas, J. Zyss, M. Blanchard-Desce and J.-M. Lehn, *Mol. Cryst. Liq. Cryst. Sci. Technol., Sec. B* **1992**, *2*, 53; b) F. Meyers, J. L. Brédas and J. Zyss, *J. Am. Chem. Soc.* **1992**, *114*, 2914.

[8.91]   a) S. R. Marder, D. N. Beratan and L.-T. Cheng, *Science* **1991**, *252*, 103; b) S. R. Marder, J. W. Perry, G. Bourhill, C. B. Gorman, B. G. Tieman and K. Mansour, *Science* **1993**, *261*, 186.

[8.92]   S. Palacin, M. Blanchard-Desce, J.-M. Lehn and A. Barraud, *Thin Solid Films* **1989**, *178*, 387.

[8.93]   a) S. Palacin, *Thin Solid Films* **1989**, *178*, 327; b) V. Dentan, M. Blanchard-Desce, S. Palacin, I. Ledoux, A. Barraud, J.-M. Lehn and J. Zyss, *Thin Solid Films* **1992**, *210/211*, 221.

[8.94]   a) S. Tomaru, S. Zembutsu, M. Kawachi and M. Kobayashi, *J. Chem. Soc. Chem. Commun.* **1984**, 1207; b) E. Kelderman, L. Derhaeg, G. J. T. Heesink, W. Verboom, J. F. J. Engbersen, N. F. van Hulst, A. Persoons and D. N. Reinhoudt, *Angew. Chem.* **1992**, *104*, 1107; *Angew. Chem. Int. Ed. Engl.* **1992**, *31*, 1075.

[8.95]   a) G. Mignani, F. Leising, R. Meyrueix and H. Samson, *Tetrahedron Lett.* **1990**, *31*, 4743; b) S. Yitzchaik, G. Berkovic and V. Krongauz, *Chem. Mater.* **1990**, *2*, 162; c) D. R. Kanis, M. A. Ratner and T. J. Marks, *J. Am. Chem. Soc.* **1992**, *114*, 10338; W. M. Laidlaw, R. G. Denning, T. Verbiest, E. Chauchard and A. Persoons, *Nature* **1993**, *363*, 58.

[8.96]   J. Friedrich and D. Haarer, *Angew. Chem.* **1984**, *96*, 96; *Angew. Chem. Int. Ed. Engl.* **1984**, *23*, 113.

[8.97]   D. Haarer and S. Silbey, *Physics Today* May **1990**, 58.

[8.98]   a) U. P. Wild, A. Rebane and A. Renn, *Adv. Mater.* **1991**, *3*, 453; b) U. P. Wild and A. Renn, in *Photochromism*, (Eds.: H. Dürr and H. Bouas-Laurent), Elsevier, Amsterdam, **1990**, p. 930.

[8.99]   a) C. Bräuchle, *Angew. Chem.* **1992**, *104*, 431; *Angew. Chem. Int. Ed. Engl.* **1992**, *31*, 426; b) T. Basché and W. E. Moerner, *Nature* **1992**, *355*, 335.

[8.100]   U. P. Wild, S. Bernet, B. Kohler and A. Renn, *Pure Appl. Chem.* **1992**, *64*, 1335.

[8.101]   I. Renge, *Mol. Cryst. Liq. Cryst.* **1992**, *217*, 121; *J. Opt. Soc. Am. B* **1992**, *9*, 719.

[8.102]   U. P. Wild and A. Renn, *J. Mol. Electr.* **1991**, *7*, 1.

[8.103]   C. Bräuchle, N. Hampp and D. Oesterhelt, *Adv. Mater.* **1991**, *3*, 420.

[8.104]   a) A. F. Garito and A. J. Heeger, *Acc. Chem. Res.* **1974**, *7*, 232; b) A. D. Yoffe, *Chem. Soc. Rev.* **1976**, *5*, 51; c) J. H. Perlstein, *Angew. Chem.* **1977**, *89*, 534; *Angew. Chem. Int. Ed. Engl.* **1977**, *16*, 519; d) M. L. Khidekel and E. I. Zhilyaeva, *Synthetic Metals* **1981**, *4*, 1; e) D. Bloor, *Chem. Brit.* **1983**, 725.

[8.105]   a) *The Physics and Chemistry of Low Dimensional Solids*, (Ed.: L. Alcácer), Redel, Dordrecht, 1980; b) *Extended Linear Chain Compounds*, (Ed.: J. S. Miller), Plenum, New York, 1981; c) *Conjugated Polymers*, (Eds.: J. L. Bredas and R. Silbey), 1991; d) *Handbook of Conducting Polymers*, (Ed.: T. J. Skotheim), Dekker, New York, 1986; e) J.-J. André, A. Bieber and F. Gautier, *Ann. Phys.* **1976**, *1*, 145; f) T. J. Marks, *Angew. Chem.* **1990**, *102*, 886; *Angew. Chem. Int. Ed. Engl.* **1990**, *29*, 857.

[8.106]   F. Garnier, *Angew. Chem.* **1989**, *101*, 529; *Angew. Chem. Int. Ed. Engl.* **1989**, *28*, 513.

[8.107]   a) H. Meier, *Organic Semiconductors*, Verlag Chemie, Weinheim, 1974; b) J. Simon and J.-J. André, *Molecular Semiconductors*, Springer, Berlin, 1985.

[8.108]   a) C. E. D. Chidsey and R. W. Murray, *Science* **1986**, *231*, 25; b) for organic thin-film transistors, see: F. Garnier, G. Horowitz, X. Peng and D. Fichou, *Adv. Mater.* **1990**, 2, 592; J.-P. Bourgoin, M. Vandevyver, A. Barraud, G. Tremblay and P. Hesto, *Molec. Engineering* **1993**, 2, 309.

[8.109]   a) *Molecular Electronic Devices*, (Ed.: F. L. Carter), Dekker, New York, *Vol. 1*, 1982; *Vol. 2*, 1987; b) *Molecular Electronics*, (Ed.: G. J. Ashwell), Wiley, New York, 1992.

[8.110]   a) A. Barraud, *J. Chim. Phys.* **1988**, *85*, 1121; b) B. Tieke, *Adv. Mater.* **1990**, 2, 222; c) M. Dupuis and E. Clementi, in *Biological and Artificial Intelligence Systems*, (Eds.: E. Clementi and S. Chin), ESCOM, Leiden, 1988, 185.

[8.111]   D. Haarer, *Angew. Chem. Adv. Mater.* **1989**, *101*, 1576; *Angew. Chem. Int. Ed. Engl. Adv. Mater.* **1989**, *28*, 1544.

[8.112]   a) H. C. Wolf, *Nachr. Chem. Tech. Lab.* **1989**, *37*, 350; b) A. Barraud, O. Kahn and J.-P. Launay, *Sci. Tech.* **1989**, *15*, 54; c) P. Day, *Chem. Brit.* **1990**, *26*, 52; d) H. Tachibana and M. Matsumoto, *Adv. Mater.* **1993**, *5*, 796; e) J. Baker, *Chem. Britain* **1991**, 728.

[8.113]   a) H. Taube, *Angew. Chem.* **1984**, *96*, 315; *Angew. Chem. Int. Ed. Engl.* **1984**, *23*, 329; b) R. A. Marcus, *Angew. Chem.* **1993**, *105*, 1161; *Angew. Chem. Int. Ed. Engl.* **1993**, *32*, 1111.

[8.114]   a) *New J. Chem.* **1991**, *15*, No. 2/3; b) *Molecular Electronics*, (Eds.: Ch. Ziegler, W. Göpel and G. Zerbi), North Holland, Amsterdam, 1993.

[8.115]   Proceeding of the 1st European Conference on Molecular Electronics, *Molec. Cryst. Liq. Cryst. A*, Special Topics 58, **1993**, 234–236.

[8.116]   For approaches toward molecular rectifiers a) see R. M. Metzger and C. A. Panetta, *J. Chim. Phys.* **1988**, *85*, 1125; *J. Mol. Electronics* **1989**, *5*, 1; b) F. R. Ahmed, P. E. Burrows, K. J. Donovan and E. G. Wilson, *Synth. Met.* **1988**, 27, B 593; c) A. S. Martin and J. R. Sambles, *Adv. Mater.* **1993**, *5*, 580; d) D. Rong and T. E. Mallouk, *Inorg. Chem.* **1993**, *32*, 1454; e) C. Joachim and J.-P. Launay, *J. Molec. Electronics* **1990**, *6*, 37; E. G. Petrov, preprints ITP-94-31E, ITP-94-32E and ITP-

94-33E, Bogolyubov Institute for Theoretical Physics, Ukrainian National Academy of Sciences, Kiev, **1994**.

[8.117] a) K. H. Likharev and T. Claeson, *Scientif. Amer.* **1992**, *267(6)*, 50; b) R. I. Gilmanshin and P. I. Lazarev, *J. Molec. Electr.* **1988**, *4*, S93.

[8.118] P. D. Beer, *Chem. Soc. Rev.* **1989**, *18*, 409.

[8.119] a) P. D. Beer, E. L. Tite and A. Ibbotson, *J. Chem. Soc., Dalton Trans.* **1991**, 1691; b) P. D. Beer, M. G. B. Drew, C. Hazlewood, D. Hesek, J. Hodacova and S. E. Stokes, *J. Chem. Soc., Chem. Commun.* **1993**, 229.

[8.120] a) S. Akabori, Y. Habata, Y. Sakamoto, M. Sato and S. Ebine, *Bull. Chem. Soc. Japan* **1983**, *56*, 537; b) H. Plenio, H. El-Desoky and J. Heinze, *Chem. Ber.* **1993**, *126*, 2403; c) H. Plenio, D. Burth and P. Gockel, *Chem. Ber.* **1993**, *126*, 2585.

[8.121] C. D. Hall, J. H. R. Tucker and S. Y. F. Chu, *Pure Appl. Chem.* **1993**, *65*, 591.

[8.122] H. Bock, B. Hierholzer, F. Vögtle and G. Hollmann, *Angew. Chem.* **1984**, *96*, 74; *Angew. Chem. Int. Ed. Engl.* **1984**, *23*, 57.

[8.123] A. R. Bernardo, J. F. Stoddart and A. E. Kaifer, *J. Am. Chem. Soc.* **1992**, *114*, 10624.

[8.124] a) F. L. Dickert and A. Haunschild, *Adv. Mater.* **1993**, *5*, 887; b) P. Bäuerle and S. Scheib, *Adv. Mater.* **1993**, *5*, 848; c) M. J. Marsella and T. M. Swager, *J. Am. Chem. Soc.* **1993**, *115*, 12214; T. M. Swager and M. J. Marsella, *Adv. Mater.* **1994**, *6*, 595.

[8.125] W. Simon and U. E. Spichiger, *Intern. Lab.* September **1991**, 35; H. Ti Tien, *Adv. Mater.* **1990**, *2*, 316.

[8.126] a) D. N. Reinhoudt and E. J. R. Sudhölter, *Adv. Mater.* **1990**, *2*, 23; b) P. L. H. M. Cobben, R. J. M. Egberink, J. G. Bomer, P. Bergveld, W. Verboom and D. N. Reinhoudt, *J. Am. Chem. Soc.* **1992**, *114*, 10573.

[8.127] G. Hafeman, J. W. Parce and H. M. McConnell, *Science* **1988**, *240*, 1182.

[8.128] L. L. Miller, *Mol. Cryst. Liq. Cryst.* **1988**, *160*, 297.

[8.129] J. Heinze, *Angew. Chem.* **1993**, *105*, 1337; *Angew. Chem. Int. Ed. Engl.* **1993**, *32*, 1268.

[8.130] a) S. Roffia, R. Casadei, F. Paolucci, C. Paradisi, C. A. Bignozzi and F. Scandola, *J. Electroanal. Chem.* **1991**, *302*, 157; b) S. Roffia, M. Marcaccio, C. Paradisi, F. Paolucci, V. Balzani, G. Denti, S. Serroni and S. Campagna, *Inorg. Chem.* **1993**, *32*, 3003; c) M. Baumgarten, W. Huber and K. Müllen, *Adv. Phys. Org. Chem.* **1993**, *28*, 1.

[8.131] A. H. Alberts, J.-M. Lehn and D. Parker, *J. Chem. Soc. Dalton Trans.* **1985**, 2311.

[8.132] D. Astruc, *New J. Chem.* **1992**, *16*, 305; *Acc. Chem. Res.* **1986**, *19*, 377.

[8.133] a) J. L. Fillaut and D. Astruc, *J. Chem. Soc., Chem. Commun.* **1993**, 1320; b) F. Moulines, L. Djakovitch, R. Boese, B. Gloaguen, W. Thiel, J.-L. Fillaut, M.-H. Delville and D. Astruc, *Angew. Chem.* **1993**, *105*, 1132; *Angew. Chem. Int. Ed. Engl.* **1993**, *32*, 1075.

[8.134] B. Steiger and L. Walder, *Helv. Chim. Acta* **1992**, *75*, 90.

[8.135] A. De Blas, G. De Santis, L. Fabbrizzi, M. Licchelli, A. M. Manotti Lanfredi, P. Pallavicini, A. Poggi and F. Ugozzoli, *Inorg. Chem.* **1993**, *32*, 106.

[8.136] a) M. T. Pope and A. Müller, *Angew. Chem.* **1991**, *103*, 56; *Angew. Chem. Int. Ed. Engl.* **1991**, *30*, 34; b) J.-P. Launay, *J. Inorg. Nucl. Chem.* **1976**, *38*, 807.

[8.137] A. V. Xavier, *J. Inorg. Biochem.* **1986**, *28*, 239.

[8.138] D. Lexa, P. Maillard, M. Momenteau and J.-M. Savéant, *J. Phys. Chem.* **1987**, *91*, 1951.

[8.139] L. Hammarström, M. Almgren, J. Lind, G. Merényi, T. Norrby and B. Akermark, *J. Phys. Chem.* **1993**, *97*, 10083.

[8.140] T. S. Arrhenius, M. Blanchard-Desce, M. Dvolaitzky, J.-M. Lehn and J. Malthête, *Proc. Natl. Acad. Sci. USA* **1986**, *83*, 5355; M. Blanchard-Desce, T. S. Arrhenius and J.-M. Lehn, *Bull. Soc. Chim. Fr.* **1993**, *130*, 266.

[8.141]   a) G. Ourisson and Y. Kakatani, in *Carotenoids: Chemistry and Biology*, (Ed.: N. R. Krinsky), Plenum, New York, **1990**, p. 237; b) Y. Nakatani, T. Lazrak, A. Milon, G. Wolff and G. Ourisson, in *General and Applied Aspects of Halophilic Organisms*, (Ed.: F. Rodriguez-Valero), Plenum, New York, **1991**, p. 207.

[8.142]   L. B.-Å. Johansson, M. Blanchard-Desce, M. Almgren and J.-M. Lehn, *J. Phys. Chem.* **1989**, *93*, 6751.

[8.143]   S.-i. Kugimiya, T. Lazrak, M. Blanchard-Desce and J.-M. Lehn, *J. Chem. Soc., Chem. Commun.* **1991**, 1179.

[8.144]   a) A. Slama-Schwok, M. Blanchard-Desce and J.-M. Lehn, *J. Phys. Chem.* **1992**, *96*, 10559; b) L. Hammarström, M. Almgren and T. Norrby, *J. Phys. Chem.* **1992**, *96*, 5017.

[8.145]   a) M. Blanchard-Desce, T. S. Arrhenius, J.-M. Lehn and E. Bamberg, unpublished work; b) M. Kemp, V. Mujica and M. A. Ratner, *J. Chem. Phys.* **1994**, *101*, 5172; V. Mujica, M. Kemp and M. A. Ratner, *J. Chem. Phys.* **1994**, *101*, 6849; *ibid.* **1994**, *101*, 6856.

[8.146]   a) M. J. Crossley and P. L. Burn, *J. Chem. Soc., Chem. Commun.* **1991**, 1569; b) H. L. Anderson, *Inorg. Chem.* **1994**, *33*, 972.

[8.147]   Y. Kobuke, M. Yamanishi, I. Hamachi, H. Kagawa and H. Ogoshi, *J. Chem. Soc., Chem. Commun.* **1991**, 895.

[8.148]   a) P. Bäuerle, *Adv. Mater.* **1992**, *4*, 102; b) D. Delabouglise, M. Hmyene, G. Horowitz, A. Yassar, F. Garnier, *Adv. Mater.* **1992**, *4*, 107.

[8.149]   a) J.-H. Liao, M. Benz, E. LeGoff and M. G. Kanatzidis, *Adv. Mater.* **1994**, *6*, 135; b) S. Hotta and K. Waragai, *Adv. Mater.* **1993**, *5*, 896.

[8.150]   P. W. Kenny and L. L. Miller, *J. Chem. Soc., Chem. Commun.* **1988**, 84.

[8.151]   a) D. M. Dietz, B. J. Stallman, W. S. V. Kwan, J. F. Penneau and L. L. Miller, *J. Chem. Soc., Chem. Commun.* **1990**, 367; b) V. Cammarata, C. J. Kolaskie, L. L. Miller and B. J. Stallman, *J. Chem. Soc., Chem. Commun.* **1990**, 1290.

[8.152]   T. Jørgensen, T. K. Hansen and J. Becher, *Chem. Soc. Rev.* **1994**, *23*, 41.

[8.153]   P. Kaszynski, A. C. Friedli and J. Michl, *J. Am. Chem. Soc.* **1992**, *114*, 601.

[8.154]   H. E. Zimmerman, R. K. King and M. B. Meinhardt, *J. Org. Chem.* **1992**, *57*, 5484.

[8.155]   J. Müller, K. Base, T. F. Magnera and J. Michl, *J. Am. Chem. Soc.* **1992**, 114, 9721.

[8.156]   Y. Ihara, J. Canceill and J.-M. Lehn, unpublished; with high stability constants of the order of 10^5 M^{-1} in aqueous solution.

[8.157]   J.-M. Lehn, J.-P. Vigneron, I. Bkouche-Waksman, J. Guilhem and C. Pascard, *Helv. Chim. Acta* **1992**, *75*, 1069.

[8.158]   a) A. Aviram, *J. Am. Chem. Soc.* **1988**, *110*, 5687; b) A. Farazdel, M. Dupuis, E. Clementi and A. Aviram, *J. Am. Chem. Soc.* **1990**, *112*, 4206.

[8.159]   a) J. M. Tour, R. Wu and J. S. Schumm, *J. Am. Chem. Soc.* **1992**, *113*, 7064; b) J. Nakayama and T. Fujimori, *J. Chem. Soc. Chem. Commun.* **1991**, 1614.

[8.160]   J. M. Savéant, *J. Electroanal. Chem.* **1988**, *242*, 1.

[8.161]   a) A. Heller, *Acc. Chem. Res.* **1990**, *23*, 128; b) A. Heller, *J. Phys. Chem.* **1992**, *96*, 3579.

[8.162]   M. S. Wrighton, *Science*, **1986**, *231*, 32.

[8.163]   a) S. G. Boxer, *Annu. Rev. Biophys. Biophys. Chem.* **1990**, *19*, 267; b) *Structure and Bonding 75. Long Range Electron Transfer in Biology*, Springer Verlag, New York, **1991**; c) C. C. Moser, J. M. Keske, K. Warncke, R. S. Farid and P. L. Dutton, *Nature* **1992**, *355*, 796.

[8.164]   a) D. S. Wuttke, M. J. Bjerrum, J. R. Winkler and H. B. Gray, *Science* **1992**, *256*, 1007; b) P. Siddarth and R. A. Marcus, *J. Phys. Chem.* **1993**, *97*, 13078.

[8.165]   J. Jortner, M. Bixon, H. Heitele and M. E. Michel-Beyerle, *Chem. Phys. Lett.* **1992**, *197*, 131.

[8.166]   J. Jortner and M. Bixon, *Mol. Cryst. Liq. Cryst.* **1993**, *234*, 29.

[8.167] M. Blanchard-Desce and J.-M. Lehn, work in progress.

[8.168] a) P. Seta, E. Bienvenue, A. L. Moore, T. A. Moore and D. Gust, *Electrochim. Acta* **1989**, *34*, 1723; b) A. Lamrabte, M. Momenteau, P. Maillard and P. Seta, *J. Molec. Electr.* **1990**, *6*, 145.

[8.169] a) B. Marczinke, K. J. Przibilla, M. Blanchard-Desce and J.-M. Lehn, unpublished work; b) see also: J. A. Thomas, C. J. Jones, T. A. Hamor, J. A. McCleverty, F. Mabbs, D. Collison and C. Harding, *Mol. Cryst. Liq. Cryst.* **1993**, *234*, 103.

[8.170] A. C. Benniston, V. Goulle, A. Harriman, J.-M. Lehn and B. Marczinke, *J. Phys. Chem.* **1994**, *98*, 7798.

[8.171] S. Woitellier, J.-P. Launay and C. W. Spangler, *Inorg. Chem.* **1989**, *28*, 758.

[8.172] a) A. Haim, *Progr. Inorg. Chem.* **1983**, *30*, 273; b) M. Beley, J.-P. Collin, R. Louis, B. Metz and J.-P. Sauvage, *J. Am. Chem. Soc.* **1991**, *113*, 8521.

[8.173] a) A. Haim, *Progr. Inorg. Chem.* **1983**, *30*, 273; b) X. D. Chai, T.-J. Li and J.-M. Lehn, unpublished work; X. D. Chai, Ph.D. Thesis, Jilin University, Peoples Republic of China, **1991**.

[8.174] O. Kahn, *Molecular Magnetism*, VCH, Weinheim, **1993**.

[8.175] a) J. S. Miller, A. J. Epstein and M. W. Reiff, *Acc. Chem. Res.* **1988**, *21*, 114; b) *Science* **1988**, *240*, 40; c) J. S. Miller, A. J. Epstein and M. W. Reiff, *Chem. Rev.* **1988**, *88*, 201; d) J. S. Miller and A. J. Epstein, *Angew. Chem.* **1994**, *106*, 399; *Angew. Chem. Int. Ed. Engl.* **1994**, *33*, 385.

[8.176] H. Iwamura, *Adv. Phys. Org. Chem.* **1990**, *26*, 179.

[8.177] a) D. A. Dougherty, *Acc. Chem. Res.* **1991**, *24*, 88; b) D. A. Dougherty, S. J. Jacobs, S. K. Silverman, M. M. Murray, D. A. Shultz, A. P. West, Jr. and J. A. Clites, *Mol. Cryst. Liq. Cryst.* **1993**, *232*, 289.

[8.178] O. Kahn, *Structure Bonding* **1987**, *68*, 89.

[8.179] K. Nakatani, P. Bergerat, E. Codjovi, C. Mathonière, Y. Pei and O. Kahn, *Inorg. Chem.* **1991**, *30*, 3977; H. O. Stumpf, Y. Pei, C. Michaut, O. Kahn, J.-P. Renard and L. Ouahab, *Chem. Mater.* **1994**, *6*, 257.

[8.180] a) C. M. Armstrong, *Quart. Rev. Biophys.* **1975**, *7*, 179; b) W. A. Catterall, *Science* **1989**, *242*, 50; c) B. Hille, *Ion Channels of Excitable Membranes*, Sinauer, Sunderland, MA, **1984**; d) *Ion Channels 2*, (Ed.: T. Narahashi), Plenum, New York, **1990**; e) for the role of protein structural changes see: D. Oesterhelt, J. Tittor and E. Bamberg, *J. Bioenerg. Biomembr.* **1992**, *24*, 181.

[8.181] a) F. Stevens, *Science* **1984**, *225*, 1346; b) N. Unwin, *Chemica Scripta* **1987**, *27 B*, 47; c) A. Maelicke, *TIBS* **1988**, 199; d) C. Miller, *Science* **1991**, *252*, 1092.

[8.182] J. Simon, M. K. Engel and C. Soulié, *New J. Chem.* **1992**, *16*, 287.

[8.183] a) J. D. Lear, Z. R. Wasserman and W. F. DeGrado, *Science* **1988**, *240*, 1177; b) D. Wade, A. Boman, B. Wåhlin, C. M. Drain, D. Andreu, H. G. Boman and R. B. Merrifield, *Proc. Natl. Acad. Sci. USA* **1990**, *87*, 4761; c) K. S. Åkerfeldt, J. D. Lear, Z. R. Wasserman, L. A. Chung and W. F. DeGrado, *Acc. Chem. Res.* **1993**, *26*, 191.

[8.184] a) S. Oiki, W. Danho and M. Montal, *Proc. Natl. Acad. Sci. USA* **1988**, *85*, 2393; b) M. Montal, in [8.180d] p. 1; c) A. Grove, M. Mutter, J. E. Rivier and M. Montal, *J. Am. Chem. Soc.* **1993**, *115*, 5919.

[8.185] C. I. Stankovic and S. L. Schreiber, *Chemtracts–Org. Chem.* **1991**, *4*, 1.

[8.186] a) M. R. Ghadiri, J. R. Granja, R. A. Milligan, D. E. McRee and N. Khazanovich, *Nature* **1993**, *366*, 324; b) N. Khazanovich, J. R. Granja, D. E. McRee, R. A. Milligan and M. R. Ghadiri, *J. Am. Chem. Soc.* **1994**, *116*, 6011, c) X. Sun and G. P. Lorenzi, *Helv. Chim. Acta* **1994**, *77*, 1520; L. Tomasic and G. P. Lorenzi, *ibid.*, **1987**, *70*, 1012.

[8.187]   a) B. Lotz, F. Colonna-Cesari, F. Heitz and G. Spach, *J. Mol. Biol.* **1976**, *106*, 915; b) R. N. Reusch and H. L. Sadoff, *Proc. Natl. Acad. Sci. USA* **1988**, *85*, 4176.

[8.188]   a) G. G. Cross, T. M. Fyles, T. D. James and M. Zojaji, *Synlett* **1993**, 449; b) T. M. Fyles, T. D. James and K. C. Kaye, *J. Am. Chem. Soc.* **1993**, *115*, 12315.

[8.189]   Y. Kobuke, K. Ueda and M. Sokabe, *J. Am. Chem. Soc.* **1992**, *114*, 7618.

[8.190]   a) E. Neher, *Angew. Chem.* **1992**, *104*, 837; *Angew. Chem. Int. Ed. Engl.* **1992**, *31*, 824; b) B. Sakmann, *Angew. Chem.* **1992**, *104*, 844; *Angew. Chem. Int. Ed. Engl.* **1992**, *31*, 830.

[8.191]   a) F. G. Riddell, S. Arumugam, P. J. Brophy, B. G. Cox, M. C. H. Payne and T. E. Southon, *J. Am. Chem. Soc.* **1988**, *110*, 734; b) D. C. Shungu and R. W. Briggs, *J. Magn. Res.* **1988**, *77*, 491, and references therein.

[8.192]   A. R. Waldeck, A. J. Lennon, B. E. Chapman and P. W. Kuchel, *J. Chem. Soc. Faraday Trans.* **1993**, *89*, 2807.

[8.193]   J.-P. Behr, C. J. Burrows, R. J. Heng and J.-M. Lehn, *Tetrahedron Lett.* **1985**, *26*, 215; J.-M. Lehn and P. G. Potvin, *Can. J. Chem.* **1988**, *66*, 195.

[8.194]   a) J.-M. Lehn, J. Malthête and A.-M. Levelut, *J. Chem. Soc. Chem. Commun.* **1985**, 1794; b) J. Malthête, A. -M. Levelut and J.-M. Lehn, *ibid.* **1992**, 1434.

[8.195]   a) C. Mertesdorf and H. Ringsdorf, *Mol. Cryst. Liq. Cryst.* **1989**, *5*, 1757; b) C. Mertesdorf, H. Ringsdorf and J. Stumpe, *Liq. Cryst.* **1991**, *9*, 337; c) S. H. J. Idziak, N. C. Maliszewskyj, G. B. M. Vaughan, P. Heiney, C. Mertesdorf, H. Ringsdorf, J. P. McCauley, Jr. and A. B. Smith, III, *J. Chem. Soc. Chem. Commun.* **1992**, 98.

[8.196]   a) V. Percec, G. Johansson, J. A. Heck, G. Ungar and S. V. Batty, *J. Chem. Soc. Perkin Trans 1* **1993**, 1411; b) V. Percec, J. A. Heck, D. Tomazos and G. Ungar, *J. Chem. Soc. Perkin Trans 2* **1993**, 2381; c) T. Komori and S. Shinkai, *Chem. Lett.* **1993**, 1455.

[8.197]   M. Armand, *Adv. Mater.* **1990**, 2, 278.

[8.198]   N. Kobayashi and A. B. P. Lever, *J. Am. Chem. Soc.* **1987**, *109*, 7433.

[8.199]   a) O. E. Sielcken, J. Schram, R. J. M. Nolte, J. Schoonman and W. Drenth, *J. Chem. Soc. Chem. Commun.* **1988**, 108; b) O. E. Sielcken, L. A. van de Kuil, W. Drenth, J. Schoonman and R. J. M. Nolte, *J. Am. Chem. Soc.* **1990**, *112*, 3086.

[8.200]   E. Yilmazer, A. Gürek, A. Gül and Ö. Bekâroglu, *Helv. Chim. Acta* **1988**, *71*, 1616.

[8.201]   T. Toupance, V. Ahsen and J. Simon, *J. Am. Chem. Soc.* **1994**, *116*, 5352.

[8.202]   N. Voyer, *J. Am. Chem. Soc.* **1991**, *113*, 1818.

[8.203]   J. Malthête, D. Poupinet, R. Vilanove and J.-M. Lehn, *J. Chem. Soc. Chem. Commun.* **1989**, 1016.

[8.204]   a) Y. Ishikawa, T. Kunitake, T. Matsuda, T. Otsuka and S. Shinkai, *J. Chem. Soc. Chem. Commun.* **1989**, 736; b) S. Ozeki, T. Ikegawa, S. Inokuma and T. Kuwamura, *Langmuir* **1989**, *5*, 222; c) C. Mertesdorf, T. Plesnivy, H. Ringsdorf and P. A. Suci, *Langmuir* **1992**, *8*, 253.

[8.205]   L. Jullien and J.-M. Lehn, *Tetrahedron Lett.* **1988**, 29, 3803; *J. Incl. Pheno. Molec. Reco. Chem.* **1992**, *12*, 55.

[8.206]   J. Canceill, L. Jullien, L. Lacombe and J.-M. Lehn, *Helv. Chim. Acta* **1992**, *75*, 791.

[8.207]   L. Jullien, T. Lazrak, J. Canceill, L. Lacombe and J.-M. Lehn, *J. Chem. Soc., Perkin Trans 2* **1993**, 1011.

[8.208]   M. J. Pregel, L. Jullien and J.-M. Lehn, *Angew. Chem.*, **1992**, *104*, 1695; *Angew. Chem. Int. Ed. Engl.* **1992**, *31*, 1637.

[8.209]   a) R. W. Holz, *Ann. N. Y. Acad. Sci.*, **1974**, *235*, 469; b) J. Bolard, *Biochim. Biophys. Acta* **1986**, *864*, 257.

[8.210]   T. M. Fyles, K. C. Kaye, T. D. James and D. W. M. Smiley, *Tetrahedron Lett.* **1990**, *31*, 1233.

[8.211] J. Rebek, Jr., *Acc. Chem. Res.* **1984**, *17*, 258.

[8.212] G. Gagnaire, G. Gellon and J.-L. Pierre, *Tetrahedron Lett.* **1988**, *29*, 933.

[8.213] P. Mitchell, *Angew. Chem.* **1979**, *91*, 733; *Angew. Chem. Int. Ed. Engl.* **1979**, *18*, 718.

[8.214] *Proton Transfer Reactions*, (Eds.: E. F. Caldin and V. Gold), Wiley, New York, 1975.

[8.215] *L'actualité chimique*, special issue, No. 1, January–February, **1991**.

[8.216] *Chem. Phys.* special issue, No. 2, **1989**.

[8.217] a) *Disc. Faraday Soc.* **1965**, *Vol. 39*; b) E. Kosower and D. Huppert, *Ann. Rev. Phys. Chem.* **1986**, *37*, 127.

[8.218] G. Zundel and M. Eckert, *J. Mol. Struct.* **1989**, *200*, 73; B. Brezinski, G. Zundel and R. Krämer, *Chem. Phys. Lett.* **1989**, *157*, 512.

[8.219] G. Zundel, *J. Membrane Sci.* **1982**, *11*, 249.

[8.220] F. Aguilar-Parrilla, G. Scherer, H.-H. Limbach, M. de la Concepción Foces-Foces, F. Hernández Cano, J. A. S. Smith, C. Toiron and J. Elguero, *J. Am. Chem. Soc.* **1992**, *114*, 9657.

[8.221] H.-H. Limbach, G. Scherer, L. Meschede, F. Aguilar-Parrilla, B. Wehrle, J. Braun, Ch. Hoelger, H. Benedict, G. Buntkowsky, W. P. Fehlhammer, J. Elguero, J. A. S. Smith and B. Chaudret in *Ultrafast Reaction Dynamics and Solvent Effects, Experimental and Theoretical Aspects*, (Eds.: Y. Gauduel and P. J. Rossky), American Institute of Physics, 1993.

[8.222] a) J. Hennig and H.-H. Limbach, *J. Am. Chem. Soc.* **1984**, *106*, 292; b) K. M. Merz and C. H. Reynolds, *J. Chem. Soc., Chem. Commun.* **1988**, 90, and references therein.

[8.223] R. C. Haddon and F. H. Stillinger, in [8.109a], p. 19.

[8.224] a) T. Mitani, *Mol. Cryst. Liq. Cryst.* **1989**, *171*, 343; b) T. Inabe, *New J. Chem.* **1991**, *15*, 129.

[8.225] M. T. Reetz, S. Höger and K. Harms, *Angew. Chem.* **1994**, *106*, 193; *Angew. Chem. Int. Ed. Engl.* **1994**, *33*, 181.

[8.226] M. G. Kuzmin, in [A.28], p. 279.

[8.227] M. Gutman, D. Huppert and E. Pines, *J. Am. Chem. Soc.* **1981**, *103*, 3709.

[8.228] M. J. Politi and J. H. Fendler, *J. Am. Chem. Soc.* **1984**, *106*, 265.

[8.229] *Photochromism; Molecules and Systems*, (Eds.: H. Dürr and H. Bouas-Laurent), Elsevier, Amsterdam, 1990.

[8.230] a) P. Borowicz, A. Grabowska, R. Wortmann and W. Liptay, *J. Luminescence* **1992**, *52*, 265; b) A. U. Acuña, A. Costela and J. M. Muñoz, *J. Phys. Chem.* **1986**, *90*, 2807.

[8.231] a) A.E. Chichibabin, B. M. Kuindzhi and S. W. Benewolenskaja, *Ber.* **1925**, *58*, 1580; b) R. Hardwick, H. S. Mosher and P. Passailaigue, *Trans. Faraday Soc.* **1960**, *56*, 44.

[8.232] a) G. Wettermark, *J. Am. Chem. Soc.* **1962**, *84*, 3658; b) H. Sixl and R. Warta, *Chem. Phys.* **1985**, *94*, 147; c) for another case, see: J. D. Margerum and R. G. Brault, *J. Am. Chem. Soc.* **1966**, *88*, 4733.

[8.233] Y. Eichen, J.-M. Lehn, M. Scherl, D. Haarer, R. Casalegno, A. Corval, K. Kuldova and H. P. Trommsdorff, *J. Chem. Soc., Chem. Commun.* **1995**, in press.

[8.234] J. Tessié, B. Gabriel and M. Prats, *TIBS* **1993**, 243.

[8.235] a) P. Haberfield, *J. Am. Chem. Soc.* **1987**, *109*, 6177 and 6178; b) M. Irie, *J. Am. Chem. Soc.* **1983**, *105*, 2078.

[8.236] a) C. J. Jalink, A. H. Huizer and C. A. G. O. Varma, *J. Chem. Soc. Faraday Trans.* **1992**, *88*, 2655; b) O. V. Chranina, F. P. Czerniakowski and G. S. Denisov, *J. Mol. Struct.* **1988**, *177*, 309.

[8.237] a) *Sensors, A Comprehensive Survey*, (Eds.: W. Göpel, J. Hesse and J. N. Zemel), VCH, Weinheim; b) *Vol. 2/3, Chemical and Biochemical Sensors*, (Eds.: W. Göpel,

T. A. Jones, M. Kleitz, I. Lundströn and T. Seiyama), VCH, Weinheim, **1991, 1992**; c) W. Göpel, *Sensors and Actuators B* **1994**, *18–19*, 1.

[8.238]   a) U. Weimar, S. Vaihinger, K. D. Schierbaum and W. Göpel, *Chemical Sensor Technology*, Kodansha, Tokyo, **1991**, *3*, 51; b) A. Hierlemann, U. Weimar, G. Kraus, G. Gauglitz and W. Göpel, *Sensors and Materials* **1995**, in press.

[8.239]   O. Kahn and J.-P. Launay, *Chemtronics* **1988**, *3*, 140.

[8.240]   a) U. Kölle, *Angew. Chem.* **1991**, *103*, 970; *Angew. Chem. Int. Ed. Engl.* **1991**, *30*, 956; b) M. Sano and H. Taube, *J. Am. Chem. Soc.* **1991**, *113*, 2327; *Inorg. Chem.* **1994**, *33*, 705.

[8.241]   a) H. Bolvin, O. Kahn and B. Vekhter, *New J. Chem.* **1991**, *15*, 889; b) J. Zarembowitch and O. Kahn, *New J. Chem.* **1991**, *15*, 181; c) J. Kröber, E. Codjovi, O. Kahn, F. Grolière and C. Jay, *J. Am. Chem. Soc.* **1993**, *115*, 9810.

[8.242]   A. P. de Silva, H. Q. N. Gunaratne and C. P. McCoy, *Nature* **1993**, *364*, 42.

[8.243]   A. Goldbeter and Y.-X. Li, in *Cell to Cell Signalling: From Experiments to Theoretical Models*, Academic Press, New York, **1989**, p. 415.

[8.244]   B. L. Feringa, W. F. Jager and B. de Lange, *Tetrahedron* **1993**, 49, 8267.

[8.245]   M. Irie, *Mol. Crystl. Liq. Cryst.* **1993**, *227*, 263.

[8.246]   T. Saika, T. Lyoda, K. Honda and T. Shimidzu, *J. Chem. Soc., Chem. Commun.* **1992**, 591.

[8.247]   a) J. Daub, J. Salbeck, T. Knöchel, C. Fischer, H. Kunkely and K. M. Rapp, *Angew. Chem.* **1989**, *101*, 1541; *Angew. Chem. Int. Ed. Engl.* **1989**, *28*, 1494; b) J. Daub, C. Fischer, J. Salbeck and K. Ulrich, *Adv. Mater.* **1990**, 2, 366.

[8.248]   T. Iyoda, T. Saika, K. Honda and T. Shimidzu, *Tetrahedron Lett.* **1989**, *30*, 5429.

[8.249]   A. K. Newell and J. H. P. Utley, *J. Chem. Soc. Chem. Commun.* **1992**, 800.

[8.250]   J. Achatz, C. Fischer, J. Salbeck and J. Daub, *J. Chem. Soc. Chem. Commun.* **1991**, 504.

[8.251]   S. L. Gilat, S. H. Kawai and J.-M. Lehn, *J. Chem. Soc. Chem. Commun.* **1993**, 1439.

[8.252]   S. H. Kawai, S. L. Gilat and J.-M. Lehn, *J. Chem. Soc. Chem. Commun.* **1994**, 1011.

[8.253]   H. Tachibana, T. Nakamura, M. Matsumoto, H. Komizu, E. Man, H. Niino, A. Yabe and Y. Kawabata, *J. Am. Chem. Soc.* **1989**, *111*, 3080.

[8.254]   M. P. O'Neil, M. P. Niemczyk, W. A. Svec, D. Gosztola, G. L. Gaines III and M. R. Wasielewski, *Science* **1992**, *257*, 63.

[8.255]   J. M. Hammerstad-Pedersen, Y. I. Kharkats, P. Sommer-Larsen and J. Ulstrup, *Adv. Mater. Optics Electr.* **1992**, *1*, 147.

[8.256]   V. Goulle, A. Harriman and J.-M. Lehn, *J. Chem. Soc. Chem. Commun.* **1993**, 1034.

[8.257]   J. H. Burroughes, D. D. C. Bradley, A. R. Brown, R. N. Marks, K. Mackay, R. H. Friend, P. L. Burns and A. B. Holmes, *Nature* **1990**, *347*, 539.

[8.258]   N. S. Hush, A. T. Wong, G. B. Bacskay and J. R. Reimers, *J. Am. Chem. Soc.* **1990**, *112*, 4192.

[8.259]   J. Salbeck, V. N. Komissarov, V.I. Minkin and J. Daub, *Angew. Chem.* **1992**, *104*, 1498; *Angew. Chem. Int. Ed. Engl.* **1992**, *31*, 1498.

[8.260]   J.-M. Lehn, J.-P. Sauvage, J. Simon, R. Ziessel, C. Piccinni-Leopardi, G. Germain, J. P. Declercq and M. Van Meerssche, *Nouv. J. Chim.* **1983**, *7*, 413.

[8.261]   J.-P. Gisselbrecht, M. Gross, J.-M. Lehn, J.-P. Sauvage, R. Ziessel, C. Piccinni-Leopardi, J. M. Arrieta, G. Germain and M. Van Meerssche, *Nouv. J. Chim.* **1984**, *8*, 661.

[8.262]   T. Saji, K. Hoshino and S. Aoyagui, *J. Chem. Soc. Chem. Commun.* **1985**, 865.

[8.263]   J. C. Medina, I. Gay, Z. Chen, L. Echegoyen and G. W. Gokel, *J. Am. Chem. Soc.* **1991**, *113*, 365.

[8.264]   M. Irie and M. Kato, *J. Am. Chem. Soc.* **1985**, *107*, 1024.

[8.265] S. M. Fatah-ur Rahman and K. Fukunishi, *J. Chem. Soc. Chem. Commun.* **1992**, 1740.

[8.266] S. M. Fatah-ur Rahman, K. Fukunishi, M. Kuwabara, H. Yamanaka and M. Nomura, *Bull. Chem. Soc. Japan* **1993**, *66*, 1461.

[8.267] S. R. Adams, J. P. Y. Kao, G. Grynkiewicz, A. Minta and R. Y. Tsien, *J. Am. Chem. Soc.* **1988**, *110*, 3212.

[8.268] a) J. H. Kaplan and G. C. R. Ellis-Davies, *Proc. Nat. Acad. Sci. USA* **1988**, *85*, 6571; b) G. C. R. Ellis-Davies and J. H. Kaplan, *Proc. Natl. Acad. Sci. USA* **1994**, *91*, 187.

[8.269] R. Warmuth, E. Grell, J.-M. Lehn, J. W. Bats and G. Quinkert, *Helv. Chim. Acta* **1991**, *74*, 671.

[8.270] R. Warmuth, E. Grell and J.-M. Lehn, *Soc. Gen. Physiol. Ser.* **1991**, *46*, 437.

[8.271] E. Grell and R. Warmuth, *Pure Appl. Chem.* **1993**, *65*, 373.

[8.272] J.-P. Souchez, B. Dietrich and J.-M. Lehn, work in progress.

[8.273] a) J. H. Kaplan, B. Forbush III and J. F. Hoffmann, *Biochemistry* **1978**, *17*, 1929; b) A. M. Gurney and A. Lester, *Physiol. Rev.* **1987**, *67*, 583.

[8.274] I. J. Colton and R. J. Kazlauskas, *J. Org. Chem.* **1992**, *57*, 7005.

[8.275] L. Zelikovitch, L. Libman and A. Shanzer, *Nature* **1995**, in press.

[8.276] a) J. Rebek, Jr., and L. Marshall, *J. Am. Chem. Soc.* **1983**, *105*, 6668; b) M. Inouye, T. Konishi and K. Isagawa, *J. Am. Chem. Soc.* **1993**, *115*, 8091; c) Y. Kobuke, Y. Sumida, M. Hayashi and H. Ogoshi, *Angew. Chem.* **1991**, *103*, 1513; *Angew. Chem. Int. Ed. Engl.* **1991**, *30*, 1496.

[8.277] a) M.-a. Haga, T.-a. Ano, K. Kano and S. Yamabe, *Inorg. Chem.* **1991**, *30*, 3843; b) W. L. Mock and J. Pierpont, *J. Chem. Soc., Chem. Commun.* **1990**, 1509.

[8.278] a) K. E. Drexler, *Nanosystems*, Wiley, New York, **1992**; b) G. M. Fahy, *Clinical Chem.* **1993**, *39*, 2011.

[8.279] a) C. Roussel, A. Lidén, M. Chanon, J. Metzger and J. Sandström, *J. Am. Chem. Soc.* **1976**, *98*, 2847; F. Cozzi, A. Guenzi, C. A. Johnson and K. Mislow, *J. Am. Chem. Soc.* **1981**, *103*, 957; H. Iwamura, *J. Mol. Struct.* **1985**, *126*, 401; b) T. R. Kelly, M. C. Bowyer, K. V. Bhaskar, D. Bebbington, A. Garcia, F. Lang, M. H. Kim and M. P. Jette, *J. Am. Chem. Soc.* **1994**, *116*, 3657; c) G. Fischer, *Angew. Chem.* **1994**, *106*, 1415; *Angew. Chem. Int. Ed. Engl.* **1994**, *33*, 1415; d) P. Timmerman, W. Verboom, F. C. J. M. van Veggel, J. P. M. van Duynhoven and D. N. Reinhoudt, *Angew. Chem.* **1994**, *106*, 2437; *Angew. Chem. Int. Ed. Engl.* **1994**, *33*, 2345.

[8.280] G. Schill, *Catenanes, Rotaxanes and Knots*, Academic Press, New York, **1971**.

[8.281] C. O. Dietrich-Buchecker and J.-P. Sauvage, *Chem. Rev.* **1987**, *87*, 795; *Tetrahedron* **1990**, *46*, 503.

[8.282] a) J.-P. Sauvage, *Acc. Chem. Res.* **1990**, *23*, 319; b) J.-P. Sauvage and C. Dietrich-Buchecker, in [A.34], Vol. 2, **1991**, p. 195; c) J.-C. Chambron, C. Dietrich-Buchecker and J.-P. Sauvage, in [A.33], p. 131.

[8.283] D. Philp and J. F. Stoddart, *Synlett* **1991**, 445.

[8.284] For a pentacatenane see: D. B. Amabilino, P. R. Ashton, A. S. Reder, N. Spencer and J. F. Stoddart, *Angew. Chem.* **1994**, *106*, 1316; *Angew. Chem. Int. Ed. Engl.* **1994**, *33*, 1286.

[8.285] a) P. R. Ashton, T. T. Goodnow, A. E. Kaifer, M. V. Reddington, A. M. Z. Slawin, N. Spencer, J. F. Stoddart, C. Vicent and D. J. Williams, *Angew. Chem.* **1989**, *101*, 1404; *Angew. Chem. Int. Ed. Engl.* **1989**, *28*, 1396; b) P. R. Ashton, C. L. Brown, E. J. T. Chrystal, K. P. Parry, M. Pietraszkiewicz, N. Spencer and J. F. Stoddart, *Angew. Chem.* **1991**, *103*, 1058; *Angew. Chem. Int. Ed. Engl.* **1991**, *30*, 1042.

[8.286] a) P. L. Anelli, N. Spencer and J. F. Stoddart, *J. Am. Chem. Soc.* **1991**, *113*, 5131; b) P. R. Ashton, D. Philp, N. Spencer and J. F. Stoddart, *J. Chem. Soc., Chem. Com-

mun. **1992**, 1124; c) D. B. Amabilino, P. R. Ashton, A. S. Reder, N. Spencer and J. F. Stoddart, *Angew. Chem.* **1994**, *106*, 450; *Angew. Chem. Int. Ed. Engl.* **1994**, *33*, 433.

[8.287] A. C. Benniston and A. Harriman, *Angew. Chem.* **1993**, *105*, 1553; *Angew. Chem. Int. Ed. Engl.* **1993**, *32*, 1459.

[8.288] F. Vögtle, W. M. Müller, U. Müller, M. Bauer and K. Rissanen, *Angew. Chem.* **1993**, *105*, 1356; *Angew. Chem. Int. Ed. Engl.* **1993**, *32*, 1295.

[8.289] a) R. Ballardini, V. Balzani, M. T. Gandolfi, L. Prodi, M. Venturi, D. Philp, H. G. Ricketts and J. F. Stoddart, *Angew. Chem.* **1993**, *105*, 1362; *Angew. Chem. Int. Ed. Engl.* **1993**, *32*, 1301 b) R. A. Bissell, E. Córdova, A. E. Kaifer and J. F. Stoddart, *Nature* **1994**, *369*, 133.

[8.290] J. Huguet and M. Vert, in *Microdomains in Polymer Solutions*, (Ed.: P. Dubin), Plenum, New York, **1985**, p. 51; D. Vallin, J. Huguet and M. Vert, *Polym. J.* **1982**, *12*, 113.

[8.291] H. Menzel, *Nachr. Chem. Tech. Lab.* **1991**, *39*, 636.

[8.292] a) Y. Osada and J. Gong, *Prog. Polym. Sci.* **1993**, *18*, 187; b) *Responsive gels*, (Ed.: K. Dusek), *Adv. Polym. Sci.* **1993**, Vol. *110*; c) Y. Osada and S. B. Ross-Murphy, *Scientific Amer.* **1993**, *268*, May p. 42.

[8.293] a) T. Yanagida, Y. Harada and A. Ishijima, *TIBS* **1993**, *18*, 319; b) I. Rayment and H. M. Holden, *TIBS* **1994**, *19*, 129; c) for the observation of reptation in actin molecules, see: E. Sackmann, J. Käs and H. Strey, *Adv. Mater.* **1994**, *6*, 507.

[8.294] D. W. Urry, *Angew. Chem.* **1993**, *105*, 859; *Angew. Chem. Int. Ed. Engl.* **1993**, *32*, 819.

[8.295] D. M. Eigler, C. P. Lutz and W. E. Rudge, *Nature* **1991**, *352*, 600.

[8.296] H.-J. Galla, *Angew. Chem.* **1992**, *104*, 47; *Angew. Chem. Int. Ed. Engl.* **1992**, *31*, 45.

[8.297] E. Di Mauro and C. P. Hollenberg, *Adv. Mater.* **1993**, *5*, 384.

[8.298] H.-D. Wiemhöfer and K. Cammann, in [8.237b], **1991**, Vol. *2*, Part *I*, p. 159.

[8.299] P. C. Jurs and T. L. Isenhour, *Chemical Applications of Pattern Recognition*, Wiley, New York, **1975**.

[8.300] J. Zupan and J. Gasteiger, *Neural Networks for Chemists*, VCH, Weinheim, **1993**.

[8.301] L. A. Zadeh, *Information and Control* **1965**, *8*, 338; *Synthese* **1975**, *30*, 407.

Chapter 9

[9.1] J.-M. Lehn, A. Rigault, J. Siegel, J. Harrowfield, B. Chevrier and D. Moras, *Proc. Natl. Acad. Sci. USA* **1987**, *84*, 2565.

[9.2] J. S. Lindsey, *New J. Chem.* **1991**, *15*, 153.

[9.3] G. M. Whitesides, J. P. Mathias and C. T. Seto, *Science* **1991**, *254*, 1312.

[9.4] M. Eigen, *Naturwiss.* **1971**, *33a*, 465.

[9.5] A. L. Lehninger, *Biochemistry*, 2nd ed., Worth, New York, **1975**, Chap. 36.

[9.6] B. Alberts, D. Bray, J. Lewis, M. Raff, K. Roberts and J. D. Watson, *Molecular Biology of the Cell*, Garland, New York, **1983**, p. 121–126; B. Hess and A. Mikhailov, *Science* **1994**, *264*, 223.

[9.7] F. Cramer, *Chaos and Order, The Complex Structure of Living Systems*, VCH, Weinheim, **1993**, Chap. 7.

[9.8] a) *Self-Organizing Systems. The Emergence of Order*, (Ed.: F. E. Yates), Plenum, New York, **1987**; b) F. E. Yates, Preface of [9.8a]; c) R. Landauer, in [9.8a] p. 435;

d) H. Haken, *Synergetics*, Springer, Berlin, **1978**; *Synergetics, Chaos, Order, Self-organization*, (Ed.: M. Bushev), World Scientific, London, **1994**; e) G. Nicolis, I. Prigogine, *Self-organization in non-equilibrium systems*, Wiley, New York, **1977**.

[9.9]   a) H. W. Kroto, *Angew. Chem.* **1992**, *104*, 113; *Angew. Chem. Int. Ed. Engl.* **1992**, *31*, 111; b) T. W. Ebbesen, *Annu. Rev. Mater. Sci.* **1994**, *24*, 235.

[9.10]   M. Simard, D. Su and J. D. Wuest, *J. Am. Chem. Soc.* **1991**, *113*, 4696.

[9.11]   *Understanding Self-Assembly and Organization in Liquid Crystals*, (Eds.: E. P. Raynes and N. Boden), *Phil. Trans. R. Soc. Lond.* A **1993**, *344*, p. 305–440.

[9.12]   D. H. Busch, *J. Incl. Phenom. Molec. Recogn. Chem.* **1992**, *12*, 389.

[9.13]   a) S. Anderson, H.K L Anderson and J. K. M. Sanders, *Accounts Chem. Res.* **1993**, *26*, 469; b) R. Hoss and F. Vögtle, *Angew. Chem.* **1994**, *106*, 389; *Angew. Chem. Int. Ed. Engl.* **1994**, *33*, 375.

[9.14]   a) J. Monod, J.-P. Changeux and F. Jacob, *J. Mol. Biol.* **1963**, *6*, 306; b) G. Gagnaire, A. Jeunet and J.-L. Pierre, *Tetrahedron Lett.* **1991**, *32*, 2021; c) J. C. Rodriguez-Ubis, O. Juanes and E. Brunet, *Tetrahedron Lett.* **1994**, *35*, 1295.

[9.15]   a) B. Perlmutter-Hayman, *Acc. Chem. Res.* **1986**, *19*, 90; b) T. G. Traylor, M. J. Mitchel, J. P. Ciccone and S. Nelson, *J. Am. Chem. Soc.* **1982**, *104*, 4986; c) I. Tabushi, S.-I. Kugimiya, M. G. Kinnaird and T. Sazaki, *J. Am. Chem. Soc.* **1985**, *107*, 4192; d) J. Rebek, Jr., T. Costello, L. Marschall, R. Wattley, R. C. Gadwood and K. Onan, *J. Am. Chem. Soc.* **1985**, *107*, 7481.

[9.16]   a) D. Jalabert, J. B. Robert, H. Roux-Buisson, J.-P. Kintzinger, J.-M. Lehn, R. Zinzius, D. Canet and P. Tekely, *Europhys. Lett.* **1991**, *15*, 435; b) E. W. Bastiaan and C. MacLean, *NMR Basic Principles and Progress* **1990**, *25*, 17; c) A. Maliniak, A. Laaksonen and J. Korppi-Tommola, *J. Am. Chem. Soc.* **1990**, *112*, 86; d) G. A. Jeffrey, in [A. 18] **1984**, *Vol. 1*, 135; e) for water structure in solutions, see: J. L. Finney and A. K. Soper, *Chem. Soc. Rev.* **1994**, *23*, 1.

[9.17]   a) A. Klug, *Angew. Chem.* **1983**, *95*, 579; *Angew. Chem. Int. Ed. Engl.* **1983**, *22*, 565; b) W. B. Wood, in [9.8a] p. 133.

[9.18]   F.-U. Hartl, R. Hlodan and T. Langer, *TIBS* **1994**, *19*, 20.

[9.19]   a) Y. S. Babu, J. S. Sack, T. J. Greenhough, C. E. Bugg, A. R. Means and W. J. Cook, *Nature* **1985**, *315*, 37; b) R. H. Kretsinger, S. E. Rudnick and L. J. Weissman, *J. Inorg. Biochem.* **1986**, *28*, 289; c) S. Forsén, H. J. Vogel and T. Drakenberg, *Calcium and Cell Function*, (Ed.: W. Y. Cheung), Academic Press, Orlando, **1986**, *6*, 113; d) S. R. Martin, A. Andersson Teleman, P. M. Bayley, T. Drakenberg and S. Forsen, *Eur. J. Biochem.* **1985**, *151*, 543; e) A. Deville, P. Laszlo and D. J. Nelson, *J. Theoret. Biol.* **1985**, *112*, 157.

[9.20]   a) H. Krautscheid, D. Fenske, G. Baum and M. Semmelmann, *Angew. Chem.* **1993**, *105*, 1364; *Angew. Chem. Int. Ed. Engl.* **1993**, *32*, 1303; b) D. Fenske ad H. Krautscheid, *Angew. Chem.* **1990**, *102*, 1513; *Angew. Chem. Int. Ed. Engl.* **1990**, *29*, 1452; c) F. M. Mulder, T. A. Stegink, R. C. Thiel, L. J. de Jongh and G. Schmid, *Nature* **1994**, *367*, 716; d) M. N. Vargaftik, I. I. Moiseev, D. I. Kochubey and K. I. Zamaraev, *Faraday Discuss.* **1991**, *92*, 13.

[9.21]   For small metal particles and colloids, see: a) *Faraday Disc.* **1991**, *Vol. 92*; b) G. Schmid, *Chem. Rev.* **1992**, *92*, 1709; c) M. T. Reetz and W. Helbig, *J. Am. Chem. Soc.* **1994**, *116*, 7401.

[9.22]   a) A. Müller, R. Rohlfing, J. Döring and M. Penk, *Angew. Chem.* **1991**, *103*, 575; *Angew. Chem. Int. Ed. Engl.* **1991**, *30*, 588; b) *Molec. Engineering*, Special issue: *Polyoxometallates*, (Eds.: M. T. Pope and A. Müller), **1993**, *3*, Nos. 1–3.

[9.23]   S. J. Lippard, *Angew. Chem.* **1988**, *100*, 353; *Angew. Chem. Int. Ed. Engl.* **1988**, *27*, 344.

[9.24]   G. Süss-Fink, *Angew. Chem.* **1991**, *103*, 73; *Angew. Chem. Int. Ed. Engl.* **1991**, *30*, 72.

[9.25]   B. K. Teo and H. Zhang, *Proc. Natl. Acad. Sci. USA* **1991**, *88*, 5067.

[9.26]   G. S. H. Leen, K. J. Fisher, D. C. Craig, M. L. Scudder and I. G. Dance, *J. Am. Chem. Soc.* **1990**, *112*, 6435.

[9.27]   A. J. Amaroso, L. H. Gade, B. F. G. Johnson, J. Lewis, P. R. Raithby and W.-T. Wong, *Angew. Chem.* **1991**, *103*, 102; *Angew. Chem. Int. Ed. Engl.* **1991**, *31*, 107.

[9.28]   K. Sakai and K. Matsumoto, *J. Am. Chem. Soc.* **1989**, *111*, 3074.

[9.29]   R. H. Cayton, M. H. Chisholm, J. C. Huffman and E. B. Lobkovsky, *J. Am. Chem. Soc.* **1991**, *113*, 8709.

[9.30]   B. Köhler, R. Kirmse, R. Richter, J. Sieler and E. Hayer, *Z. Anorg. Allg. Chem.* **1986**, *537*, 133.

[9.31]   G. Süss-Fink, J.-L. Wolfender, F. Neumann and H. Stoeckli-Evans, *Angew. Chem.* **1990**, *102*, 447; *Angew. Chem. Int. Ed. Engl.* **1990**, *29*, 429.

[9.32]   D. P. Smith, E. Baralt, B. Morales, M. M. Olmstead, M.F. Maestre and R. H. Fish, *J. Am. Chem. Soc.* **1992**, *114*, 10647

[9.33]   S. Rüttimann, G. Bernardinelli and A. F. Williams, *Angew. Chem.* **1993**, *105*, 432; *Angew. Chem. Int. Ed. Engl.* **1993**, *32*, 392.

[9.34]   P. Chaudhuri, I. Karpenstein, M. Winter, M. Lengen, C. Butzlaff, E. Bill, A. X. Trautwein, U. Flörke and H.-J. Haupt, *Inorg. Chem.* **1993**, *32*, 888.

[9.35]   a) G. Newton, I. Haiduc, R. B. King and C. Silvestru, *J. Chem. Soc., Chem. Commun.* **1993**, 1229; b) I. Haiduc and C. Silvestru, to be published.

[9.36]   S. Gambarotta, C. Floriani, A. Chiesi-Villa and C. Guastini, *J. Chem. Soc., Chem. Commun.* **1983**, 1156.

[9.37]   J. A. J. Jarvis, B. T. Kilbourn, R. Pearce, M. F. Lappert, *J. Chem. Soc. Chem. Commun.* **1973**, 475.

[9.38]   A. N. Nesmeyanov, Yu. T. Struchkov, N. N. Sedova, V. G. Andrianov, Yu. V. Volgin and V. A. Sazonova, *J. Organomet. Chem.* **1977**, *137*, 217.

[9.39]   a) H. Hartl, F. Mahdjour-Hassan-Abadi, *Angew. Chem.* **1984**, *96*, 359; *Angew. Chem. Int. Ed. Engl.* **1984**, *23*, 378; b) F. Mahdjour-Hassan-Abadi, H. Hartl, J. Fuchs, *Angew. Chem.* **1984**, *96*, 497; *Angew. Chem. Int. Ed. Engl.* **1984**, *23*, 514.

[9.40]   J. Lorberth, M. El-Essawi, W. Massa and L. Labib, *Angew. Chem.* **1988**, *100*, 1194; *Angew. Chem. Int. Ed. Engl.* **1988**, *27*, 1160; A. Caneschi, D. Gatteschi, J. Laugier, P. Rey, R. Sessoli and C. Zanchini, *J. Am. Chem. Soc.* **1988**, *110*, 2795.

[9.41]   I. G. Dance, M. L. Scudder and R. Secomb, *Inorg. Chem.* **1985**, *24*, 1201.

[9.42]   O. Poncelet, L. G. Hubert-Pfalzgraf, J.-C. Daran and R. Astier, *J. Chem. Soc. Chem. Commun.* **1989**, 1846.

[9.43]   K. L. Taft, C. D. Delfs, G. C. Papaefthymiou, S. Foner, D. Gatteschi and S. J. Lippard, *J. Am. Chem. Soc.*, **1994**, *116*, 823.

[9.44]   a) M. Fujita, J. Yazaki and K. Ogura, *J. Am. Chem. Soc.* **1990**, *112*, 5645; b) P. J. Stang and V. Zhdankin, *J. Am. Chem. Soc.* **1993**, *115*, 9808; P. J. Stang and K. Chen, *J. Am. Chem. Soc.* **1995**, *117*, 1667.

[9.45]   M.-T. Youinou, N. Rahmouni, J. Fischer and J. A. Osborn, *Angew. Chem.* **1992**, *104*, 771; *Angew. Chem. Int. Ed. Engl.* **1992**, *31*, 733.

[9.46]   J. R. Bradbury, J. L. Hampton, D. P. Martone and A. W. Maverick, *Inorg. Chem.* **1989**, *28*, 2392.

[9.47]   S. K. Mandal, L. K. Thompson, M. J. Newlands, E. J. Gabe and F. L. Lee, *J. Chem. Soc. Chem. Commun.* **1989**, 744.

[9.48]   M. Fujita, S. Nagao, M. Iida, K. Ogata and K. Ogura, *J. Am. Chem. Soc.* **1993**, *115*, 1574.

[9.49]   A. W. Maverick, M. L. Ivie, J. H. Waggenspack and F. R. Fronczek, *Inorg. Chem.* **1990**, *29*, 2403.

[9.50]   M. Fujita, J. Yazaki and K. Ogura, *Tetrahedron Lett.* **1991**, 5589.

[9.51]   a) R. W. Saalfrank; A. Stark, M. Bremer and H.-U. Hummel, *Angew. Chem.* **1990**, *102*, 292; *Angew. Chem. Int. Ed. Engl.* **1990**, *29*, 311; b) R. W. Saalfrank, B. Hörner, D. Stalke and J. Salbeck, *Angew. Chem.* **1993**, *105*, 1223; *Angew. Chem. Int. Ed. Engl.* **1993**, *32*, 1179; c) R. W. Saalfrank, R. Burak, A. Breit, D. Stalke, R. Herbst-Irmer, J. Daub, M. Porsch, E. Bill, M. Müther and A. X. Trautwein, *Angew. Chem.* **1994**, *106*, 1697; *Angew. Chem. Int. Ed. Engl.* **1994**, *33*, 1621.

[9.52]   R. W. Saalfrank and R. Burak, in *Advances in the Use of Synthons in Organic Chemistry, Vol. 1*, JAI Press, Greenwich, Conn., **1993**, 103.

[9.53]   M. Fujita, F. Ibukuro, H. Hagih and K. Ogura, *Nature* **1994**, *367*, 720.

[9.54]   a) E. C. Constable and A. M. W. Cargill Thompson, *J. Chem. Soc., Chem. Commun.* **1992**, 617; b) E. C. Constable, A. M. W. Cargill Thompson and D. A. Tocher, in [A.28], p. 219.

[9.55]   M. M. Harding, U. Koert, J.-M. Lehn, A. Marquis-Rigault, C. Piguet and J. Siegel, *Helv. Chim. Acta* **1991**, *74*, 594.

[9.56]   a) For an early case of a complex related to the helicate formed by **128** see [8.260]; b) related structural features are found in some other dinuclear metal complexes; see for instance: G. Struckmeier, U. Thewalt and J. H. Fuhrhop, *J. Am. Chem. Soc.* **1976**, *98*, 278; D. Wester and G. J. Palenik, *J. Chem. Soc. Chem. Commun.* **1975**, 74; G. C. Van Stein, G. Van Koten, K. Vrieze, C. Brevard and A. L. Spek, *J. Am. Chem. Soc.* **1984**, *106*, 4486.

[9.57]   J.-M. Lehn and A. Rigault, *Angew. Chem.* **1988**, *100*, 1121; *Angew. Chem. Int. Ed. Engl.* **1988**, *27*, 1095.

[9.58]   J. Harrowfield, J.-M. Lehn, B. Chevrier and D. Moras, unpublished results.

[9.59]   T. M. Garrett, U. Koert, J.-M. Lehn, A. Rigault, D. Meyer and J. Fischer, *J. Chem. Soc., Chem. Commun.* **1990**, 557.

[9.60]   Y. He and J.-M. Lehn, *Chem. J. Chinese Univ.* **1990**, *6*, 184.

[9.61]   M. T. Youinou, R. Ziessel and J.-M. Lehn, *Inorg. Chem.* **1991**, *30*, 2144.

[9.62]   T. Nabeshima, T. Inaba, N. Furukawa, T. Hosoya and Y. Yano, *Inorg. Chem.* **1993**, *32*, 1407.

[9.63]   D. Funeriu and J.-M. Lehn, unpublished results

[9.64]   A. Pfeil and J.-M. Lehn, *J. Chem. Soc. Chem. Commun.* **1992**, 838.

[9.65]   T. M. Garrett, U. Koert and J.-M. Lehn, *J. Phys. Org. Chem.* **1992**, *5*, 529.

[9.66]   U. Koert, M. M. Harding and J.-M. Lehn, *Nature* **1990**, *346*, 339.

[9.67]   B. Schoentjes and J.-M. Lehn, *Helv. Chim. Acta* **1995**, *78*, 1; B. Schoentjes, Doctorat ès Sciences, Université Louis Pasteur, Strasbourg, **1993**.

[9.68]   W. Zarges, J. Hall, J.-M. Lehn and C. Bolm, *Helv. Chim. Acta* **1991**, *74*, 1843.

[9.69]   a) R. Glaser, *Chirality* **1993**, *5*, 272; *Struct. Chem.* **1990**, *2*, 479; b) F. A. L. Anet, S. S. Miura, J. Siegel, and K. Mislow, *J. Am. Chem. Soc.* **1983**, *105*, 1419; c) K. Mislow, *Croat. Chim. Acta* **1985**, *58*, 353; d) M. Cinquini, F. Cozzi, F. Sannicolo and A. Sironi, *J. Am. Chem. Soc.* **1988**, *110*, 4363; e) K. Mislow, *Bull. Soc. Chim. Fr.* **1994**, *131*, 534.

[9.70]   a) K. T. Potts, *Bull. Soc. Chim. Belg.* **1990**, *99*, 741; b) K. T. Potts, K. A. Gheysen Raiford and M. Keshavarz-K, *J. Am. Chem. Soc.* **1993**, *115*, 2793.

[9.71]   E. C. Constable, R. Chotalia and D. A. Tocher, *J. Chem. Soc., Chem. Commun.* **1992**, 771.

[9.72]   a) E. C. Constable, *Tetrahedron* **1992**, *48*, 10013; b) *Progress Inorg. Chem.* **1994**, *42*, 67.

[9.73]   E. C. Constable, M. D. Ward and D. A. Tocher, *J. Am. Chem. Soc.* **1990**, *112*, 1256; E. C. Constable and R. Chotalia, *J. Chem. Soc. Chem. Commun.* **1992**, 64.

[9.74]   K. T. Potts, M. Keshavarz-K, F. S. Tham, H. D. Abruña and C. R. Arana, *Inorg. Chem.* **1993**, *32*, 4422.

[9.75]   a) K. T. Potts, M. Keshavarz-K, F. S. Tham, H. D. Abruña and C. Arana, *Inorg. Chem.* **1993**, *32*, 4436; b) K. T. Potts, M. Keshavarz-K, F. S. Tham, H. D. Abruña and C. Arana, *Inorg. Chem.* **1993**, *32*, 4450.

[9.76]   D. M. Walba, *Tetrahedron* **1985**, *41*, 3161.

[9.77]   a) C. O. Dietrich-Buchecker, J. Guilhem, C. Pascard and J.-P. Sauvage, *Angew. Chem.* **1990**, *102*, 1202; *Angew. Chem. Int. Ed. Engl.* **1990**, *29*, 1154; b) J.-F. Nierengarten, C. O. Dietrich-Buchecker and J.-P. Sauvage, *J. Am. Chem. Soc.* **1994**, *116*, 375.

[9.78]   a) C. Piguet, G. Bernardinelli and A. F. Williams, *Inorg. Chem.* **1989**, *28*, 2920; b) O. J. Gelling, F. van Bolhuis and B. L. Feringa, *J. Chem. Soc., Chem. Commun.* **1991**, 917.

[9.79]   a) T. W. Bell and H. Jousselin, *Nature* **1994**, *367*, 441; E. C. Constable, A. J. Edwards, P. R. Raithby and J. V. Walker, *Angew. Chem.* **1993**, *105*, 1486; *Angew. Chem. Int. Ed. Engl.* **1993**, *32*, 1465; c) A. Marquis and J.-M. Lehn, unpublished work; d) R. F. Carina, G. Bernardinelli and A. F. Williams, *Angew. Chem.* **1993**, *105*, 1483; *Angew. Chem. Int. Ed. Engl.* **1993**, *32*, 1463.

[9.80]   J. D. Crane and J.-P. Sauvage, *New J. Chem.* **1992**, *16*, 649.

[9.81]   a) J. Hall, B. Hasenknopf and J.-M. Lehn, unpublished work; b) B. Hasenknopf and J.-M. Lehn, unpublished work.

[9.82]   a) R. C. Scarrow, D. L. White and K. N. Raymond, *J. Am. Chem. Soc.* **1985**, *107*, 6540; b) B. R. Serr, K. A. Andersen, C. M. Elliott and O. P. Anderson, *Inorg. Chem.* **1988**, *27*, 4499; c) J. Libman, Y. Tor and A. Shanzer, *J. Am. Chem. Soc.* **1987**, *109*, 5880; A. Shanzer, J. Libman, Y. Tor and H. Gottlieb, in *Transport Through Membranes: Carriers, Channels and Pumps*, (Eds.: A. Pullman et al.), Kluwer, Dordrecht, **1988**, p. 57.

[9.83]   A. F. Williams, C. Piguet and G. Bernardinelli, *Angew. Chem.* **1991**, *103*, 1530; *Angew. Chem. Int. Ed. Engl.* **1991**, *30*, 1490.

[9.84]   C. Piguet, J. C. G. Bünzli, G. Bernardinelli, G. Hopfgartner and A. F. Williams, *J. Am. Chem. Soc.* **1993**, *115*, 8197.

[9.85]   R. Krämer, J. -M. Lehn, A. DeCian and J. Fischer, *Angew. Chem.* **1993**, *105*, 764; *Angew. Chem. Int. Ed. Engl.* **1993**, *32*, 703.

[9.86]   A trinuclear triple helical cuprous complex of a non-chelating ligand has been reported: K. T. Potts, C. P. Horwitz, A. Fessak, M. Keshavarz-K, K. E. Nash and P. J. Toscano, *J. Am. Chem. Soc.* **1993**, *115*, 10444.

[9.87]   P. Braunstein, B. Oswald, A. Tiripicchio and M. Tiripicchio-Gamellini, *Angew. Chem.* **1990**, *102*, 1206; *Angew. Chem. Int. Ed. Engl.* **1990**, *29*, 1140.

[9.88]   a) M. R. Ghadiri, C. Soares and C. Choi, *J. Am. Chem. Soc.* **1992**, *114*, 825; b) M. Lieberman and T. Sazaki, *J. Am. Chem. Soc.* **1991**, *113*, 1470; c) M. R. Ghadiri and M. A. Case, *Angew. Chem.* **1993**, *105*, 1663; *Angew. Chem. Int. Ed. Engl.* **1993**, *32*, 1594.

[9.89]   M. R. Ghadiri, C. Soares and C. Choi, *J. Am. Chem. Soc.* **1992**, *114*, 4000.

[9.90]   For the use of organic templates for the assembly of model proteins, see: I Ernest, S. Vuilleumier, H. Fritz and M. Mutter, *Tetrahedron Lett.* **1990**, *31*, 4015; M. Mutter and S. Vuilleumier, *Angew. Chem.* **1989**, *101*, 551; *Angew. Chem. Int. Ed. Engl.* **1989**, *28*, 535; T. Sasaki and E. T. Kaiser, *J. Am. Chem. Soc.* **1989**, *111*, 380.

[9.91]   a) K. P. Meurer and F. Vögtle, *Topics Curr. Chem.* **1985**, *127*, 1 and references therein; b) R. Bishop and I. G. Dance, *ibid.* **1988**, *149*, 137.

[9.92]   a) K. Mislow, D. Gust, P. Finacchiaro and R. J. Boettcher, *Topics Curr. Chem.* 1974, 47, 1; b) R. J. M. Nolte, *Chem. Soc. Rev.* 1994, 23, 11.

[9.93]   P. Baxter, J.-M. Lehn, A. DeCian and J. Fischer, *Angew. Chem.*, 1993, 105, 92; *Angew. Chem. Int. Ed. Engl.*, 1993, 32, 69.

[9.94]   a) E. Leize, A. Van Dorsselaer, R. Krämer and J.-M. Lehn, *J. Chem. Soc., Chem. Commun.* 1993, 990; b) A. Marquis and J.-M. Lehn, work in progress.

[9.95]   a) G. Hopfgartner, C. Piguet, J. D. Henion and A. F. Williams, *Helv. Chim. Acta* 1993, 76, 1759; b) A. F. Williams, C. Piguet and R. F. Carina, p. 409 in [A.40].

[9.96]   P. N. W. Baxter and J.-M. Lehn, unpublished results.

[9.97]   a) H. Sleiman, P. N. W. Baxter, J.-M. Lehn and K. Rissanen, *J. Chem. Soc., Chem. Commun.* 1995, in press; b) G. S. Hanan, C. R. Arana, J.-M. Lehn and D. Fenske, *Angew. Chem.* 1995, in press.

[9.98]   O. Heyke, K. Wänmark, J. Thomas and J.-M. Lehn, unpublished results.

[9.99]   a) C. M. Drain and J.-M. Lehn, *J. Chem. Soc. Chem. Commun.* 1994, 2313; b) C. Von Borczyskowski, U. Rempel, M. Lindrum and A. Kern, *Mol. Cryst. Liq. Cryst.* 1993, 234, 97.

[9.100] P. N. W. Baxter, J.-M. Lehn, J. Fischer and M.-T. Youinou, *Angew. Chem.* 1994, 106, 2432; *Angew. Chem. Int. Ed. Engl.* 1994, 33, 2284.

[9.101] G. R. Desiraju, *Crystal Engineering. The Design of Organic Solids*, Elsevier, Amsterdam, 1989.

[9.102] a) R. Taylor and O. Kennard, *Acc. Chem. Res.* 1984, 17, 320; b) O. Kennard and W. N. Hunter, *Angew. Chem.* 1991, 103, 1280; *Angew. Chem. Int. Ed. Engl.* 1991, 30, 1254.

[9.103] For a recent overview of structure and hydrogen bonding in the organic solid state, see: *Chem. Materials, Special Issue* 1994, 6, No. 8.

[9.104] G. A. Jeffrey and W. Saenger, *Hydrogen Bonding in Biologial Structures*, Springer, Berlin, 1991.

[9.105] J. Bernstein, M. C. Etter and L. Leiserowitz, in *Structure Correlation*, (Eds.: J. D. Dunitz and H. B. Bürgi), VCH, Weinheim, 1993.

[9.106] J. Pranata, S. G. Wierschke and W. L. Jorgensen, *J. Am. Chem. Soc.* 1991, 113, 2810.

[9.107] J. Hunziker, H.-J. Roth, M. Böhringer, A. Giger, U. Diederichsen, M. Göbel, R Krishnan, B. Jaun, C. Leumann and A. Eschenmoser, *Helv. Chim. Acta* 1993, 76, 259.

[9.108] W. Guschlbauer, J.-F. Chantot and D. Thiele, *J. Biomol. Struct. and Dyn.* 1990, 8, 491.

[9.109] S. Bonazzi, M. M. De Morais, A. Garbesi, G. Gottarelli, P. Mariani and G. P. Spada, *Liquid Crystals* 1991, 10, 495.

[9.110] R. Jin, K. J. Breslauer, R. A. Jones and B. L. Gaffney, *Science* 1990, 250, 543.

[9.111] F. Seela and K. Mersmann, *Helv. Chim. Acta* 1993, 76, 1435.

[9.112] S. Bonazzi, M. Capobianco, M. M. De Morais, A. Garbesi, G. Gottarelli, P. Mariani, M. G. Ponzi Bossi, G. P. Spada and L. Tondelli, *J. Am. Chem. Soc.* 1991, 113, 5809.

[9.113] a) C. H. Kang, X. Zhang, R. Ratliff, R. Moyzis and A. Rich, *Nature* 1992, 356, 126; b) G. Laughlan, A. I. H. Murchie, D. G. Norman, M. H. Moore, P. C. E. Moody, D. M. J. Lilley and B. Luisi, *Science* 1994, 265, 520.

[9.114] a) D. Sen and W. Gilbert, *Nature* 1988, 334, 364; b) W. I. Sundquist and A. Klug, *Nature* 1989, 342, 825.

[9.115] F. Ciuchi, G. Di Nicola, H. Franz, G. Gottarelli, P. Mariani, M. G. Ponzi Bossi and G. P. Spada, *J. Am. Chem. Soc.* 1994, 116, 7064.

[9.116] a) M. C. Etter, Z. Urbañczyk-Lipkowska, D. A. Jahn and J. S. Frye, *J. Am. Chem. Soc.* 1986, 108, 5871; M. C. Etter, D. L. Parker, S. R. Ruberu, T. W. Panunto and D. Britton, *J. Incl. Phenom. and Mol. Recogn. Chem.* 1990, 8, 395; b) D. J. Duchamp, R. E. Marsh, *Acta Cryst.* **1969**, *B25*, 5; c) J. Yang, J. L. Marendaz, S. J. Geib and A. D. Hamilton, *Tetrahedron Lett.* **1994**, 35, 3665.

[9.117]   a) Y. Ducharme and J. D. Wuest, *J. Org. Chem.* **1988**, *53*, 5787; b) For cyclotrimerisation see: S. C. Zimmerman and B. F. Duerr, *J. Org. Chem.* **1992**, *57*, 2215.

[9.118]   For a 2+2 cyclic heterodimer, see: J. Yang, E. Fan, S. J. Geib and A. D. Hamilton, *J. Am. Chem. Soc.* **1993**, *115*, 5314.

[9.119]   J.-M. Lehn, M. Mascal, A. DeCian and J. Fischer, *J. Chem. Soc., Chem. Commun.* **1990**, 479.

[9.120]   a) J. A. Zerkowski, C. T. Seto, D. A. Wierda and G. M. Whitesides, *J. Am. Chem. Soc.* **1990**, *112*, 9025; b) see also, J. C. MacDonald and G. M. Whitesides, *Chem. Rev.* **1994**, *94*, 2383.

[9.121]   J. A. Zerkowski, C. T. Seto and G. M. Whitesides, *J. Am. Chem. Soc.* **1992**, *114*, 5473.

[9.122]   J.-M. Lehn, M. Mascal, A. DeCian and J. Fischer, *J. Chem. Soc. Perkin Trans. 2* **1992**, 461.

[9.123]   A. Marsh, E. G. Nolen, K. M. Gardinier and J.-M. Lehn, *Tetrahedron Lett.* **1993**, *35*, 397.

[9.124]   V. Prelog and G. Helmchen, *Angew. Chem.* **1982**, 94, 614; *Angew. Chem. Int. Ed. Engl.* **1982**, *21*, 567.

[9.125]   A. Marsh, M. Silvestri and J.-M. Lehn, unpublished work.

[9.126]   N. Branda, S. Erickson, U. Hoffmann, A. Marsh and J.-M. Lehn, work in progress.

[9.127]   P. E. Nielsen, M. Egholm, R. H. Berg and O. Buchardt, *Science* **1991**, *254*, 1497.

[9.128]   B. Hyrup, M. Egholm, P. E. Nielsen, P. Wittung, B. Nordén and O. Buchardt, *J. Am. Chem. Soc.* **1994**, *116*, 7964.

[9.129]   M. S. Kim and G. W. Gokel, *J. Chem. Soc., Chem. Commun.* **1987**, 1686.

[9.130]   a) R. P. Bonar-Law and J. K. M. Sanders, *Tetrahedron Lett.* **1993**, *34*, 1677; b) R. Wyler, J. de Mendoza and J. Rebek, Jr., *Angew. Chem.*, **1993**, *105*, 1820; *Angew. Chem. Int. Ed. Engl.* **1993**, *32*, 1699.

[9.131]   a) C. T. Seto, J. P. Mathias and G. M. Whitesides, *J. Am. Chem. Soc.* **1993**, *115*, 1321; b) C. T. Seto and G. M. Whitesides, *J. Am. Chem. Soc.* **1993**, *115*, 1330; c) J. P. Mathias, E. E. Simanek and G. M. Whitesides, *J. Am. Chem. Soc.* **1994**, *116*, 4326.

[9.132]   a) R. Wyler, J. de Mendoza and J. Rebek Jr., *Angew. Chem.* **1993**, *105*, 1820; *Angew. Chem. Int. Ed. Engl.* **1993**, *32*, 1699; b) N Branda, R. Wyler and J. Rebek, Jr., *Science* **1994**, *263*, 1267; c) N. Branda, R. M. Grotzfeld, C. Valdés and J. Rebek, Jr., *J. Am. Chem. Soc.* **1995**, *117*, 85.

[9.133]   C. M. Drain, R. Fischer, E. G. Nolen and J.-M. Lehn, *J. Chem. Soc., Chem. Commun.* **1993**, 243

[9.134]   G. Decher and J.-D. Hong, *Makromol. Chem., Macromol. Symp.* **1991**, *46*, 321.

[9.135]   A. Ulman, *Adv. Mater.* **1990**, 2, 573.

[9.136]   H. Sellers, A. Ulman, Y. Shnidman and J. E. Eilers, *J. Am. Chem. Soc.* **1993**, *115*, 9389.

[9.137]   C. D. Bain and G. M. Whitesides, *Angew. Chem.* **1989**, *101*, 522; *Angew. Chem. Int. Ed. Engl.* **1989**, *28*, 506.

[9.138]   A. Barraud, *Pour la Science* **1993**, *189*, 62.

[9.139]   F. Garnier, A. Yassar, R. Hajlaoui, G. Horowitz, F. Deloffre, B. Servet, S. Ries and P. Alnot, *J. Am. Chem. Soc.* **1993**, *115*, 8716.

[9.140]   F. M. Menger and C. A. Littau, *J. Am. Chem. Soc.* **1993**, *115*, 10083.

[9.141]   S. Zhang, T. Holmes, C. Lockshin and A. Rich, *Proc. Natl. Acad. Sci. USA* **1993**, *90*, 3334.

[9.142]   a) V. Percec, H. Jonsson and D. Tomazos, in [7.13] p. 1; b) V. Percec, J. Heck, M. Lee, G. Ungar and A. Alvarez-Castillo, *J. Mater. Chem.* **1992**, 2, 1033; c) V. Percec and G. Johansson, *J. Mater. Chem.* **1993**, *3*, 83.

[9.143] a) A.-M. Giroud-Godquin and P. M. Maitlis, *Angew. Chem.* **1991**, *103*, 370; *Angew. Chem. Int. Ed. Engl.* **1991**, *30*, 375; b) S. A. Hudson and P. M. Maitlis, *Chem. Rev.* **1993**, *93*, 861.

[9.144] a) R. Eidenschink, *Angew. Chem. Adv. Mater.* **1989**, 101, 1454; b)

[9.145] M. Ebert, R. Kleppinger, M. Soliman, M. Wolf, J. H. Wendorff, G. Lattermann and G. Staufer, *Liquid Crystals* **1990**, *7*, 553.

[9.146] S. Hoffmann and W. Witkowski, in *Mesomorphic Order in Polymers*, (Ed.: A. Blumstein), *ACS Symp. Ser.* **1978**, *74*, 78; S. Hoffmann, *Z. Chem.* **1987**, *27*, 395.

[9.147] a) U. Kumar, T. Kato and J. M. J. Fréchet, *J. Am. Chem. Soc.* **1992**, *114*, 6630; b) T. Kato, H. Kihara, T. Uryu, A. Fujishima and J. M. J. Fréchet, *Macromolecules* **1992**, *25*, 6836; c) L. J. Yu, *Liquid Crystals* **1993**, *14*, 1303; d) C. Alexander, C. P. Jariwala, C. M. Lee and A. C. Griffin, *Makromol. Chem., Macromol. Symp.* **1994**, *77*, 283.

[9.148] a) J.-H. Fuhrhop and W. Helfrich, *Chem. Rev.* **1993**, *93*, 1565; b) J.-H. Fuhrhop, S. Svenson, P. Luger and C. André, *Supramolecular Chem.* **1993**, 2, 157; c) J.-H. Fuhrhop and C. Boettcher, *J. Am. Chem. Soc.* **1990**, *112*, 1768; d) J. M. Schnur, *Science* **1993**, *262*, 1669.

[9.149] J.-M. Lehn, *Makromol. Chem., Macromol. Symp.* **1993**, *69*, 1.

[9.150] M.-J. Brienne, J. Gabard, J.-M. Lehn and I. Stibor, *J. Chem. Soc. Chem. Commun.* **1989**, 1868.

[9.151] C. Fouquey, J.-M. Lehn and A.-M. Levelut, *Adv. Mater.* **1990**, 2, 254.

[9.152] T. Gulik-Krzywicki, C. Fouquey and J.-M. Lehn, *Proc. Natl. Acad. Sci. USA* **1993**, *90*, 163.

[9.153] M.-J. Brienne, C. Fouquey, A.-M. Levelut and J.-M. Lehn, unpublished work.

[9.154] M. Kotera, J.-M. Lehn and J.-P. Vigneron, *J. Chem. Soc. Chem. Commun.* **1994**, 197.

[9.155] G. Wegner, *Thin Solid Films* **1992**, *216*, 105 and references therein.

[9.156] C. Fouquey, J.-M. Lehn, work in progress.

[9.157] M. Antonietti and S. Heinz, *Nachr. Chem. Tech. Lab.* **1992**, *40*, 308.

[9.158] a) M. Warner, in [9.11], p. 403; b) C. T. Imrie, *TRIP*, **1995**, *3*, 22.

[9.159] *Mesomorphic Order in Polymers*, (Ed.: A. Blumstein), *ACS Symp. Ser.*, No. 74, **1978**.

[9.160] a) *Protein Folding*, (Ed.: T. E. Creighton), Freeman, New York, **1992**; b) A. Sali, E. I. Shaknovich and M. Karplus, *Nature* **1994**, *369*, 248; c) M. Karplus and A. Šali, *Current Opinion in Structural Biology* **1995**, *5*, 58; d) R. L. Baldwin, *Nature* **1994**, *369*, 183.

[9.161] C. Hilger and R. Stadler, *Makromol. Chem.* **1991**, *192*, 805; *Polymer* **1991**, *32*, 17, 3244.

[9.162] C. Hilger, R. Stadler and L. L. de Lucca Freitas, *Polymer* **1990**, *31*, 818.

[9.163] R. Stadler and L. L. de Lucca Freitas, *Colloid Polym. Sci.* **1988**, *266*, 1102.

[9.164] A. I. Kitaigorodsky, *Molecular Crystals and Molecules*, Academic Press, New York, **1973**.

[9.165] a) M. D. Ward, *Pure Appl. Chem.* **1992**, *64*, 1623; b) D. Braga and F. Grepioni, *Acc. Chem. Res.* **1994**, *27*, 51.

[9.166] a) J. D. Dunitz, *Pure Appl. Chem.* **1991**, *63*, 177; b) J. D. Dunitz, in [A.37], *Vol.* 2; c) A. Gavezzotti and M. Simonetta, *Chem. Rev.* **1982**, *82*, 1.

[9.167] O. Erner and A. Eling, *J. Chem. Soc., Perkin Trans 2* **1994**, 925.

[9.168] a) X. Zhao, Y.-L. Chang, F. W. Fowler and J. W. Lauher, *J. Am. Chem. Soc.* **1990**, *112*, 6627; b) J. W. Lauher, Y.-L. Chang and F. W. Fowler, *Mol. Cryst. Liq. Cryst.* **1992**, *211*, 99; c) Y.-L. Chang, M.-A. West, F. W. Fowler and J. W. Lauher, *J. Am. Chem. Soc.* **1993**, *115*, 5991.

[9.169]   K. C. Russell, J.-M. Lehn, N. Kyritsakas, A. DeCian and J. Fischer, to be published.

[9.170]   For recent approaches, see M. C. Etter and G. M. Frankenbach, *Chem. Mater.* 1989, *1*, 10; I. Weissbuch, M. Lahav, L. Leiserowitz, G. R. Meredith, and H. Vanherzeele, *ibid.* 1989, *1*, 114; S. R. Marder, J. W. Perry and W. P. Schaefer, *Science* 1989, *245*, 626.

[9.171]   b) K. C. Russell and J.-M. Lehn, unpublished results; b) K. C. Russell, E. Leize, A. Van Dorsselaer and J.-M. Lehn, *Angew. Chem.* 1995, *107*, 244; *Angew. Chem. Int. Ed. Engl.* 1995, *34*, 209.

[9.172]   a) For an example of one-dimensional energy migration see: D. Markovitsi, F. Rigaut, M. Mouallem and J. Malthête, *Chem. Phys. Lett.* 1987, *135*, 236; see also [8.32]; b) for photoinduced charge separation in porphyrin-derived liquid crystals see: M.-A. Fox, A. J. Bard, H.-L. Pan and C.-Y. Liu, *J. Chin. Chem. Soc.* 1993, *40*, 321; c) for photoconduction in discotic liquid crystals see: D. Adam, D. Haarer, F. Closs, T. Frey, D. Funhoff, K. Siemensmeyer, P. Schuhmacher and H. Ringsdorf, *Ber. Bunsenges. Phys. Chem.* 1993, *97*, 1366.

[9.173]   R. Krämer, J.-M. Lehn and A. Marquis-Rigault, *Proc. Natl. Acad. Sci. USA* 1993, *90*, 5394.

[9.174]   a) S. Brenner and R. A. Lerner, *Proc. Natl. Acad. Sci. USA*, 1992, *89*, 5381; b) M. Famulok and J. W. Szostak, *Angew. Chem.*, 1992, *104*, 1001; *Angew. Chem. Int. Ed. Engl.* 1992, *31*, 979.

[9.175]   a) M. C. Needels, D. G. Jones, E. H. Tate, G. L. Heinkel, L. M. Kochersperger, W. J. Dower, R. W. Barrett and M. A. Gallop, *Proc. Natl. Acad. Sci. USA* 1993, *90*, 10700; b) For a recent overview see: *Bioorg. and Medic. Chem. Lett.* (Ed.: M. R. Pavia, T. K. Sawyer and W. H. Moos), 1993, *3*, 387–476; c) D. J. Kenan, D. E. Tsai and J. D. Keene, *TIBS* 1994, *19*, 57; d) R. M. Baum, *Chem. Eng. News* 1994, *February 7*, 20.

[9.176]   a) M. Sassanfar and J. W. Szostak, *Nature* 1993, *364*, 550; b) A. Borchardt and W. C. Still, *J. Am. Chem. Soc.* 1994, *116*, 373; c) for a recent approach see: K. D. Janda, C.-H. L. Lo, T. Li, C. F. Barbas III, P. Wirsching and R. A. Lerner, *Proc. Natl. Acad. Sci. USA* 1994, *91*, 2532.

[9.177]   a) R. W. Armstrong, J.-M. Beau, S. H. Cheon, W. J. Christ, H. Fujioka, W.-H. Ham, L. D. Hawkins, H. Jin, S. H. Kang, Y. Kishi, M. J. Martinelli, W. W. McWhorter, Jr., M. Mizuno, M. Nakata, A. E. Stutz, F. X. Talamas, M. Taniguchi, J. A. Tino, K. Ueda, J.-i. Uenishi, J. B. White and M. Yonaga, *J. Am. Chem. Soc.* 1989, *111*, 7525; *ibid.* 7530; b) K C. Nicolaou, *Angew. Chem.* 1993, *105*, 1462; *Angew. Chem. Int. Ed. Engl.* 1993, *32*, 1377; c) for an outloook on organic synthesis see: D. Seebach, *Angew. Chem.* 1990, *102*, 1363; *Angew. Chem. Int. Ed. Engl.* 1990, *29*, 1320.

[9.178]   V. I. Sokolov, *Russ. Chem. Rev.* 1973, *42*, 452; H. L. Frisch and E. Wasserman, *J. Am. Chem. Soc.* 1961, *83*, 3789; for a recent overview see: *Topology in Chemistry*, *New J. Chem.* Special issue 1993, *17*, 617-763; M. Suffczynski, *Polish J. Chem.* 1995, *69*, 157.

[9.179]   a) H. C. Anderson and J. K. M. Sanders, *Angew. Chem.* 1990, *102*, 1478; *Angew. Chem. Int. Ed. Engl.* 1990, *29*, 1400; b) S. Anderson, H. C. Anderson and J. K. M. Sanders, *Angew. Chem.* 1992, *104*, 921; *Angew. Chem. Int. Ed. Engl.* 1992, *31*, 907.

[9.180]   J.-C. Chambron, V. Heitz and J.-P. Sauvage, *J. Chem. Soc., Chem. Commun.* 1992, 1131.

[9.181]   a) C. Dietrich-Buchecker and J.-P. Sauvage, *New. J. Chem.* 1992, *16*, 277; b) for knots based on DNA, see: S. M. Du, B. D. Stollar and N. C. Seeman, *J. Am. Chem. Soc.* 1995, *117*, 1194.

[9.182]   A. C. Benniston and A. Harriman, *Synlett* 1993, 223.

[9.183] a) P. Beak and J. M. Zeigler, *J. Org. Chem.*, 1981, *46*, 619; b) R. G. Chapman, N. Chopra, E. D. Cochien and J. C. Sherman, *J. Am. Chem. Soc.* 1994, *116*, 369.

[9.184] D. Armspach, P. R. Ashton, C. P. Moore, N. Spencer, J. F. Stoddart, T. J. Wear and D. J. Williams, *Angew. Chem.* 1993, *105*, 944; *Angew. Chem. Int. Ed. Engl.* 1993, *32*, 854.

[9.185] H. Ogino, *New. J. Chem.* 1993, *17*, 683.

[9.186] A. Harada and M. Kamachi, *J. Chem. Soc., Chem. Commun.* 1990, 1322.

[9.187] A. Harada, J. Li and M. Kamachi, *Nature* 1992, *356*, 325; *J. Am. Chem. Soc.* 1994, *116*, 3192.

[9.188] G. Wenz and B. Keller, *Angew. Chem.* 1992, *104*, 201; *Angew. Chem. Int. Ed. Engl.* 1992, *31*, 197; G. Wenz, F. Wolf, M. Wagner and S. Kubik, *New. J. Chem.* 1993, *17*, 729.

[9.189] G. Wenz, *Angew. Chem.* 1994, *106*, 851; *Angew. Chem. Int. Ed. Engl.* 1994, *33*, 803.

[9.190] J. F. Stoddart, *Angew. Chem.* 1992, *104*, 860; *Angew. Chem. Int. Ed. Engl.* 1992, *31*, 846.

[9.191] H. W. Gibson and H. Marand, *Adv. Mater.* 1993, *5*, 11.

[9.192] C. A. Hunter, *J. Am. Chem. Soc.* 1992, *114*, 5303.

[9.193] A. Harada, J. Li and M. Kamachi, *Nature* 1993, *364*, 516.

[9.194] a) P. K. Dhal and F. H. Arnold, *J. Am. Chem. Soc.* 1991, *113*, 7417; b) C. Dallaire, K. C. Russell and J.-M. Lehn, work in progress.

[9.195] a) G. M. J. Schmidt, *Pure Appl. Chem.* 1971, *27*, 647; b) I. C. Paul and D. Y. Curtin, *Acc. Chem. Res.* 1973, *6*, 217.

[9.196] L. E. Orgel, *Nature* 1992, *358*, 203.

[9.197] a) G. von Kiedrowski, J. Helbing, B. Wlotzka, S. Jordan, M. Mathen, T. Achilles, D. Sievers, A. Terfort and B. C. Kahrs, *Nachr. Chem. Tech. Lab.* 1992, *40*, 578; b) D. Sievers and G. von Kiedrowski, *Nature* 1994, *369*, 221.

[9.198] S. Hoffmann, *Angew. Chem.* 1992, *104*, 1032; *Angew. Chem. Int. Ed. Engl.* 1992, *31*, 1013.

[9.199] a) G. von Kiedrowski, *Angew. Chem.* 1986, *98*, 932; *Angew. Chem. Int. Ed. Engl.* 1986, *25*, 932; b) for a recent case involving triple helix formation see: T. Li and K. C. Nicolaou, *Nature* 1994, *369*, 218.

[9.200] a) G. von Kiedrowski, B. Wlotzka and J. Helbing, *Angew. Chem.* 1989, *101*, 1259; *Angew. Chem. Int. Ed. Engl.* 1989, *28*, 1235; b) G. von Kiedrowski, B. Wlotzka, J. Helbing, M. Matzen and S. Jordan, *Angew. Chem.* 1991, *103*, 456, 1066; *Angew. Chem. Int. Ed. Engl.* 1991, *30*, 423, 892.

[9.201] a) W. S. Zielinski and L. E. Orgel, *Nature* 1987, *327*, 346; b) see also T. Wu and L. E. Orgel, *J. Am. Chem. Soc.* 1992, *114*, 317.

[9.202] a) T. Tjivikua, P. Ballester and J. Rebek, Jr., *J. Am. Chem. Soc.* 1990, *112*, 1249; b) J. S. Nowick, Q. Feng, T. Tjivikua, P. Ballester and J. Rebek, Jr., *J. Am. Chem. Soc.* 1991, *113*, 8831; c) M. Famulok, J. S. Nowick and J. Rebek, Jr., *Acta Chem. Scand.* 1992, *46*, 315; d) E. A. Wintner, M. M. Conn and J. Rebek Jr., *Acc. Chem. Res.* 1994, *27*, 198.

[9.203] A. Terfort and G. von Kiedrowski, *Angew. Chem.* 1992, *104*, 626; *Angew. Chem. Int. Ed. Engl.* 1992, *31*, 654.

[9.204] C. Böhler, W. Bannwarth and P. L. Luisi, *Helv. Chim. Acta* 1993, *76*, 2313.

[9.205] a) T. Achilles and G. von Kiedrowski, *Angew. Chem.*, 1993, *105*, 1225; *Angew. Chem. Int. Ed. Engl.* 1993, *32*, 1198; b) V. Rotello, J.-I. Hong and J. Rebek, Jr., *J. Am. Chem. Soc.* 1991, *113*, 9422.

[9.206] a) F. M. Menger, A. V. Eliseev and N. A. Khanjin, *J. Am. Chem. Soc.* 1994, *116*, 3613; b) M. M. Conn, E. A. Wintner and J. Rebek, Jr., *J. Am. Chem. Soc.* 1994, *116*, 8823.

[9.207] a) P. A. Bachmann, P. Walde, P. L. Luisi and J. Lang, *J. Am. Chem. Soc.* 1990, *112*, 8200; b) *ibid.* 1991, *113*, 8204; c) P. Walde, R. Wick, M. Fresta, A. Mangone and P. L. Luisi, *J. Am. Chem. Soc.* 1994, *116*, 11649.

[9.208] R. Stiller and J.-M. Lehn , work in progress.

[9.209] J. Jacques, A. Collet and S. H. Wilen, *Enantiomers, Racemates and Resolutions*, Krieger, Malabar, Florida, 1991.

[9.210] a) D. P. Cray and D. P. Mellor, *Topics Curr. Chem.* 1976, *63*, 1; L. Salem, X. Chapuisat, G. Segal, P. C. Hiberty, Ch. Minot, C. Leforestier and P. Sautet, *J. Am. Chem. Soc.* 1987, *109*, 2887; H. Wynberg and B. Feringa, *Tetrahedron* 1976, *32*, 2831; b) J. Brienne, J. Gabard, M. Leclercq, J.-M. Lehn, M. Cesario, C. Pascard, M. Chevé and G. Dutruc-Rosset, *Tetrahedron Lett.* 1994, *35*, 8157.

[9.211] S. Mann, *Nature* 1993, *365*, 499.

[9.212] a) *Supramolecular Architecture, ACS Symp. Ser. No. 499*, (Ed.: T. Bein), ACS Washington DC, 1992; see also *Chem. Eng. News* 1991, May 27, p. 24; b) *Inorganic Materials*, (Eds.: D. W. Bruce and D. O'Hare), Wiley, Chichester, 1992.

[9.213] J. Rouxel, *Adv. Synth. React. Solids* JAI Press, Greenwich, Conn., 1994, *2*, 27.

[9.214] G. D. Stucky and J. E. MacDougall, *Science* 1990, *247*, 669.

[9.215] a) P. J. Fagan, M. D. Ward and J. C. Calabrese, *J. Am. Chem. Soc.* 1989, *111*, 1698; b) *Inorganic and Organometallic Polymers with Special Properties*, (Ed.: R. M. Laine), Kluwer, Dordrecht, 1989.

[9.216] A. J. Ashe, III, W. Butler and T. R. Diephouse, *J. Am. Chem. Soc.* 1981, *103*, 207.

[9.217] M. G. Newton, R. B. King, I. Haiduc and A. Silvestru, *Inorg. Chem.* 1993, *32*, 3795.

[9.218] R. H. Cayton, M. H. Chisholm, J. C. Huffman and E. B. Lobkovsky, *Angew. Chem.* 1991, *103*, 893.

[9.219] S. J. Leob and G. K. H. Shimizu, *J. Chem. Soc., Chem. Commun.* 1993, 1395.

[9.220] A. McAuley, S. Subramanian and M. J. Zaworotko, *J. Chem. Soc., Chem. Commun.* 1992, 1321.

[9.221] B. F. Abrahams, B. F. Hoskins, J. Liu and R. Robson, *J. Am. Chem. Soc.* 1991, *113*, 3045.

[9.222] B. F. Abrahams, M. J. Hardie, B. F. Hoskins, R. Robson and G. A. Williams, *J. Am. Chem. Soc.* 1992, *114*, 10641.

[9.223] a) B. F. Hoskins and R. Robson, *J. Am. Chem. Soc.* 1989, *111*, 5962; *ibid.* 1990, *112*, 1546; b) A. K. Brimah, E. Siebel, R. D. Fischer, N. A. Davies, D. C. Apperley and R. K. Harris, *J. Organometal. Chem.* 1994, *475*, 85.

[9.224] M. Fujita, Y. J. Kwon, S. Washizu and K. Ogura, *J. Am. Chem. Soc.* 1994, *116*, 1151.

[9.225] D. A. Evans, K. A. Woerpel and M. J. Scott, *Angew. Chem.* 1992, *104*, 439; *Angew. Chem. Int. Ed. Engl.* 1992, *31*, 430.

[9.226] B. F. Abrahams, B. F. Hoskins and R. Robson, *J. Am. Chem. Soc.* 1991, *113*, 3606.

[9.227] G. De Munno, M. Julve, F. Nicolo, F. Lloret, J. Faus, R. Ruiz and E. Sinn, *Angew. Chem.* 1993, *105*, 585; *Angew. Chem. Int. Ed. Engl.* 1993, *32*, 613.

[9.228] K.-W. Kim and M. G. Kanatzidis, *J. Am. Chem. Soc.* 1992, *114*, 4878.

[9.229] V. Soghomonian, Q. Chen, R. C. Haushalter, J. Zubieta and C. J. O'Connor, *Science* 1993, *259*, 1596.

[9.230] H. O. Stumpf, L. Ouahab, Y. Pei, D. Grandjean and O. Kahn, *Science* 1993, *261*, 447.

[9.231] O. Ermer and L. Lindenberg, *Helv. Chim. Acta* 1991, *74*, 825.

[9.232] S. Mann and F. C. Meldrum, *Adv. Mater.* 1991, *3*, 316; F. C. Meldrum, V. J. Wade, L. Nimmo, B. R. Heywood and S. Mann, *Nature* 1991, *349*, 684.

[9.233]   a) M. P. Byrn, C. J. Curtis, S. I. Khan, P. A. Sawin, R. Tsurumi and C. E. Strouse, *J. Am. Chem. Soc.* **1990**, *112*, 1865; b) M. P. Byrn, C. J. Curtis, I. Goldberg, T. Huang, Y. Hsiou, S. I. Khan, P. A. Sawin, S. K. Tendick, A. Terzis, C. E. Strouse, *Mol. Cryst. Liq. Cryst.* **1992**, *211*, 135.

[9.234]   S. Iijima, *Nature* **1991**, *354*, 56; D. Ugarte, *Nature*, **1992**, *359*, 707; S. Iijima and T. Ichihashi, *Nature* **1993**, *363*, 603.

[9.235]   For general features of molecular materials see for instance: J. Simon, J.-J. André and A. Skoulios *Nouv. J. Chim.* **1986**, *10*, 9836 and references therein.

[9.236]   D. J. O'Shannessy, L. I. Andersson and K. Mosbach, *J. Mol. Recogn.* **1989**, *2*, 1.

[9.237]   C. Bamford and K. Al-Lamee, *J. Chem. Soc., Chem. Commun.* **1993**, 1580.

[9.238]   L. Jullien, J.-M. Lehn and V. Marchi, work in progress.

[9.239]   S. I. Stupp, S. Son, H. C. Lin and L. S. Li, *Science* **1993**, *259*, 59.

[9.240]   M. Djabourov, *Polym. Int.* **1991**, *25*, 135.

[9.241]   As illustration, see for instance: A. Thierry, C. Straupé, B. Lotz and J. C. Wittmann, *Polym. Commun.* **1990**, *31*, 299.

[9.242]   a) K. Hanabusa, J. Tange, Y. Taguchi, T. Koyama and H. Shirai, *J. Chem. Soc. Chem. Commun.* **1993**, 390; b) K. Hanabusa, T. Miki, Y. Taguchi, T. Koyama and H. Shirai, *J. Chem. Soc., Chem. Commun.* **1993**, 1382.

[9.243]   M. Mascal and J.-M. Lehn, unpublished observations.

[9.244]   a) E. Sackmann, *Macromol. Chem. Phys.* **1994**, *195*, 7; b) E. Sackmann, *FEBS Letters* **1994**, *346*, 3.

[9.245]   P. J. Fagan and M. D. Ward, *Scientific Amer.* **1992**, *267(1)*, 48.

[9.246]   B. R. Heywood and S. Mann, *Adv. Mater.* **1994**, *6*, 9.

[9.247]   J. C. Wittmann and B. Lotz, *Prog. Polym. Sci.* **1990**, *15*, 909.

[9.248]   J. C. Wittmann and P. Smith, *Nature* **1991**, *352*, 414.

[9.249]   C. B. Aakeröy and K. R. Seddon, *Chem. Soc. Rev.* **1993**, *22*, 397.

[9.250]   a) *Molecular Engineering and Structure Design, Isr. J. Chem.* **1985**, *25*, special issue; b) J. S. Moore and S. Lee, *Chemistry and Industry* **1994**, 556.

[9.251]   a) *Fundamentals of Adhesion*, (Ed.: L.-H. Lee), Plenum Press, New York, **1991**; b) J. N. Israelachvili, *Intermolecular and Surface Forces*, Academic Press, New York, **1985**; c) K. Kendall, *Science* **1994**, *263*, 1720.

[9.252]   a) A. H. Heuer, D. J. Fink, V. J. Laraia, J. L. Arias, P. D. Calvert, K. Kendall, G. L. Messing, J. Blackwell, P. C. Rieke, D. H. Thompson, A. P. Wheeler, A. Veis and A. I. Caplan, *Science* **1992**, *255*, 1098; b) Y. Zhang and N. C. Seeman, *J. Am. Chem. Soc.* **1994**, *116*, 1661; D. M. J. Lilley and R. M. Clegg, *Quarterly Rev. Biophys.* **1993**, *26*, 131; c) K. Nagayama, *Nanobiology* **1992**, *1*, 25.

[9.253]   S. A. MacDonald, C. G. Willson and J. M. J. Fréchet, *Acc. Chem. Res.* **1994**, *27*, 151; b) J. L. Wilbur, A. Kumar, E. Kim and G. M. Whitesides, *Adv. Mater.* **1994**, *6*, 600.

[9.254]   a) A. J. Bard, *Integrated Chemical Systems: A Chemical Approach to Nanotechnology*, Wiley, New York, **1994**; b) C. R. Martin, *Science* **1994**, *266*, 1961.

[9.255]   G. Binnig and H. Rohrer, *Angew. Chem.* **1987**, *99*, 622; *Angew. Chem. Int. Ed. Engl.* **1987**, *26*, 606.

[9.256]   a) P. K. Hansma, V. B. Elings, O. Marti and C. E. Bracker, *Science* **1988**, *242*, 209; b) M. Radmacher, R. W. Tillmann, M. Fritz and H. E. Gaub, *Science* **1992**, *257*, 1900.

[9.257]   J. A. Stroscio and D. M. Eigler, *Science* **1991**, *254*, 1319.

[9.258]   a) H. Fuchs, *J. Molec. Struct.* **1993**, *292*, 29; b) J. Frommer, *Angew. Chem.* **1992**, *104*, 1325; *Angew. Chem. Int. Ed. Engl.* **1992**, *31*, 1298.

[9.259]   a) P. Avouris, *Acc. Chem. Res.* **1994**, *27*, 159; b) P. Avouris, in *Highlights in Condensed Matter Physics and Future Prospects*, (Ed.: L. Esaki), Plenum Press, New York, **1991**.

[9.260]   D. M. Eigler and E. K. Schweizer, *Nature* 1990, *344*, 524.
[9.261]   M. F. Crommie, C. P. Lutz and D. M. Eigler, *Science* 1993, *262*, 218.
[9.262]   P. Zeppenfeld, C. P. Lutz, D. M. Eigler, *Ultramicroscopy* 1992, *42–44 (Pt. A)*, 128.
[9.263]   I.-W. Lyo and P. Avouris, *Science* 1989, *245*, 1369.
[9.264]   P. Bedrossian, D. M. Chen, K. Mortensen and J. A. Golovchenko, *Nature* 1989, *342*, 258.
[9.265]   J. M. R. Weaver, L. M. Walpita and H. K. Wickramasinghe, *Nature* 1989, *342*, 783.
[9.266]   a) R. Berndt, R. Gaisch, J. K. Gimzewski, B. Reihl, R. R. Schlittler, W. D. Schneider and M. Tschudy, *Science* 1993, *262*, 1425; b) for optical near-field microscopy, see: T. Basché, *Angew. Chem.* 1994, *106*, 1805; *Angew. Chem. Int. Ed. Engl.* 1994, *33*, 1723.
[9.267]   G. Nunes, Jr. and M. R. Freeman, *Science* 1993, *262*, 1029.
[9.268]   A. Sato and Y. Tsukamoto, *Adv. Mater.* 1994, *6*, 79.
[9.269]   M. Pomerantz, A. Aviram, R. A. McCorkle, L. Li and A. G. Schrott, *Science* 1992, *255*, 1115.
[9.270]   Y.-H. Tsao, D. F. Evans, H. Wennerström, *Science* 1993, *262*, 547.
[9.271]   F. Creuzet, G. Ryschenkow and H. Arribart, *J. Adhesion* 1992, *40*, 15.
[9.272]   J. H. Hoh, J. P. Cleveland, C. B. Prater, J.-P. Revel and P. K. Hansma, *J. Am. Chem. Soc.* 1992, *114*, 4917.
[9.273]   X.-Y. Lin, F. Creuzet and H. Arribart, *J. Phys. Chem.* 1993, *97*, 7272.
[9.274]   P. A. Christensen, *Chem. Soc. Rev.* 1992, *21*, 197.
[9.275]   a) F.-R. Fan and A. J. Bard, *J. Electrochem. Soc.* 1989, *136*, 3216; b) R. Christoph, H. Siegenthaler, H. Rohrer and H. Wiese, *Electrochim. Acta* 1989, *34*, 1011.
[9.276]   H. D. Abruña, J. H. Schott, J. E. Hudson, S. R. Snyder and H. S. White, *Comments Inorg. Chem.* 1994, *15*, 171.
[9.277]   a) E.-L. Florin, V. T. Moy, H. E. Gaub, *Science* 1994, *264*, 415; b) F. Pincet, E. Perez, G. Bryant, L. Lebeau and C. Mioskowski; *Phys. Rev. Lett.* 1994, *73*, 2780; c) G. U. Lee, L. A. Chrisey and R. J. Colton, *Science* 1994, *266*, 771; d) W. M. Heckl and J. F. Holzrichter, *Nonlinear Optics* 1992, *1*, 53.
[9.278]   K. Lieberman, S. Harush, A. Lewis and R. Kopelman, *Science* 1990, *247*, 59.
[9.279]   R. P. Feynman, *Eng. and Sci.*, 1960, *23*, 22; see also: *Science* 1991, *254*, 1300.

Chapter 10

[10.1]   F. E. Yates, Epilogue in [9.8a], p. 617.
[10.2]   T. D. Schneider, *Nanotechnology* 1994, *5*, 1 and references therein.
[10.3]   For a discussion of the information content of nucleotide sequences, see: T. D. Schneider, G. O. Stormo, L. Gold and A. Ehrenfeucht, *J. Mol. Biol.* 1986, *188*, 415.
[10.4]   a) N. Guihery, G. Durand and M.-B. Lepetit, *Chem. Phys.* 1994, *183*, 45; b) N. Guihery, G. Durand and M.-B. Lepetit and J.-P. Malrieu, *Chem. Phys.* 1994, *183*, 61.
[10.5]   a) M. A. Reed, *Scientific Amer.* 1993, *268(1)*, 98; see for instance the electrode grid pictured on p. 103; ion grids such as **167** are however much more compact; b) C. Weisbuch and B. Vinter, *Quantum Semiconductor Structures*, Academic Press, Boston, 1991.
[10.6]   M. Sundaram, S. A. Chalmers, P. F. Hopkins and A. C. Gossard, *Science* 1991, *254*, 1326.
[10.7]   C. M. Fisher, M. Burghard, S. Roth and K. von Klitzing, *Europhys. Lett.* in press.

[10.8]   a) M. Conrad, *Commun. Assoc. Comp. Mach.* **1985**, *28*, 464; b) M. Conrad, *Nanobiology* **1993**, *2*, 5.

[10.9]   One may note that folding and structure generation in biological macromolecules could occur through parallel search with formation of local arrangements which then interact with each other giving larger structured areas, that in turn will interact and so on, in a convergent fashion until the final superstructure is reached. For instance, protein folding [9.160], as a self-organizing process, would present characteristics of parallel computing.

[10.10]   a) I. M. Klotz, D. W. Darnall and N. R. Langerman, in *The Proteins*, (Eds.: H. Neurath and R. L. Hill), *Vol. 3*, Academic Press, **1975**, 293; b) J. Monod, J. Wyman and J.-P. Changeux, *J. Mol. Biol.* **1965**, *12*, 88; c) D. S. Goodsell and A. J. Olson, *TIBS* **1993**, *18*, 65.

[10.11]   a) R. Günther, B. Schapiro and P. Wagner, *Chaos, Solitons and Fractals* **1994**, *4*, 635; b) see in M. Gell-Mann, *The Quark and the Jaguar*, Little, Brown, **1992**.

[10.12]   For a discussion and evaluation of the structural complexity of molecules on the basis of their graphs, see: S. H. Bertz, *J. Am. Chem. Soc.* **1981**, *103*, 3599; *Bull. Math. Biol.* **1983**, *45*, 849; in *Chemical Applications of Topology and Graph Theory*, (Ed.: R. B. King), *Studies in Physical and Theoretical Chemistry* **1983**, *28*, 206.

[10.13]   a) G. Yagil, *J. theor. Biol.* **1985**, *112*, 1; b) G. Yagil, in *1992 Lecturers in Complex Systems* (Eds.: L. Nadel and D. Stein), SFI Studies in the Sciences of Complexity, Lect. Vol. V, Addison-Wesley, **1993**.

[10.14]   Two species have a single interaction. A third interacts with the first two and also perturbs their interaction. Thus, levels of increasing complexity consist of (1) the isolated species; (2) two species, their interaction and the perturbation of each species by the interaction; (3) three or more species, their interactions with each other, the perturbation of each by any other one and the perturbation of the interaction between any pair by each other species (2nd order interaction). A set of N interacting species S_i comprises $N(N-1)/2$ direct, first order I_{ij} interactions between S_i and S_j and $N(N-1)(N-2)/2$ perturbations of I_{ij} by S_k, i.e. 2nd order interactions, up to $N!/2$ perturbations of $(N-1)$th order.

[10.15]   H. T. Odum, *Science* **1988**, *242*, 1132.

[10.16]   B. Berger, P. W. Shor, L. Tucker-Kellogg and J. King, *Proc. Natl. Acad. Sci. USA* **1994**, *91*, 7732.

[10.17]   H. A. Simon, *Proc. Amer. Phil. Soc.* **1962**, *106*, 467.

[10.18]   J. R. Platt, *J. Theoret. Biol.* **1961**, *1*, 342.

[10.19]   See for instance, J.-L. Lions, *Compt. Rend. Acad. Sci., Sér. gén.*, **1993**, *10*, 305 and references therein.

[10.20]   a) D. Ruelle, *Chaotic Evolution of Strange Attractors*, Lezioni Lincee, Cambridge University Press, Cambridge, **1989**; b) *La Science du désordre, La Recherche*, special issue, **1991**, *22*, No. 232.

[10.21]   Th. Mann, *Der Zauberberg*, Fischer, Frankfurt, **1924**; see also in [10.22].

[10.22]   M. Eigen, *Stufen zum Leben*, Piper, München, **1987**.

[10.23]   A. Kornberg, *Biochemistry* **1987**, *26*, 6888.

[10.24]   See ref. [9.5], p. 3.

[10.25]   G. Whitesides, *Angew. Chem.* **1990**, *102*, 1247; *Angew. Chem. Int. Ed. Engl.* **1990**, *29*, 1209.

[10.26]   J.-M. Lehn, *Alchimères*, a text with photographs by Tromeur, edn. Rémanences, Strasbourg, **1991**.

[10.27]   Auguste Rodin, *La Main de Dieu*, Musée Rodin, Paris, France.

[10.28]   M. Berthelot, *Chimie Organique Fondée sur la Synthèse*, Mallet-Bachelier, Paris, **1860**, *2*, 811.

[10.29]   The original sentence (transcribed into modern italian) is: "*In effetti l'uomo non si varia dalli animali se non nella accidentale, col quale esso si dimostra essere cosa divina, perche dove la nature finisce il produrre le sue spezie l'omo quivi comincia colle cose naturali a fare, coll'aiutorio d'essa natura infinite spezie . . .*", Leonardo da Vinci, in *Disegni Anatomici* nella Biblioteca Reale di Windsor (*Anatomic Drawings* in the Royal Library of Windsor) n. *72 verso* (earlier numerotation 19030 *verso* (B13 V)). Translated into english: "In fact man does not vary from the animals except in what is accidental, and it is in this that he shows himself to be a divine thing, for where nature finishes producing its species there man begins with natural things to make with the aid of this nature an infinite number of species . . .", in *The Literary Works of Leonardo da Vinci*, compiled and edited from the original manuscripts by J.-P. Richter, commentary by C. Pedretti, Phaidon, Oxford, 1977 p. 102. In this sentence "accidental" designates the capability of experimentation (A. Vezzosi). Replacing "fare" by "creare" in the shortened version matches more faithfully the thoughts of Leonardo da Vinci (G. Caglioti). See also: A. M. Brizio, *Leonardo pittore*, in *Leonardo* (Ed.: L. Reti), Mondadori, Milano, 1974, p. 24. I wish to thank Professor G. Caglioti, (Politecnico di Milano), for his help in elucidating these literary matters and for a stimulating correspondence, and Professor A. Vezzosi (Museo Ideale Leonardo da Vinci, Vinci) for providing relevant information.

Appendix

Monographs, Edited Books, Special Issues

A.1   M. Hiraoka *Crown Compounds. Their Characteristics and Applications*, Kodansha, Tokyo, **1978**.
A.2   M. Dobler, *Ionophores and their Structures*, J. Wiley, New York, **1981**.
A.3   G. W. Gokel and S. H. Korzeniowski, *Macrocyclic Polyether Syntheses*, Springer, Berlin, **1982**.
A.4   J.-M. Lehn, *Chemia Supramolekularna*, translated and supplemented by J. Lipkowski, M. Pietraszkiewicz, K. Suwinska, T. Koscielski, M. Kozbial, Polskiej Akad., Nauk., Warsaw, **1985**; 2nd ed., **1993**.
A.5   H. Dugas, *Bioorganic Chemistry*, Springer, Heidelberg, **1989**.
A.6   C. D. Gutsche, *Calixarenes*, RSC, Cambridge, **1989**.
A.7   L. F. Lindoy, *The Chemistry of Macrocyclic Ligand Complexes*, Cambridge University Press, Cambridge, **1989**.
A.8   F. Vögtle, *Supramolekulare Chemie*, Teubner, Stuttgart, **1989**; Supramolecular Chemistry, Wiley, New York, **1991**.
A.9   F. Vögtle, *Cyclophan-Chemie*, Teubner, Stuttgart, **1990**.
A.10 V. Balzani and F. Scandola, *Supramolecular Photochemistry*, Ellis Horwood, New York, **1991**.
A.11 F. Diederich, *Cyclophanes*, RSC, Cambridge, **1991**.
A.12 B. Dietrich, P. Viout and J.-M. Lehn, *Aspects de la Chimie des Composés Macrocycliques*, InterEditions/CNRS, Paris, **1991**; *Macrocyclic Chemistry*, VCH, Weinheim, **1993**.
A.13 G. Gokel, *Crown Ethers and Cryptands*, RSC, Cambridge, **1991**.
A.14 J. S. Bradshaw, K. E. Krakowiak and R. M. Izatt, *Aza-Crown Macrocycles*, Wiley, New York, **1993**.
A.15 *Synthetic Multidentate Macrocyclic Compounds*, (Eds.: R. M. Izatt and J. J. Christensen), Academic Press, New York, **1978**.
A.16 *Coordination Chemistry of Macrocyclic Compounds*, (Ed.: G. A. Melson), Plenum Press, New York, **1979**.
A.17 *Progress in Macrocyclic Chemistry*, (Eds.: R. M. Izatt and J. J. Christensen), *Vol. 1*, **1979**; *Vol. 2*, **1981**, *Vol. 3*, **1987**, Wiley, New York.
A.18 *Inclusion Compounds*, (Eds.: J. L. Atwood, J. E. D. Davies and D. D. MacNicol), *Vols. 1–3*, **1984**; *Vols. 4, 5*, **1991**, Academic Press, London.
A.19 *Design and Synthesis of Organic Molecules Based on Molecular Recognition*, (Ed.: G. van Binst), Springer, Heidelberg, **1986**.

A.20   *Supramolecular Photochemistry*, (Ed.: V. Balzani), Reidel, Dordrecht, **1987**.

A.21   *Crown Ethers and Analogs*, (Ed.: S. Patai and Z. Rappoport), J. Wiley, New York, **1989**.

A.22   *Cation Binding by Macrocycles*, (Ed.: Y. Inoue and G. W. Gokel), Marcel Dekker, New York, **1990**.

A.23   *Calixarenes: a Versatile Class of Macrocyclic Compounds*, (Eds.: J. Vicens and V. Böhmer), Kluwer, Dordrecht, **1991**.

A.24   *Frontiers in Supramolecular Organic Chemistry and Photochemistry*, (Eds.: H.-J. Schneider and H. Dürr), VCH, Weinheim, **1991**.

A.25   *Inclusion Aspects of Membrane Chemistry*, (Eds.: T. Osa and J. L. Atwood), Kluwer, Dordrecht, **1991**.

A.26   *Advances in Supramolecular Chemistry*, (Ed.: G. W. Gokel), JAI Press, Greenwich, *Vol. 1*, **1990**; *Vol. 2*, **1992**; *Vol. 3*, **1993**.

A.27   *Crown Compounds: Toward Future Applications*, (Ed.: S. R. Cooper), VCH, Weinheim, **1992**.

A.28   *Supramolecular Chemistry*, (Eds.: V. Balzani and L. De Cola), Kluwer, Dordrecht, **1992**.

A.29   *Structure and Bonding*, Springer, Berlin, *Vol. 16*, **1973**.

A.30   *Host-Guest Complex Chemistry, Topics Curr. Chem. Vol. 98*, **1981**; *Vol. 101*, **1982**; *Vol. 121*, **1984**.

A.31   *Biomimetic and Bioorganic Chemistry, Topics Curr. Chem. Vol. 128*, **1985**; *Vol. 132*, **1986**; *Vol. 136*, **1986**.

A.32   *Molecular Inclusion and Molecular Recognition, Topics Curr. Chem. Vol. 140*, **1987**; *Vol. 149*, **1988**.

A.33   *Supramolecular Chemistry, Topics Curr. Chem. Vol. 165*, **1993**.

A.34   *Bioorganic Chemistry Frontiers*, Springer, Heidelberg, *Vol. 1*, **1990**; *Vol. 2*, **1991**.

A.35   *Supramolecular Chemistry, J. Coord. Chem.* Special Issue, (Eds.: T. Iwamoto, J. Lipkowski and J. F. Biernat), **1992**, *1-3*.

A.36   *Supramolecular Chemistry, Rec. Trav. Chim. Pays-Bas*, Special Issue, (Ed.: D. N. Reinhoudt), **1993**, *112*, no. 6.

A.37   *Computational Approaches in Supramolecular Chemistry*, (Ed.: G. Wipff), Kluwer, Dordrecht, **1994**.

A.38   D. J. Cram and J. M. Cram, *Container Molecules and Their Guests*, RSC, Cambridge, **1994**.

A.39   *Perspectives in Supramolecular Chemistry, Vol. 1*, (Ed.: J. P. Behr), **1994**; *Vol. 2*, (Ed.: G. Desiraju), in preparation; *Vol. 3*, (Ed.: A. D. Hamilton), in preparation, Wiley, Chichester.

A.40   *Transition Metals in Supramolecular Chemistry*, (Eds.: L. Fabbrizzi and A. Poggi), Kluwer, Dordrecht, **1994**.

A.41   J.-H. Fuhrhop and J. König, *Membranes and Molecular Assemblies: The Synkinetic Approach*, RSC, Cambridge, **1994**.

A.42   *Molecular Recognition, Tetrahedron*, Symposium-in-print No. 56 (Ed.: A. D. Hamilton), **1995**, *51*, No. 2.

A.43   *An Introduction to Molecular Electronics* (Eds.: M. C. Petty, M. R. Bryce and D. Bloor), Arnold, London, **1995**.

Illustration Acknowledgements

The following figures, prepared by the author, have been previously published.

Fig. 5   *Pure Appl. Chem.* **1980**, *52*, 2441.
Fig. 8   *Pure Appl. Chem.* **1979**, *12*, 1305.
Fig. 9   *J. Chem. Soc., Chem. Commun.* **1993**, 1819.
Fig. 19   *Supramolecular Photochemistry* (Ed.: V. Balzani), Reidel, Dortrecht, **1987**, p. 1.
Fig. 22   *J. Chem. Soc., Chem. Commun.* **1991**, 1179
Fig. 34   *Helv. Chim. Acta* **1991**, *74*, 1843.
Fig. 37   *J. Chem. Soc. Perkin Trans. 2* **1992**, 461.
Fig. 41   *Proc. Natl. Acad. Sci. USA* **1993**, *90*, 163.
Fig. 43   *Makromol. Chem., Macromol. Symp.* **1993**, *69*, 1.
Fig. 44   *Makromol. Chem., Macromol. Symp.* **1993**, *69*, 1.
Fig. 45   *J. Chem. Soc., Chem Commun.* **1990**, 479.
Fig. 46   *J. Chem. Soc., Chem Commun.* **1990**, 479.
Fig. 47   *J. Chem. Soc. Perkin Trans. 2* **1992**, 461.
Fig. 48   *Proc. Natl. Acad. Sci. USA* **1993**, *90*, 5394.
Fig. 49   *Proc. Natl. Acad. Sci. USA* **1993**, *90*, 5394.

Index